Partial Order Concepts in Applied Sciences

Marco Fattore • Rainer Bruggemann
Editors

Partial Order Concepts in Applied Sciences

Springer

Editors
Marco Fattore
Department of Statistics and Quantitative Methods
University of Milan "Milano-Bicocca"
Milan, Italy

Rainer Bruggemann
Leibniz—Institute of Freshwater Ecology and Inland Fisheries
Berlin, Germany

ISBN 978-3-319-45419-1 ISBN 978-3-319-45421-4 (eBook)
DOI 10.1007/978-3-319-45421-4

Library of Congress Control Number: 2016955833

Printed on acid-free paper

This Springer imprint is published by Springer Nature
The registered company is Springer International Publishing AG
The registered company address is: Gewerbestrasse 11, 6330 Cham, Switzerland

Preface

On 9th and 10th April 2015, the 11th International workshop on partial orders in applied sciences was held in Florence, Italy. Scientists from several countries gathered on the Tuscanian hills to share and discuss ideas on Partial Order Theory and its applications to open problems in different research fields. This book is the synthesis of what has been presented at the workshop. It collects papers of different nature (theoretical, applied, computational), pertaining to many different topics, from indicator construction and cluster analysis on multi-indicator systems to socio-economic, environmental, and genetic applications.

The theory of partially ordered sets (posets, for short) is a well-established and developed branch of discrete mathematics. Differently from classical analysis and linear algebra, it is not part of the standard toolbox of statisticians and applied scientists yet. However, in the last 15 years, interest in applications of poset tools to different research areas has been increasing. The reasons for that are quite clear. The description of the world around us involves increasingly complex multidimensional data structures, made up of weakly interdependent variables, often of an ordinal kind. Main examples can be found in the socio-economic analysis of well-being or in many multi-criteria decision problems in the environmental sciences. Usually, one wants to reduce complexity, making synthetic information emerge in the form of indicators or of rankings of statistical units. The use of analytical tools designed for highly correlated numerical variables (as in classical dimensionality reduction procedures) is well at odds with synthesizing this kind of data systems. It is the existence of essential (i.e., not due to noise in the data) incomparabilities among statistical units to be of central importance. Incomparabilities are an intrinsic feature of complex data and methods to deal with them are to be worked out, in order to develop data treatment tools capable of simplifying complexity, while preserving its fundamental and meaningful features. Partial order theory is designed precisely to deal with this kind of problems, being a mathematical theory solely based on algebraic relations, i.e., here on the concepts of comparability and incomparability among "entities" (statistical units, multidimensional profiles . . .). The description of complex concepts in several different research fields imports notions and tools from poset theory into the daily practice of applied scholars. The theory of partially ordered sets, however, is not only a way to enlarge the analytical toolbox of data analysts; primarily, it is a way to conceptualize the problems and address them consistently with their nature. From this point of view, the effort of spreading "poset language" across disciplines has also a cultural and epistemological goal, as a contribution towards the development of new paradigms in data analysis and interpretation. Perhaps, this is the most challenging, and the hardest, issue, i.e., to introduce the scientific community to a new look on new (and old) problems.

The series of partial order workshops is an attempt in this perspective, as are the research results "condensed" in the book. It is a pleasure to see how new results and new applications of the posetic language are continuously worked out, and how poset theory is reaching new scientific areas, as witnessed by the contents of this volume, which cover an astonishing variety of methods and topics. The book is organized into five parts. Part I is devoted to methodological advances and collects papers of a more theoretical or technical nature, oriented to the analysis of multi-indicator systems. Part II pertains to applications of poset theory and posetic techniques to socio-economic analysis; multidimensional poverty, social participation, and development are the main topics. In Part III, poset theory is applied to environmental sciences, namely to biodiversity, sustainability, and risk management in the chemical industry. Part IV investigates the use of posetic tools in new application fields, like genetics, botanics, and engineering. Finally, part V presents some recent developments in software tools for posetic analysis.

The book is of interest both for scholars working on the specific topics touched upon in the volume and for all who wants to be gently introduced, by examples, to new ideas and perspectives in complex data analysis. We hope it may inspire other scholars to undertake and tread innovative research paths in their own application fields.

Milan, Italy Marco Fattore
Berlin, Germany Rainer Bruggemann

Contents

Contributors

Robert Ackermann Technische Universität Berlin, Berlin, Germany

Ghanima Al-Sharrah Department of Chemical Engineering, Kuwait University, Safat, Kuwait

Paola Annoni Economic Analysis Unit, European Commission-Directorate General for Regional and Urban Policy, Brussels, Belgium

Alberto Arcagni Department of Statistics and Quantitative Methods, University of Milano-Bicocca, Milan, Italy

Vanessa Bach Technische Universität Berlin, Berlin, Germany

Michael Bariotakis Department of Biology, University of Crete, Heraklion, Greece

Hans-Georg Bartel Department of Chemistry, Humboldt University Berlin, Berlin, Germany

Tugce Beycan Department of Sociology, University of Neuchâtel, Neuchâtel, Switzerland

Krunoslav Brčić-Kostić Institute Ruđer Bošković, Zagreb, Croatia

Rainer Bruggemann Department of Ecohydrology, Leibniz—Institute of Freshwater Ecology and Inland Fisheries, Berlin, Germany

Lars Carlsen Awareness Center, Roskilde, Denmark

Elias Castanas Laboratory of Experimental Endocrinology, School of Medicine, University of Crete, Heraklion, Greece

Matteo Corsi Department of Economics and Business Studies, University of Genoa, Genoa, Italy

Stefania Della Queva ISTAT—Italian National Institute of Statistics, Rome, Italy

Enrico di Bella Department of Economics and Business Studies, University of Genoa, Genoa, Italy

Marco Fattore Department of Statistics and Quantitative Methods, University of Milano-Bicocca, Milan, Italy

Matthias Finkbeiner Technische Universität Berlin, Berlin, Germany

Frauke Fuhrmann Technical University of Applied Sciences Wildau, Wildau, Germany

Giulio Guarini Dipartimento di Economia e impresa, Università degli studi della Tuscia, Viterbo, Italy

Elvis Hernández-Perdomo University of Hull, Hull, UK

Alexander Hilckmann Technische Universität Berlin, Berlin, Germany

Adalbert Kerber Department of Mathematics, University of Bayreuth, Bayreuth, Germany

Peter Koppatz Sudile Gbr, Potsdam, Germany

Technical University of Applied Sciences Wildau, Wildau, Germany

Danae Laina Department of Biology, University of Crete, Heraklion, Greece

Lucia Leporatti Department of Economics and Business Studies, University of Genoa, Genoa, Italy

Christos Lionis Clinic of Social and Family Medicine, School of Medicine, University of Crete, Heraklion, Greece

Petar T. Mitrikeski Institute Ruđer Bošković & Institute for Research and Development of Sustainable Ecosystems, Zagreb, Croatia

Hans-Joachim Mucha Weierstrass Institute of Applied Analysis and Stochastics, Berlin, Germany

Johnathan Mun U.S. Naval Postgraduate School, Monterey, CA, USA

Jan W. Owsiński Systems Research Institute, Polish Academy of Science, Warszawa, Poland

Stergios Pirintsos Department of Biology, University of Crete, Heraklion, Greece

Botanical Garden, University of Crete, Rethymnon, Greece

Nancy Y. Quintero Laboratorio de Química Teórica, Universidad de Pamplona, Pamplona, Colombia

Universidad de Antioquia, Antioquia, Colombia

Universidad Pontificia Bolivariana, Antioquia, Colombia

Universidad Católica de Oriente, Antioquia, Colombia

Guillermo Restrepo Laboratorio de Química Teórica, Universidad de Pamplona, Pamplona, Colombia

Bioinformatics Group, Department of Computer Science, Universität Leipzig, Leipzig, Germany

Claudio M. Rocco Facultad de Ingeniería, Universidad Central de Venezuela, Caracas, Venezuela

Margit Scholl Technical University of Applied Sciences Wildau, Wildau, Germany

Patricia Seitz University of Texas Medical Branch, Galveston, TX, USA

William Seitz Department of Marine Sciences, Texas A&M University at Galveston, Galveston, TX, USA

Christian Suter Department of Sociology, University of Neuchâtel, Neuchâtel, Switzerland

Part I
Theoretical and Methodological Advances

Endowing Posets with Flesh: If, Why and How?

Jan W. Owsiński

1 Introduction

Resulting from numerous empirical studies are multidimensional data, say x_i, $i = 1, \ldots, n$, $x_i = \{x_{ik}\}$, with i's denoting observations, or objects, and $k = 1, \ldots, m$, denoting the descriptive variables, features, or criteria. We would often like to put these data into a ranking type of a structure, i.e. to order the items i. This means, implicitly, obtaining a sequence of "ranks" $\{o_i\}$, corresponding to $i = 1, \ldots, n$, where o_i are natural numbers ranging from 1 to at most n. (We assume that x_{ik} are the values of measurements regarding certain criteria, numbered k, $k \in K = \{1, \ldots, m\}$, and that in all cases "the more, the better". None of these assumptions limits the generality of the considerations.) Yet, the sheer multiplicity of dimensions prohibits, as a rule, a straightforward ordering of the data items. This is the obvious consequence of the situations, in which $x_{ik} > x_{jk}$ for some i, j and a definite subset of $k \in K$, while $x_{ik'} < x_{jk'}$ for the same i, j and another subset of $k' \in K$.

Thus, we very often stop at the result of analysis, being a poset, encompassing all the situations, where $x_{ik} \geq x_{jk}$ for all $k \in K$, and leaving out all the other ones, its illustration being constituted by the respective Hasse diagram.

There are—indeed numerous—situations, though, in which we would like to go beyond the poset "skeleton" and endow it with "flesh", up to construction of a complete order, perhaps with some additional characterisation. (A feasible alternative might be a kind of information, resulting from the poset processing, that is effectively "close enough" to the actual ranking.)

We argue here that such extension of a poset may be legitimate, and a shorthand analysis is provided of why and how one could go about it, based on the essential properties of the analytical tasks in general.

J.W. Owsiński (✉)
Systems Research Institute, Polish Academy of Science, Newelska 6, 01-447 Warszawa, Poland
e-mail: owsinski@ibspan.waw.pl

M. Fattore, R. Bruggemann (eds.), *Partial Order Concepts in Applied Sciences*,
DOI 10.1007/978-3-319-45421-4_1

2 Why Not?

There exist serious reservations, concerning going beyond the poset structure as a result of the ordering-oriented study for the given set of data items $\{x_i\}$.

The primary one is that the empirical data do not contain any other information than that corresponding to the poset obtained. If we go further away from this point, first of all by adding (i,j) edges that do not exist on the original Hasse diagram, and especially toward the complete order, then we are unfaithful to the data. There is, actually, an often justified suspicion of manipulation, motivated by "political" interests, behind the operations, leading from the original data-based poset to some complete order. This suspicion may, of course, be well founded.

The second reservation refers to the fact that while forming a poset from the initial $\{x_i\}$ data is straightforward and unambiguous, virtually all approaches meant to go beyond it either involve subjectivity, or have to refer to data that may have little to do with the original empirical data used in the study.

It is largely in view of these two types of reservations that the technique of counting the consistent linear extensions for a poset is advocated, which, even if still arbitrary, appears to be a possibly neutral operation, based only on the relation between the given poset and the structure of the entire lattice.

3 But Perhaps...

On the other hand, though, there are quite obvious, and, at that, quite numerous and diverse, reasons for insisting on complementing the posets to completeness, or at least somehow transforming it in a definite direction and manner. These result from the considerations, associated with the aspects, roughly illustrated in Fig. 1.

3.1 The Purpose and the Utility

First is the sheer utility: it may be so that the very objective of the endeavour, from which the data originate, includes the determination of a (possibly) complete order, for quite practical purposes. Lack of such a structure may mean a failure and a loss in economic or social terms.[1]

This argument involves a much broader background, involving such notions as: *a problem, an image (model, theory, perception) of the problem, the need to deal with it (to resolve it), the need to cognise it (to identify its structure and mechanism), the*

[1]The issue whether this loss is not worth avoiding a wrong decision, based on the "inadequate" extension of the poset, or vice versa, is exactly the subject matter of this note.

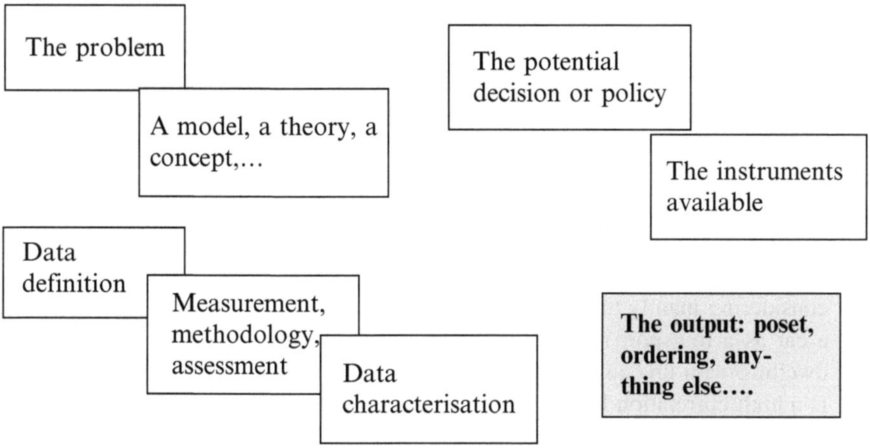

Fig. 1 The environment of the studies, leading to data structures, including posets

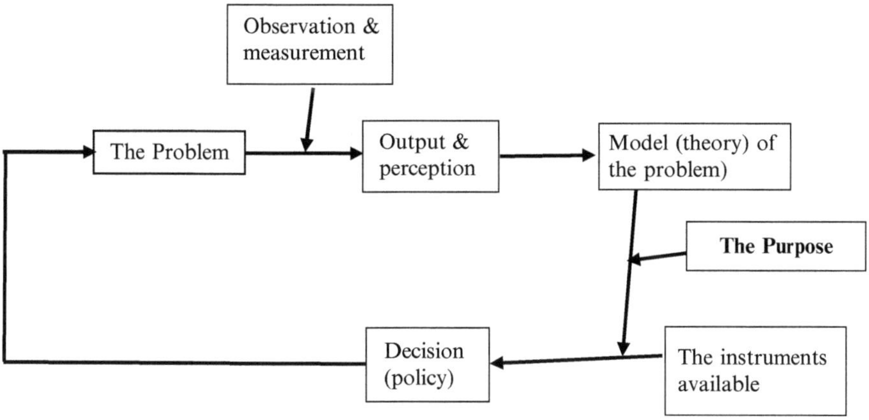

Fig. 2 The classical decision-making loop, in which the notions referred to appear

need to apply definite means to resolve the problem, based on the cognition of the problem, on the *purpose* (the *objective*), and the *instruments* we operate. All these enter the classical decision-making loop of Fig. 2.

If our *purpose* does not involve (imply) a decision or a policy, an action regarding the problem, then our cognition may be the last step in the procedure, and we might not need anything more than a poset, in case we compare some objects or states, and especially then, when not so much the values x_{ik} are important, as sheer binary relations between them.

This last remark is quite telling. A simplest pertinent situation is outlined in the frame of Example 1 (which will be continued further on, through addition of consecutive aspects).

So, if there is a purpose, requiring action, based on a decision or policy, not only ordering may be required, but also measurement of quantities x_i and their transformations (mappings). An illustration of the exposure to a situation with an explicit structure of purpose and instruments is provided below, with a hint to an important proposition.

Assume we deal with two dimensions of "wealth": $k=1$—income, and $k=2$—"usable wealth", meaning lump assessment of the value of assets, considering mainly their utility and only secondarily their market value (e.g. a car as a transport means, not as a certain saleable good; ownership of a dwelling with its equipment being the primary instance). Even though there is a high correlation between the two dimensions, there are numerous cases when households with lower incomes dispose of an ampler "usable wealth".[2] This is especially important when such situations occur close to the border of the derived deprivation function, $D(.)$, namely near $D(x)=0$, whether we speak of $x_{.1}$ or of $x_{.2}$. If the authorities dispose of only one instrument, *the general subsidy*, then a single "reaction" (decision) function $S(D_1,D_2)$ has to be developed, meaning, in fact, appropriate weighing (implicit or explicit) of the two dimensions. Now, assume that the authorities can deploy a second instrument, say, a *non-transferable allowance for housing costs.* We deal with $S_1(D_1,D_2)$, $S_2(D_1,D_2)$. The fundamental question is: how are the two pairs of dimensions interrelated? Most conveniently, the dimensions k would correspond directly to the instruments. If such a correspondence existed, even in the form of a demonstrable correlation, then the task of the authorities would be straightforward, and no additional analysis, beyond (two) unidimensional rankings, would be necessary.

Thus, it is obvious that when we dispose of more than one instrument, then there exists a room for finding plausible structures other than a single ranking. This is the case in many situations, where, in fact, not just a single ordering is required, but, rather, some structure, obtained from the poset, which corresponds, on the one hand, to the "measurements" made, and on the other—to the ultimate purpose of the exercise, which need not be unidimensional.

[2]This is very often the case on the micro-scale in countries with low or ineffective broadly conceived property taxation, but it is also, in a way, present at the national level—some countries, featuring high personal incomes of their citizens, rank much lower in terms of bank deposits, value of property owned, etc.

3.2 The Data Themselves

Then, the second broad motivation to go beyond the raw poset obtained from the data collected comes from the data themselves, as if in a paradox. And there are, indeed, several aspects or sources, for this kind of motivation:

1. The data which served to set up the poset are often, if not always, charged with definite uncertainty, coming from various origins. A trivial illustration for this fact is provided in Table 1.
2. There exists a definite "model" or "theory", which was, at least to some extent, the background for the study considered, the "model" or "theory" resulting, at the minimum level, in (a) the selection and specification of criteria (variables), $k = 1, \ldots, m$, and (b) the ways in which they are measured, or "evaluated". Thus, these criteria or variables are not some jack-out-of-the-box entities, about which we know nothing, and which cannot be subject to any operations—analytical or manipulative. They are an internally consistent fragment of perception of the reality, on which we might wish to act, basing upon the results of the study at hand.
3. Accompanying the "model" or "theory" there are empirical data, which are parallel to and intertwined with those having formed the basis for formulating the model or theory, but also for undertaking the study under consideration. These data offer a "logic" on their own, through statistical or otherwise relations between them, but, as well, through their characteristics in terms of uncertainty. Here, also, belongs the issue, which is very often formulated as one of main reproaches against the "pure poset" approach—namely the absence of scale of values of the variables (criteria), whenever they are not binary or strictly nominal.

Figure 3 shows the notions related both to data themselves (metadata, complementary data, etc.) and to the broader environment, partly introduced in connection with Fig. 2. The difference between Figs. 2 and 3 consists also in the fact that Fig. 3 presents more of the actual "data processing" aspect than of the general "thought framework", oriented at the problem and the potential solution to it (a "policy" or a "decision"), with respect to which the poset or another structure, resulting from the study, constitutes only an instrument.

4 So What?

None of the above listed aspects can be simply shrugged away. Depending upon the case, these aspects intervene in various manners and with various importance. This chapter is not meant to provide any definite methodological proposals, nor solutions—the problems touched upon differ so widely in their structure and character that dozens of theories and methodologies may not suffice to encompass all the situations, the technical variants put apart. Thus, we shall concentrate on some types of situations, and forward comments related to them.

Table 1 Some examples of the sources and character of uncertainty in data specification

Problem area	Criteria/variables (exemplary)	Importance of uncertainty[a]	Type of uncertainty[b]	Possible treatment of uncertainty
Welfare assessment; questionnaire based	Income level	Moderate	Statistical	Other data sources
	Dwelling ownership and state	Low	Formal	Legal assessment
	Car ownership	Low	Formal	Other data sources
	Nutritional norms	High	Vagueness	Other data sources
	Home appliance ownership	Moderate	Vagueness	Consistency
	Vacationing	High	Vagueness	?
Chemical compounds; hazard assessment	Toxicity	Moderate	Statistical	Biomedical check
	Persistence	Moderate	Statistical	Empirical data
	Ubiquity	Moderate	Statistical	Empirical data
	Removal cost	High	Vagueness	Market data
Job candidates	Education	Low	Formal	Questioning
	Skills	Moderate	Vagueness	Questioning
	Experience	Moderate	Vagueness	Credential checks
	Social relations	High	Vagueness	Drama test
Battery	Price	None	–	–
	Performance	High	Statistical	Testing
	Duration	High	Statistical	Testing

[a]Importance of uncertainty is a "product" of the lack of precise and well-justified knowledge of definite data item (number(s)) and the significance of this uncertainty for the exercise considered ("the ultimate purpose of the study")
[b]Only indicative—mixtures of uncertainty types are very frequent

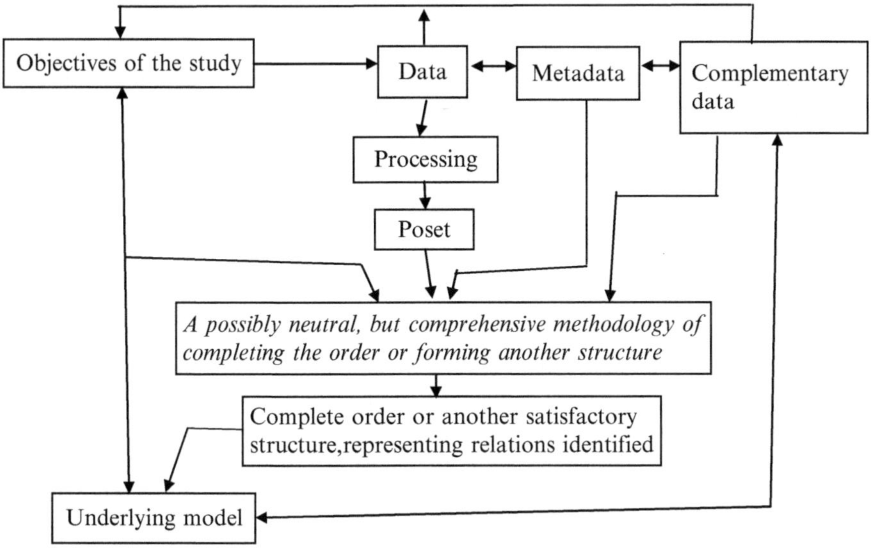

Fig. 3 Scheme of interrelations between the components of the study, to be considered when analysing the ordering of observations

In the light of the above, it is obvious that an approach, that tries to go beyond the "pure poset" result, must first of all account for the aspects mentioned, in order to avoid the trap of technical correctness that overlooks the actual issues for the sake of such technical correctness.

4.1 The Purpose and the Policy Instruments

As said, if our purpose is just "to know", and there is no possibility of developing and verifying a true-to-life model of the phenomenon, obtaining a poset may indeed be the justified terminal result. The situation may turn out to be similarly straightforward, although leading to different results (e.g. single dimension orders), when we explicitly account for more than one "resulting dimensions" (instruments to be applied). For this, though, additional assumptions have to be satisfied, first of all concerning relations between the criteria of assessment and the instruments envisaged. There may exist cases, when the presence of multiple instruments, which would correspond to potential (or "required") multiple rankings, even if not leading to ultimate simplification in the form of single-dimensional complete orders, may allow for "disentangling" of the poset, in the sense of obtaining more than one structure, each of the resulting structures being closer to the complete order than the original poset, without any additional operations. Yet, there might also exist situations, in which the relations mentioned lead to more complex issues and potential structures, and the initial problem remains unsolved.

The analysis of these potential situations is not only beyond the scope of the present note, but may also turn out to be highly complicated, at least for more general cases. Still, the possibility of simplification should be kept in mind when designing the respective study and when proceeding with analysis of data. This issue is quite closely associated with the subsequent ones, which refer to the consideration of a "model" or "theory", standing behind the study.

4.2 The "Model" Behind the Study

Whenever a study, including data collection, is undertaken, there must exist some concept, underlying the very launching of the study, as well as the potential "decision" or "policy", possibly together with the instruments considered. While this is obvious, the scope and the degree of precision of such models range extremely broadly: from situations, where very little is known or assumed, up to those, where definite, well-grounded hypotheses are being verified against a broader knowledge of the respective domain. Yet, even in the former situations, it cannot be accepted that our entire knowledge consists in saying "there is a problem": this knowledge, actually, led to the specification of the objects, variables, methods of measurement, etc. It also contributed, in a vast majority of cases, to some evaluation framework, that is—variables turned into criteria, along with respective scales.

Even if we neglect the models involving definite hypotheses (e.g. "the poor can be classified into two classes, class A, for which . . . , and class B, . . . "),[3] and the aspect of instruments/policies, there must exist some "minimum specification", originating from the most primitive perception of a given problem. This minimum specification almost certainly involves some concept either directly involving comparability or leading to a possibility of comparison. Let us consider this on an example from Table 1. Assume, namely, that we consider the case of batteries, and we deal with four, otherwise quite the same, batteries, as exemplified in Table 2.

It is obvious that we are not capable of simply ordering these objects, but, on the other hand, we can formulate and ask questions, helping to come at least closer to the ultimate linear order. We should keep in mind that we deal with one-at-a-time situation: it would be different, if we were buying batteries sequentially. Then,the optimum decisions could take the form of a closed-loop strategy, based on the results from the preceding decision, like in Fig. 2 (with the incremental objective function, based on cost of duration over time).

The types of questions, and potential answers, make a part of usual decision-making process, and are, indeed, applied both in the proper context of the poset-based analysis (like in the proposal from Tagliabue et al. 2015) and, more generally, in the context of multicriteria decision-making (like in Kaliszewski and Miroforidis 2014, or Kaliszewski et al. 2014).

[3]This aspect is elaborated at length in the papers by Fattore (2008, 2015).

Table 2 An illustrative example: four kinds of batteries

Battery	Price per unit	Expected durability	Question 1	Question 2
A	10	2 years (quite surely)	Is additional certainty worth 2? Is my time horizon really 2 years?	Can I measure this additional certainty (or difference of certainties)?
B	8	2 years (rather surely)		
C	5	1 year (quite surely)	Is this certainty equal to that for A?	Is it worth the same as A?
D	2	Unknown (unknown)	What if I make an experiment and buy 2 for 4?	Do I have time/wish for experimenting?

4.3 The Statistical Features of the Data

It is also frequent that an important aspect of the study is constituted by the statistical—in the popular sense—characterisation of the objects to be ordered. This usually means that we deal with the numbers of observations, or occurrences, of a given object. This has an obvious relation to the model or theory that we may have. We often interpret such statistics as some reflections of probabilities, resulting from the "inner" working of the process or system. This aspect, again, can by no means be overlooked, even if our goal is just to (somehow) order the objects.

Thus, if the statistics reflect some reality, inherent to the system at hand, and there are significant differences among the numbers of occurrences, then not all linear orderings of the objects considered should be taken as equally probable. Even though not a straightforward exercise (additional assumptions have to be made), the statistics ought to be used to determine the probabilities ("weights") of the particular linear extensions, in the approaches as described and analysed in Lerche et al. (2003), Patil and Taillie (2005), or De Loof et al. (2008). The same applies to the counting approach.

Thus, it is not so that we would introduce the "weights" by some subjectively designed back door—they are a direct reflection of the data, coming from the same study, having the very same degree of "legitimacy", and a similar, or even higher, level of reliability (up to a truly well-based statistical analysis of the distribution of particular "paths"—extensions—through the entire set of possible states).

Let us also indicate that the "statistical" aspect to a problem or study may entail a plethora of different problem structures, calling for entirely different approaches, even if in virtually each of them a poset structure might be obtained. So, in particular, we may deal with unique, separate cases (like, e.g. in characterisation of a set of chemical compounds, or the set of tender offers), or with a sample, and possibly even an entire population, in which certain "states" appear, their numbers of appearances often widely differing, while other ones do not appear, or have very

Table 3 An example of feasible pupil assessments for a class of 25 pupils (grades from 1 to 5 in four domains: numbers of pupils)

Assessments and corresponding numbers of pupils		
5555: 3		
5455: 1	5554: 1	5545: 2
5445: 4	5454: 2	
5444: 1		
		5535: 2
5335: 4	5433: 2	5534: 1
4335: 1		
3434: 1		

low ("exceptional") frequencies. As an illustration, we give a realistic, though quite stylised, example of a scholarly classification at lower grades of the primary school in Poland.[4]

Thus, assume pupils are classified with respect to four broadly conceived domains: 1. Behaviour, social attitude, cooperation; 2. Humanities; 3. Sciences; 4. Physical and technical exercises. The assessments are made on the 5-point scale: 5—very good, 4—good, 3—sufficient, 2—insufficient, 1—very bad. In a class of 25 pupils, we may have the "statistics" of assessments as in Table 3.

First, there are far less objects than possible states (25 vs. 625, or, rather, actually, in terms of states, 13 vs. 625, see Fig. 4). Then, most importantly, there is a clear "statistical" nature to such exercise. Actually, if we have, say, 12 binary variables and a sample or population of, say, 10,000 items, there is definitely a high probability that some of the possible 2^{12} states shall be "empty", or "close to empty", while other ones might group quite a number of items.

There are several regularities, well known for Polish teachers, parents and children, as well, appearing in the data. One of them is the "shortness" or "flatness" of the distribution. Another, known also in other countries, is the relation of marks for humanities and sciences. Here, for 14 pupils the assessments are equal, for three the marks for humanities are lower than for sciences, and for eight—vice versa. A hint for linear extensions, indeed.

Actually, in order to specify all the linear extensions in this case, we need only altogether 22 evaluation "states", as listed in Table 4 below.

Note that both the "a priori" model or theory and the direct implications of the "statistics" of the empirical results constitute a different basis for potential processing of the results, including poset extensions, than those mentioned in regard to the mutual or "absolute" importance of criteria, or variables in Sect. 4.2, and also those from Bruggemann and Carlsen (2015).

[4]Due to the discussion at the 11th workshop on Partial Order in Applied Sciences, Florence, 9–10 April 2015, the author can confirm that the stylised facts, reflected in this example, are also characteristic for many other educational systems.

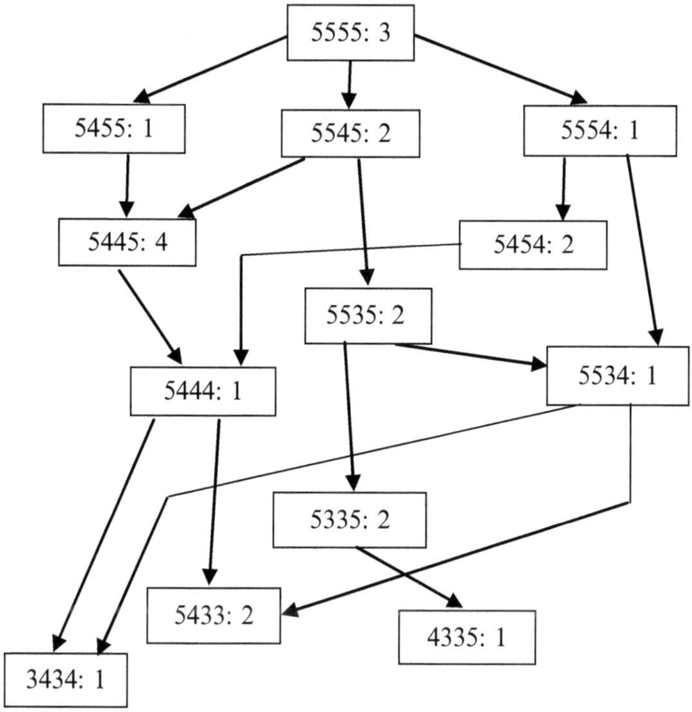

Fig. 4 Illustration of the partial order for the data of Table 3

Table 4 Assessment "states" from Table 3, complemented with the missing ones for full extensions

Assessments and the corresponding numbers of pupils				
		5555: 3		
5455: 1		5554: 1		5545: 2
5445: 4		5544: 0	5454: 2	5535: 2
		5444: 1		5534: 1
5435: 0	5345: 0	5443: 0	5434: 0	4534: 0
5335: 4			5433: 2	4434: 0
4335: 1		4444: 0		
		3444: 0		
		3434: 1		

4.4 *Representations of Uncertainty: Just One Hint*

Another important aspect of the perception outlined is that of direct representation of uncertainty. This representation might be a statistical or probabilistic, given sufficient knowledge and adequate sample or population. Yet, in the situation we deal here with we can hardly afford such assumptions (were this not the case, we would not have to face the issue that we are trying to resolve). In such cases one of

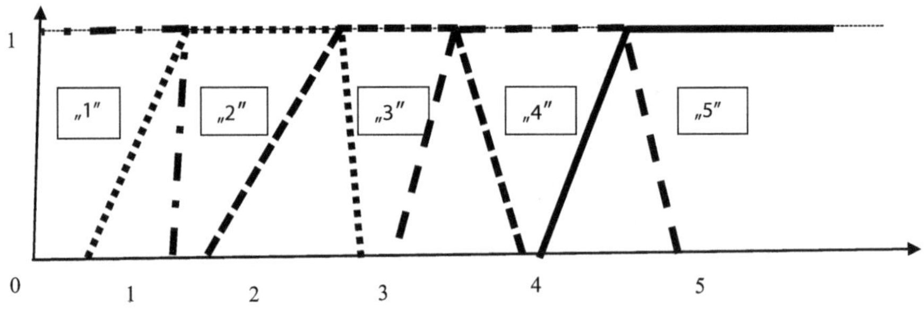

Fig. 5 Potential fuzzy "definitions" of the marks for humanities, sciences, and exercises

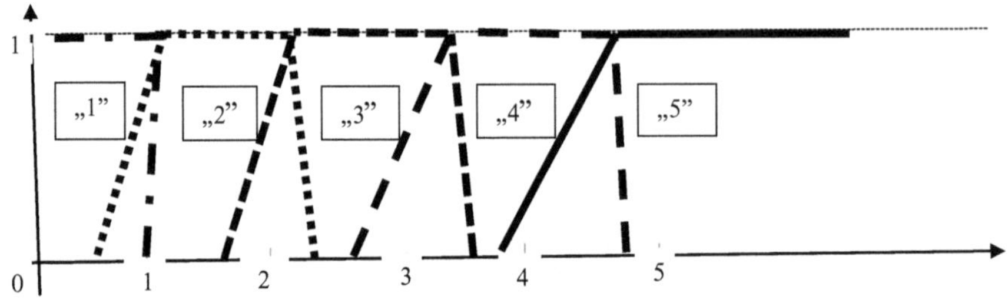

Fig. 6 Potential fuzzy "definitions" of the marks for behaviour and cooperation

the ways out is to use fuzzy set-based representation that we shall illustrate for the example of the previous section.

Thus, we can assume that the actual meaning of the assessment marks for humanities, sciences and exercises is as shown in Fig. 5, while for behaviour and cooperation—as in Fig. 6. The meaning of these definitions of marks is that "there is no precise statement of a mark **X** corresponding to the level of quality = **X**, for at least two reasons: first, the marks are strongly discrete, while the actual assessment concerns a continuum, second: various aspects (criteria) of assessment (effort, diligence, skill, knowledge, . . .); so, the actual mark **X** corresponds to a fuzzy number **X***". We assume consistency in these quasi-definitions (e.g. for two neighbouring "marks" only one can be equal to 1 at any point) although this may not be necessary at all.

What may be the consequences of such a character of data for the ordering? Comparing fuzzy numbers is possible (see, e.g. Brunelli and Mezei 2013) even if—like many operations over fuzzy sets—quite heavily loaded with arbitrariness. So, it would be possible, in principle, to employ such data in both linear extensions and the counting approach (the issue of scalability left apart), but either under very stringent conditions (e.g. all the fuzzy sets, representing numbers, have the same form, and all of the comparisons take, therefore, standard forms) or in a very simple manner (e.g. fractional weights).

On the other hand, though, it would be quite feasible to aggregate such numbers (e.g. like in clustering), and also make a straightforward ordering. The operations on fuzzy sets, representing variable values, and their collections, representing objects, would have to involve a high degree of arbitrariness, but such a possibility exists, and is indeed being made use of.

4.5 Back to the Essential Issue

Thus, the basic question arises: How much do we, in fact, sacrifice, either way? How much is worth the loss of quantitative measurements? disregard of the known or hypothesised interrelations? etc. And vice versa: How big is the risk of introducing arbitrariness into the "raw" result, and can we measure the degree of arbitrariness? And so on.

Some of the pertinent questions shall remain unanswered, or answers would only be very superficial. This means simply trying out various techniques, allowing for the possibly effective extension of the poset towards the complete order, and, eventually, those that might lead straight towards the complete order from the initial data, while violating as little as possible the formal requirements, and taking into account to the maximum extent the data available and the associated knowledge. This means, also, that we should try to find the answers to such questions as:

- Can we establish conditions for equivalence between the various procedures, related to poset, based on the output from definite data, these procedures aiming at obtaining the complete order?
- If not, what would be the results of comparison of the obtained complete orders, including the properties of the respective structures, as related to the initial poset?

This involves, in particular, the issue of:

- Potential parallelism and differences between the counting of linear extensions consistent with the poset (assumption of equal probabilities!) and the solution to the relational programming problem (see Owsiński 2011), the former possibly enriched with the assessment of different probabilities of the extensions, the latter being a way to obtain the Kemeny-median-like structure from the data.

5 A Real-Life Example Without Constructive Conclusions

5.1 The Project Outline

In 2011–2014 a modelling project was carried out at the Systems Research Institute, Polish Academy of Sciences, in collaboration with the Institute of Geography and Spatial Organization of the Academy, on the development and implementation of forecasting models for the variables, characterising the socio-economic situation of the capital province of Masovia (for an ampler description, see Czapiewski et al. 2016).

The list of modelling domains was specified by the commissioning institution, Masovian Bureau of Regional Planning. Altogether some 70 indicators were developed, along with the respective models—usually relatively simple econometric models. All this had to be done for each of the 314 municipalities of the province and for 15 years. The biggest was, of course, the demographic model, producing at each run some half a million numbers (municipalities, age groups, sexes, etc.) (Table 5).

For a number of domains, model developers were asked by the commissioning agency to provide the "synthetic indicator", given the presence of, say, three to five indicators, oriented at definite phenomena. This was done in some cases, but, generally, the reaction of the respective model authors was negative ("how do we put variables X and Y together, if they express completely different phenomena/processes?"). Yet, it must be admitted that these same authors made often quite arbitrary choices when selecting variables and developing their particular models. The question would therefore be quite justified: how does one compare these two kinds of arbitrariness, and if one is to be accepted, why the other one should not be?

5.2 An Illustrative Case

In one of such several cases of "need to develop a synthetic indicator", in domain no. 9: Technical infrastructure, there were three variables, corresponding to the shares of inhabitants of the municipalities (in %), with access to water supply system, sewage system, and water treatment plant. So, all values ranged between 0 and 100. The author of the model in question declined providing an integrative indicator of the "state of technical infrastructure".

Yet, it was easily shown that in the population of 314 municipalities of the province there are some—otherwise absolutely obvious—regularities, such as:

- Highly stable and mostly significant correlations between all three variables.
- These correlations needing a correction in view of the 100 % upper bound on values.
- High correlation with population density.
- In view of the latter, the "synthetic indicator" had to be corrected for density/urban character.

Altogether, the conclusions and the results from this microstudy were as follows:

1. The three variables could be replaced relatively safely by one (in just few cases doubts might have arisen), most handily, their average.

Table 5 The list of domains and indicators of the project illustrated

No.	Domain name	Indicators	Comments
1	Population	Population totals, according to age and sex, feminisation, share of post-productive population, ratio of working age to nonworking age population	Model types for municipality types; assumptions concerning birth rate, mortality and migration scenarios
2	Social capital	Relative number of NGOs, of sports clubs members, of cultural and art groups members; also a synthetic indicator	Three variables, treated as proxies, and an "artificial" synthetic measure
3	Financial standing of self-governments	Revenues of local authorities, own and total, expenditures, investments; auxiliary: average employment per business, past ratios of expenditures to revenues	Model types elaborated for particular municipality types
4	Propensity of municipalities to invest	Investment-related expenditures (alternative to above), budget debt, current budget surplus, propensity to invest	A specially devised model, with assumed interaction with the user
5	Intellectual capital	University graduates, university students, companies with foreign share, a synthetic indicator	Three basic variables and an "artificial" synthetic indicator
6	Labour market absorption	Demand for labour (from GDP and productivity), auxiliary: productivity	Simple model based on variables from other domains
7	Social exclusion	Indicator based on: elderly, transport-wise accessibility, university educated, jobless; Gini-like measure of income inequality	Two entirely different indicators
8	Quality of life	Synthetic indicator, based on variables from domains 3, 5, 7, 9, 15, 16 and 17	An arbitrary relative indicator, based on seven variables
9	Technical infrastructure	Shares of inhabitants with access to water supply, to sewage system and to water treatment plant	Models depend upon the type of municipality and the levels attained until now
10	External investments	Magnitude of external investments—value per capita	Model based on variables from other domains
11	Innovativeness	Indicators based on intellectual capital, magnitude structure of companies, municipal and company investments	The last component is omitted in the first indicator

(continued)

Table 5 (continued)

No.	Domain name	Indicators	Comments
12	Information society	Indicators, based on innovativeness (domain 11) and internet presence in schools	Two indicators, differing by schools considered
13	Value of fixed assets	Fixed assets of public bodies, companies, total per capita, auxiliary: investments in self-governmental and private sectors	Very rough assessment is only possible, except for the formal figures
14	Entrepreneurship and employment	Number of businesses, jobs per business, jobless, total employment, share of employment in manufacturing and service, auxiliary: proxy for farm employment	Models for municipality types, differing by coefficient values
15	Transport accessibility	Expressed in numbers of people within a definite travel time outside and inside	Model based on road network and settlement system
16	Quality of environment	Based on share of green areas, farmland, population density, car number, overbuilt area share	Model partly based on variables from domains 17 and 19
17	Sustainable development	Anthropogenic pressure: car number, population density; state of sustainable development: protected areas, forests, permanent grasslands, physical plans	A number of partial models (e.g. representing car numbers), contributing to the overall indicators
18	GDP value	Global GDP dynamics, and local dynamics, based on salary distribution	A simple macroeconomic model, with six scenarios
19	Scale and rate of urbanisation	Population density, overbuilt areas, persons employed in manufacturing and service; areas under residential and non-residential structures	Some variables taken from other domains (1. Population, 14. Persons employed outside farming)

2. The proper "quality indicator" was taken as a function of the former, with an experimentally established divisor, based on population density and population number.

We never consulted this approach with the author of the model, who did not consent to participate in the exercise.

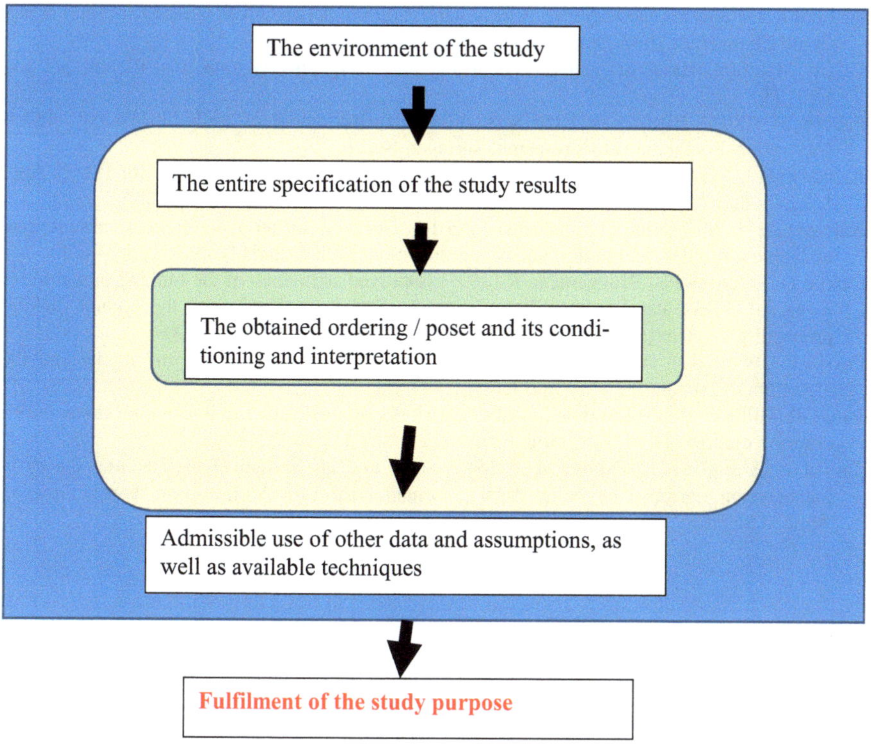

Fig. 7 General framework for the data analytic studies involving possible ordering structures

6 Some Conclusions

Figure 7 subsumes the conclusions, according to which the exercise in ordering of objects is (almost) always just a piece in a broader process or system, and should be viewed, and treated, as such. Hence, the direct results, while valid in themselves, are a part of a broader perspective, and therefore can be processed, under definite assumptions, so as to provide the feasible and necessary information.

References

Bruggemann R, Carlsen L (2015) Incomparable: what now II? Absorption of incomparabilities by a cluster method. Qual Quant. doi:10.1007/s11135-014-0076-x

Brunelli M, Mezei J (2013) How different are ranking methods for fuzzy numbers? A numerical study. Int J Approx Reason 54(5):627–639

Czapiewski K, Janc K, Owsiński JW, Śleszyński P (2016) A modeling project in Poland: the social and intellectual capital aspects. Stud Agric Econ 118(1):5–15

De Loof K, De Baets B, De Meyer H, Bruggemann R (2008) A Hitchhiker's guide to poset ranking. Comb Chem High Throughput Screen 11:734–744

Fattore M (2008) Hasse diagrams, poset theory and fuzzy poverty measures. Riv Int Sci Soz 1:63–75

Fattore M (2015) Partially ordered sets and the measurement of multidimensional ordinal deprivation, Social indicators research. Springer, Berlin

Kaliszewski I, Miroforidis J (2014) Two-sided Pareto front approximation. J Optim Theory Appl 162:845–855

Kaliszewski I, Miroforidis J, Stańczak J (2014) Decision maker's preferences, airport gate assignment problem and multiobjective optimization. Mult Criteria Decis Mak 8:84–100

Lerche D, Sørensen PB, Bruggemann R (2003) Improved estimation of the ranking probabilities in partial orders using random linear extension by approximation of the mutual ranking probability. J Chem Inf Comput Sci 43(5):1471–1480. doi:10.1021/ci0300036

Owsiński JW (2011) The bi-partial approach in clustering and ordering: the model and the algorithms. Stat Applicazioni (Special Issue):43–59

Patil GP, Taillie C (2005) Multiple indicators, partially ordered sets, and linear extensions: multi-criterion ranking and prioritization. Environ Ecol Stat 11:199–228

Tagliabue G, Barberis L, Fattore M (2015) A successful application of partial order theory in management: customer targeting. Paper presented at the 11th Workshop on "Partial Order in Applied Sciences", Florence, 9–10 April 2015

Incomparability/Inequality Measures and Clustering

Hans-Georg Bartel and Hans-Joachim Mucha

1 Some Notations and Definitions

For the following, we need some notations and definitions:

1. Let $(X, \leq)$ be a poset. The elements $\mathbf{x}_i$ of the set X are ordered m-tuples of (real) numbers: $\mathbf{x}_i = (x_{i1}, x_{i2}, \ldots, x_{im}) \in X,\ \ i = 1, 2, \ldots, |X|$. Here, the relation $\leq$ is the ordinary one between numbers.
2. Let $G = \{g_1, g_2, \ldots, g_g\}$ be a set of $g = |G|$ objects and let $X = \{\mathbf{x}_1, \mathbf{x}_2, \ldots, \mathbf{x}_g\}$ be a set of $|X| = g$ m-tuples, which is associated with a data matrix $\mathbf{X}(G, M)$ in the following way (M is a set of $m = |M|$ attributes):

$$\mathbf{X}(G, M) = \begin{pmatrix} x_{11} & \ldots & x_{1j} & \ldots & x_{1m} \\ \vdots & \vdots & \vdots & \vdots & \vdots \\ x_{i1} & \ldots & x_{ij} & \ldots & x_{im} \\ \vdots & \vdots & \vdots & \vdots & \vdots \\ x_{g1} & \ldots & x_{gj} & \ldots & x_{gm} \end{pmatrix} = \begin{pmatrix} \mathbf{x}_1 \\ \vdots \\ \mathbf{x}_i \\ \vdots \\ \mathbf{x}_g \end{pmatrix} = X^{\mathrm{T}}$$

3. Equality of the two m-tuples $\mathbf{x}_i$ and $\mathbf{x}_j$ (denoted by $\mathbf{x}_i = \mathbf{x}_j$) is fulfilled, if all elements of $\mathbf{x}_i$ and $\mathbf{x}_j$ are pairwise equal: $(x_{ih} \in \mathbf{x}_i) = (x_{jh} \in \mathbf{x}_j)$ for $h = 1, 2, \ldots, m$. Otherwise $\mathbf{x}_i$ and $\mathbf{x}_j$ are called "unequal" (denoted by $\mathbf{x}_i \neq \mathbf{x}_j$).

H.-G. Bartel
Department of Chemistry, Humboldt University Berlin, Berlin, Germany
e-mail: hg.bartel@yahoo.de

H.-J. Mucha (✉)
Weierstrass Institute of Applied Analysis and Stochastics, Mohrenstraße 39, D-10117 Berlin, Germany
e-mail: mucha@wias-berlin.de

M. Fattore, R. Bruggemann (eds.), *Partial Order Concepts in Applied Sciences*, DOI 10.1007/978-3-319-45421-4_2

Obviously, regarding this definition two m-tuples $\mathbf{x}_i$ and $\mathbf{x}_j$ can only be unequal or not. There are only two states: either "1 ≡ unequal/not equal" or "0 ≡ not unequal/equal."

4. In this publication, it is assumed that each pair $\mathbf{x}_i \neq \mathbf{x}_j$ with $i \neq j$ must differ at least in one of its elements:

$$\forall \mathbf{x}_i, \mathbf{x}_j \in X,\ i \neq j:\ \ \mathbf{x}_i \neq \mathbf{x}_j \Longleftrightarrow \exists\,(x_{ih} \in \mathbf{x}_i) \neq \left(x_{jh} \in \mathbf{x}_j\right),\ h \in \{1, 2, \ldots, m\}.$$

5. Let the $(X, \leq)$ be a poset and the ordered m-tuples $\mathbf{x}_i$ and $\mathbf{x}_j$ elements of the set X. If $x_{ih} \leq x_{jh}$ ($x_{jh} \leq x_{ih}$) is true for all pairs $\left\{x_{ih} \in \mathbf{x}_i, x_{jh} \in \mathbf{x}_j\right\}$ ($h = 1, 2, \ldots, m$), then $\mathbf{x}_i \leq \mathbf{x}_j$ ($\mathbf{x}_j \leq \mathbf{x}_i$) holds. In this case, $\mathbf{x}_i$ and $\mathbf{x}_j$ are called "comparable" (denoted by $\mathbf{x}_i \perp \mathbf{x}_j$). Otherwise, they are called "incomparable" (denoted by $\mathbf{x}_i || \mathbf{x}_j$).

   Using this definition, the question whether two m-tuples $\mathbf{x}_i$ and $\mathbf{x}_j$ are incomparable or not, obviously can only be answered with yes or no. Regarding this definition, the incomparability between two elements or objects can have only one of two states: "'1 ≡ incomparable/not comparable" or "0 ≡ not incomparable/comparable."

2 Incomparability/Inequality Measures

However, the experience obtained during the consideration of the real world shows that a detailed graduation of inequality and incomparability exists (also reflected in linguistics) and should be mathematically described.

Therefore, in a previous publication (Bartel and Mucha 2014), measures have been proposed regarding the incomparability, inequality, and combination of both. To describe these measures, we require the following numbers.

To define them, let us pairwise compare the elements of the ordered m-tuples $\mathbf{x}_i$ and $\mathbf{x}_j$: $(x_{ih} \in \mathbf{x}_i)\, \pi_\tau \left(x_{jh} \in \mathbf{x}_j\right)$ ($h = 1, 2, \ldots, m$). Here, π_τ ($\pi_\tau \in \{=, <, >\}$) denotes the result of the comparison. Therefore, the result is either $x_{ih} = x_{jh}$, $x_{ih} < x_{jh}$, or $x_{ih} > x_{jh}$, and three numbers can be determined with respect to the comparison of $\mathbf{x}_i$ with $\mathbf{x}_j$:

- $q_{ij}^{=}$, the number of the results of the comparison, in which is $x_{ih} = x_{jh}$
- $q_{ij}^{<}$, the number of the results of the comparison, in which is $x_{ih} < x_{jh}$
- $q_{ij}^{>}$, the number of the results of the comparison, in which is $x_{ih} > x_{jh}$.

Corresponding to the comparison of $\mathbf{x}_j$ with $\mathbf{x}_i$, the following relations hold for the numbers $q_{ji}^{=}$, $q_{ji}^{<}$, and $q_{ji}^{>}$:

$q_{ji}^{=} = q_{ij}^{=}$, $q_{ji}^{<} = q_{ij}^{>}$, $q_{ji}^{>} = q_{ij}^{<}$, $q_{ii}^{=} = m$, $q_{ii}^{<} = q_{ii}^{>} = 0$, and $q_{ij}^{=} + q_{ij}^{<} + q_{ij}^{>} = m$. ($m$ is the dimension of $\mathbf{x}_i$ or $\mathbf{x}_j$, respectively: $m = |\mathbf{x}_i| = |\mathbf{x}_j|$.) If it is clear which comparison is considered, these three numbers can be denoted by $q^{=}$, $q^{<}$, and $q^{>}$, respectively. So, for instance, $q^{=} + q^{<} + q^{>} = m$ can be written in general.

The three measures introduced in (Bartel and Mucha 2014) are:

(A) Measure of Incomparability $u_{ij} = u\left(\mathbf{x}_i, \mathbf{x}_j\right)$ between the m-tuples $\mathbf{x}_i$ and $\mathbf{x}_j$

$$u_{ij} = \min\left(q_{ij}^{<}, q_{ij}^{>}\right) \cdot \left[\frac{m}{2}\right]^{-1}. \tag{1}$$

(B) Measure of Inequality $v_{ij} = v\left(\mathbf{x}_i, \mathbf{x}_j\right)$ between the m-tuples $\mathbf{x}_i$ and $\mathbf{x}_j$

$$v_{ij} = \frac{m - q_{ij}^{=}}{m} = \frac{q_{ij}^{<} + q_{ij}^{>}}{m}. \tag{2}$$

(C) Distance of Combined Inequality and Incomparability $a_{ij} = a\left(\mathbf{x}_i, \mathbf{x}_j\right)$ between the m-tuples $\mathbf{x}_i$ and $\mathbf{x}_j$

$$a_{ij} = \tfrac{1}{2}\left(u_{ij} + v_{ij}\right). \quad \text{(DIST)} \tag{3}$$

In addition, a new distance measure is recommended:

(D) Distance of Combined v_{ij}-/(1$-v_{ij}$)-Weighted Incomparability and Inequality $z_{ij} = z\left(\mathbf{x}_i, \mathbf{x}_j\right)$ between the m-tuples $\mathbf{x}_i$ and $\mathbf{x}_j$

$$z_{ij} = u_{ij} w_{ij}^{(u)} + v_{ij} w_{ij}^{(v)}, \quad \text{(WDIST)} \tag{4}$$

where u_{ij} and v_{ij} are the measure of incomparability (A) and inequality (B), respectively. $w_{ij}^{(u)} = \frac{q_{ij}^{<} + q_{ij}^{>}}{m} = v_{ij}$ and $w_{ij}^{(v)} = \frac{q_{ij}^{=}}{m} = 1 - v_{ij}$ are the corresponding weights.

DIST and WDIST are special cases of the distance $z_{ij}^{\text{gen}} = u_{ij} h_{ij}^{(u)} + v_{ij} h_{ij}^{(v)}$ with generalized weighted incomparability and inequality and with $h_{ij}^{(u)} + h_{ij}^{(v)} = 1$: (C) $h_{ij}^{(u)} = h_{ij}^{(v)} = \frac{1}{2}$, (D) $h_{ij}^{(u)} = w_{ij}^{(u)}$, $h_{ij}^{(v)} = w_{ij}^{(v)}$.

The main advantage of our four proposed distance measures is that they are entirely independent of the scale of measure. This is also important from the practical point of view where usually the attributes are measured in different scales (see the application to archaeometry in Sect. 4.2). In addition, our distance measures are not affected by outliers.

3 Partitional Clustering

Usually, hierarchical cluster analysis is used when a distance/proximity matrix is given (Späth 1980). In this publication however, we consider partitional clustering based on the pairwise incomparability/inequality distance measures. The advantage of partitional clustering is often a better solution concerning the corresponding (minimum) criterion. However, there is the well-known disadvantage of iterative partitional clustering: the solution depends on the choice of the initial partition. Therefore, many different initial partitions are used to find the best partition concerning the criterion.

For the decomposition of the set G of $g = |G|$ objects in a partition of K non-empty classes (clusters) C_k, $k = 1, 2, \ldots, K$, a method of partitioning cluster analysis is used, namely Helmut Späth's exchange method TIHEXM (Späth 1985), see also (Bartel et al. 2003). This procedure appears particularly suitable, since no restrictions on the nature of the distance function exist (Bartel and Mucha 2014). Concretely, the criterionConcretely, the criterion

$$V_K = \sum_{k=1}^{K} \sum_{i \in C_k} \sum_{\substack{l \in C_k \\ l > i}} d_{il} \tag{5}$$

has to be minimized based on pairwise distances d_{il} such as defined in (1) to (4). In the language of graph theory, this method is described in detail for example in Bartel and Mucha (2014).

As a small example, the same data matrix is used as in Bartel and Mucha (2011, 2014), which was taken from the publication of Lars Carlsen and Rainer Bruggemann (2011). The 10×3 data matrix $\mathbf{X}_{\mathrm{CaBr}}(G, M)$ is considered as given in Table 1.

Figure 1 compares graphically three cluster analysis results coming from the use of the following different distance measures DIST (3), WDIST (4), and EUCLID^2. The latter stands for the squared Euclidean distance. Obviously, there are small differences of the partitions into $K = 2$ and $K = 3$ clusters in the case of DIST and WDIST. Here, only the object D_nt (at the bottom of the Hasse diagram) changes its cluster membership. On the other hand, the results based on EUCLID^2 look quite different.

Table 1 Data matrix $\mathbf{X}_{\mathrm{CaBr}}(G, M)$, taken from Carlsen and Bruggemann (2011: 126)

Name of chemical	ID	Volat	Sedim	Advec
Phenanthrene	A_ph	3	2	4
Pyrene	A_py	3	3	4
Fluoranthene	A_fl	2	3	4
Chloroform	C_ch	4	1	2
Tetrachlormethane	C_tt	4	1	3
Trichlorethene	C_tr	4	2	2
Tetrachlorethene ("Per")	C_pe	3	2	3
Atrazine	D_at	1	2	4
Nitrilotriacetic acid	D_nt	1	1	1
EDTA	D_ed	1	1	3

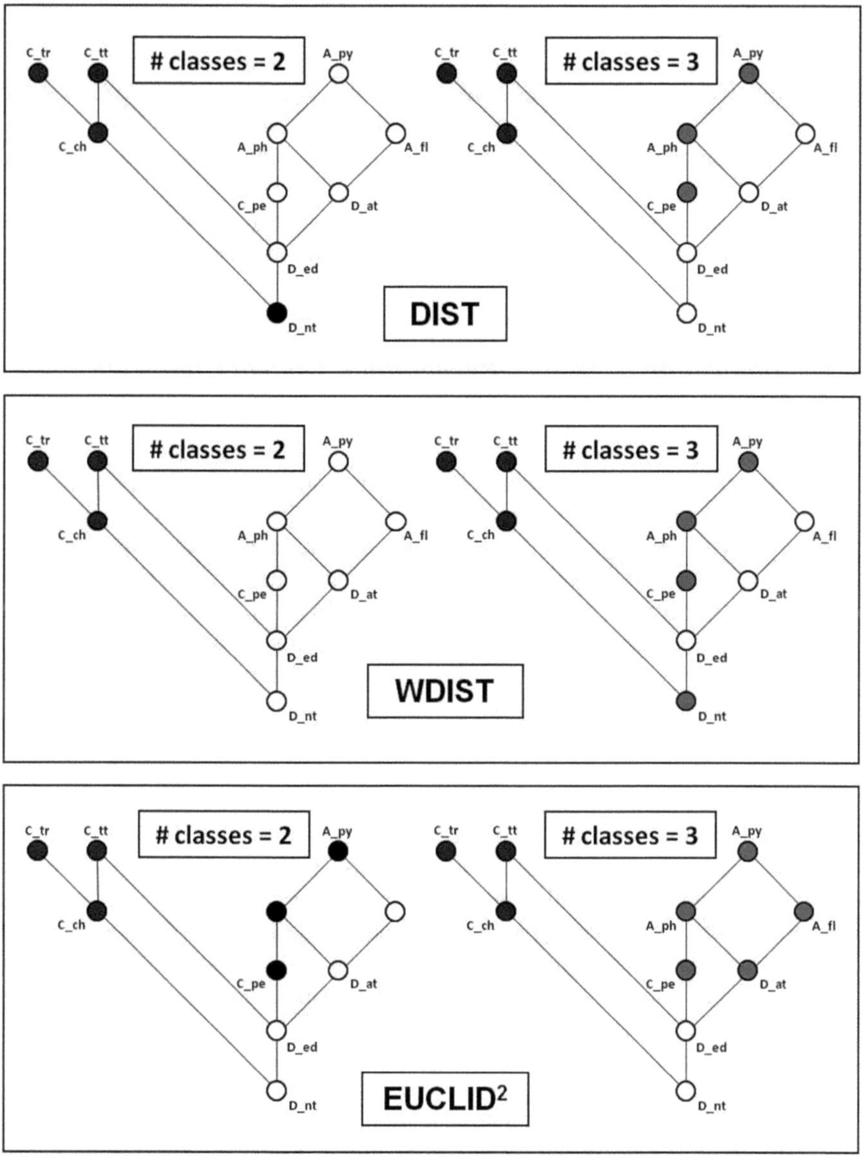

Fig. 1 The clusters found and their visualization by the Hasse diagram corresponding to the data matrix $\mathbf{X}_{\mathrm{CaBr}}$ using DIST (Bartel and Mucha 2014), WDIST, or the squared Euclidean (EUCLID^2) distances, respectively

Because of the low number of objects, and in particular the small number of attributes, a further interpretation of the results is not worthwhile.

4 Two Examples

In the following, we demonstrate the proposed clustering (5) in applications to different fields, namely pattern recognition and archaeometry. In the first, the performance of the method can be shown in comparison with a well-known statistical machine learning method. The latter concerns with a real application based on ancient archaeological artifacts. The aim is to demonstrate the applicability of our proposed distance measures used by the partitional graph clustering technique (5).

4.1 Application to Pattern Recognition: Optical Character Recognition (OCR)

The basis are normalized bitmaps of handwritten digits from a pre-printed form. From a total of 43 people, 30 contributed to the training set and different 13 ones to the test set. The resulting 32×32 bitmaps are divided into non-overlapping blocks of 4×4, and the number of on pixels is counted in each block. This generates an "aggregated bitmap" of 8×8, where each element is an integer in the range $0, \ldots, 16$. This aggregation reduces the dimensionality from 1024 ($=32 \times 32$) to 64 variables ($=8 \times 8$) and gives invariance to small distortions. The UCI web site (University of California, Irvine) is:

http://archive.ics.uci.edu/ml/datasets/Optical+Recognition+of+Handwritten+Digits

In case of such data, our incomparability/inequality measures seem to be not the best choice because we lose most of the quantitative information. First, without loss of generality, let us consider the cluster analysis of the digits "**0**" and "**1**" of the training set, where the true class membership is known. So, we are able to assess the performance of clustering based on our proposed distance measures. Figure 2 shows a visualization of the first observation of digit "**0**" and digit "**1**" in the dataset, respectively. Here, the area of the squares represents the counts of the corresponding 8×8 matrix of blocks.

Figure 3 shows a heat map of the ordered (765×765) distance matrix DIST (3) of the OCR dataset of the handwritten digits "**0**" and "**1**". Great distances were marked by the colors green to yellow, small ones by blue to red. Obviously, the region at the lower right corner reflects small within-cluster distances of the group of "**0**" digits. Thus, the group of digits "**0**" seems to be more homogeneous than the one of digits "**1**".

The confusion matrix of Table 2 summarizes the result of partitional clustering (5) using DIST (3). The error rate is really quite low. A similar good performance ($3+2=5$ errors) is obtained when using WDIST (4). For comparison, both the simple Bayesian classifier and the dual scaling classification (Mucha 2002) count only

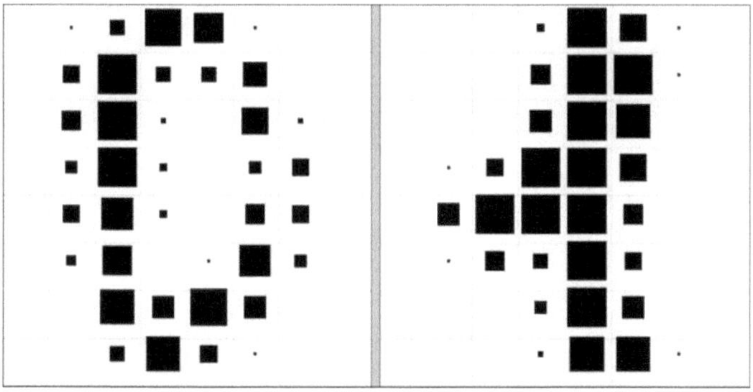

Fig. 2 OCR training data: example of the digit "**0**" and the digit "1", respectively, as a visualization of counts of (8 × 8) blocks of aggregated pixels. The square of each block represents an integer value ranging from 0 (*empty square*) to 16 (*full black square*)

one error: one digit "**0**" comes into the wrong class, for details see Mucha (2014). However, only the classification methods make use of additional information: they take the true class membership into account when estimating the classifiers. This is an important advantage over the proposed "pure" clustering method (5), and this "assistance" should be taken into account when the misclassification rates are compared, as done in Table 2.

Second, let's consider a more difficult OCR-task, the cluster analysis of the digits "**3**" and "**9**" of the training set. Figure 4 shows a visualization of the first observation of digit "**3**" and digit "**9**" in the dataset, respectively. The confusion matrix of Table 3 summarizes the result of partitional clustering using WDIST. Here, the error rate of 9.99 % is much higher compared with the simple Bayesian classifier (3.24 %) and the dual scaling classification (4.15 %), see (Mucha et al. 2014). Obviously, the main reason for the difference between the error rates is that the classification methods take into account additional information about the true class membership. And, obviously, the more difficult the classification task is the more important is the usage of this important a priori knowledge. In our clustering method (5), however, we make no use of such important information.

The partitional clustering (5) based on the distance measure DIST (3) performs similar to WDIST (4): Here, $38 + 42 = 80$ errors were counted. Using the measure of incomparability (1) or inequality (2), the cluster analysis results in $39 + 43 = 82$ errors or $40 + 44 = 84$ errors, respectively.

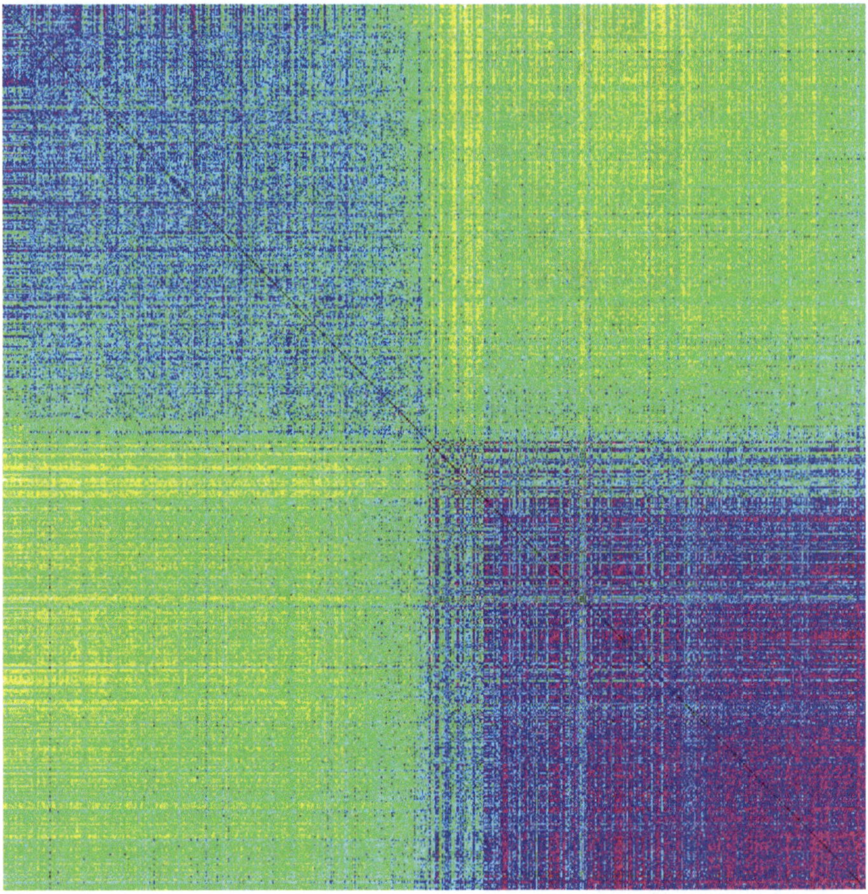

Fig. 3 OCR data: Heatplot of the ordered distance matrix DIST (3) of the digits of the two groups "digit 1" (first 376 rows and columns) and "digit 0" (last 389 rows and columns)

4.2 Application to Archaeometry: Chemical Composition of Egyptian Bronzes

Josef Riederer measured the chemical composition of 76 samples of ancient Egyptian bronze statuettes using atomic absorption spectroscopy analysis (AAS) (Riederer 1992). This is a real application of the proposed methods to archaeometry. Here, we consider only six out of the published eleven elements as attributes: Cu, Sn, Pb, Fe, Sb, and As. The reason for this is that too many measurements are below the detection thresholds of the corresponding elements.

In order to compare our distance measures (DIST, WDIST) with the well-known (squared) Euclidean ($EUCLID^2$) distance, it is necessary to transform the original data. That is because the Euclidean distance is not independent on scales of the

Table 2 OCR data of digits "**0**" and "**1**": Comparison of the results of partitional cluster analysis (rows) with the true class membership by error rates

Counts	True classes (digits)		
Cluster	0	1	Total
1	3	388	391
2	373	1	374
Total	376	389	765

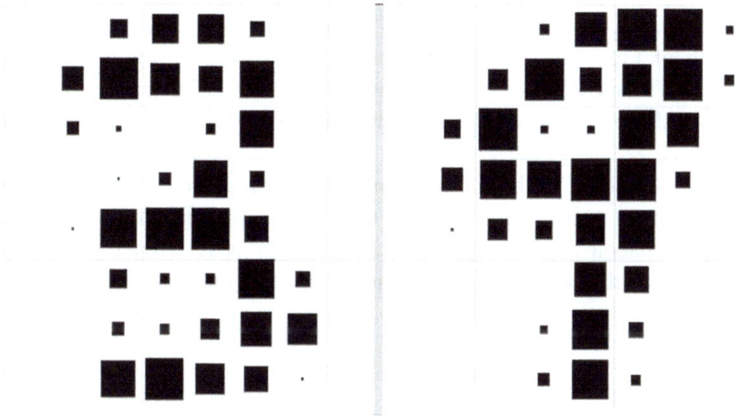

Fig. 4 OCR training data: example of the digit "**3**" and the digit "**9**", respectively, as a visualization of counts of (8×8) blocks of aggregated pixels

Table 3 OCR data of digits "**3**" and "**9**": Comparison of the results of partitional cluster analysis (rows) based on WDIST (4) with the true class membership by error rates

Counts	True classes (digits)		
Cluster	3	9	Total
1	37	342	379
2	352	40	392
Total	389	382	771

attributes. Without loss of generality, here the transformation of data values into ranks is chosen. The main advantage of our four proposed distance measures is scale independence. As already mentioned above, our proposed measures DIST and WDIST are independent on scales. Moreover, our distance measures are not affected by outliers. From a practical point of view, this is also an important advantage.

This transformation into ranks is quite simple: the measurements are replaced by their corresponding ranks $1, 2, \ldots, g$, where g is the number of objects. The mean of each of the new rank order variables becomes the same: $(g + 1)/2$. Moreover, the variance of each of the new variables becomes the same: $(g^2–1)/12$. In case of multiple values, we recommend to average the corresponding ranks. For further details concerning rank analysis in archaeometry, see Mucha et al. (2008).

Figure 5 visualizes several results of the partitional clustering method (5). Concretely, we compare the clustering into $K = 2$ clusters based on EUCLID^2, DIST, and WDIST (Fig. 5a–c). In addition, the clustering into $K = 3$ clusters based on WDIST is shown in Fig. 5d.

Figure 6 visualizes the cluster means of the clustering into $K = 2$ clusters based on EUCLID^2, DIST, and WDIST. Finally, Table 4 summarizes the visual comparison of Fig. 5 for these partitions by the corresponding confusion tables. In detail, DIST and WDIST give similar results (on the left hand side), whereas EUCLID^2 is different to both DIST (in the middle) and WDIST (on the right hand side).

Perhaps, the reason for the compositional differences (see Fig. 6) is the existence of two deposits of copper ore in Egypt. Concerning further details and additional interesting archaeological interpretation, the reader is referred to results of our ongoing research that will be published in Bartel and Mucha (2016).

5 Conclusion

In this chapter, we propose a generalization of the distance measure of incomparability/inequality that was originally introduced by Bartel and Mucha (2014). The main advantage of both proposed distance measures of incomparability/inequality is that they are entirely independent of the scale of measure. This is important especially from the practical point of view where usually the attributes are measured in different scales such as in the application to archaeometry. In addition, the proposed distance measures are not affected by outliers.

On the basis of these proposed pairwise distance measures, a partitional graph clustering technique is applied to real datasets of different fields, namely pattern recognition and archaeometry. In the first, our partitional graph clustering was compared with well-known statistical machine learning methods such as the simple Bayesian classifier. The performance of clustering is similar: the (true but unknown) classes were identified to a high degree. In clustering, only incomparability/inequality information is used. So, the good performance of cluster analysis for finding the true classes is really surprising because only the classification methods make use of the true class membership when building the classifier.

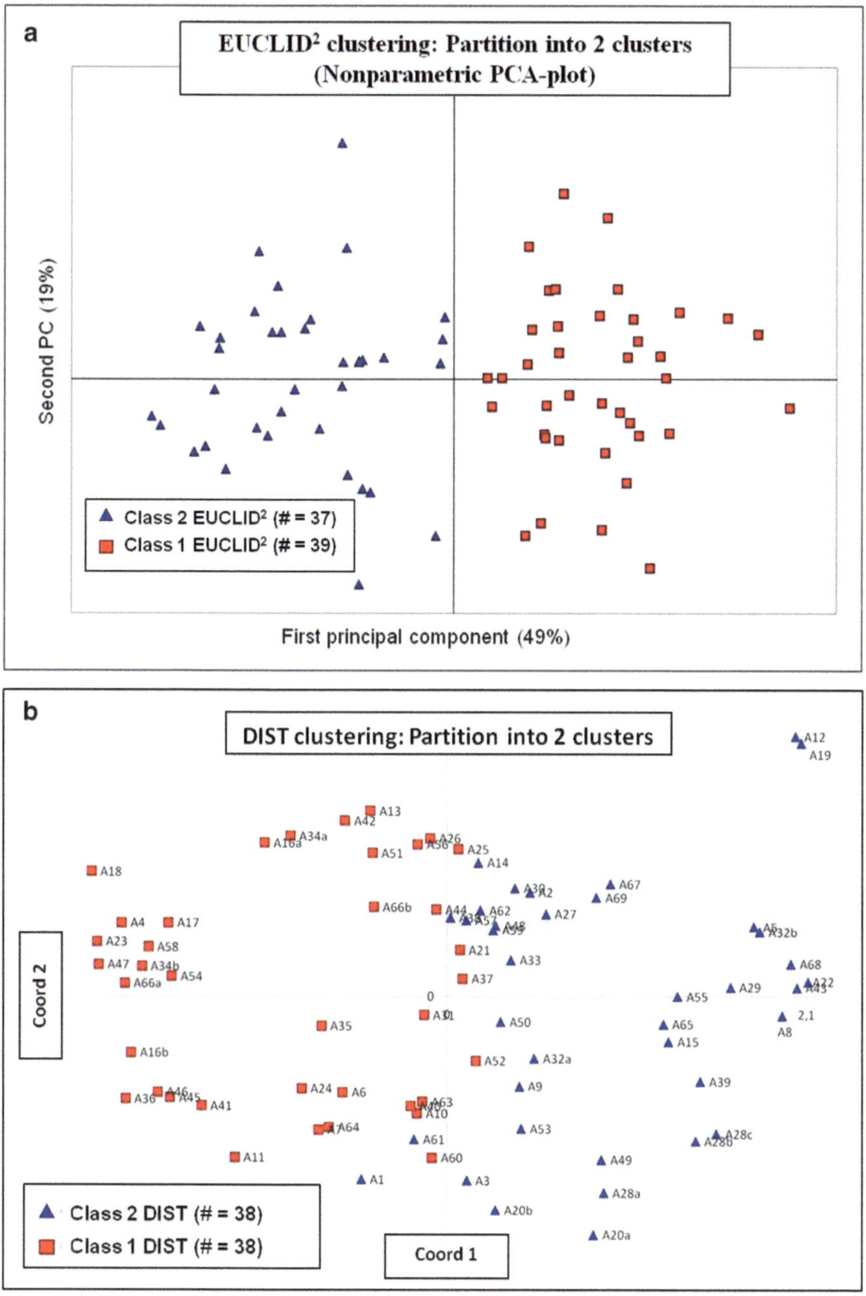

Fig. 5 (**a**) Scatterplot: visualization of the clustering result that comes from principal components analysis. (**b**) Visualization of the clustering result based on the measure DIST. This projection comes from the multidimensional scaling of the corresponding distance matrix

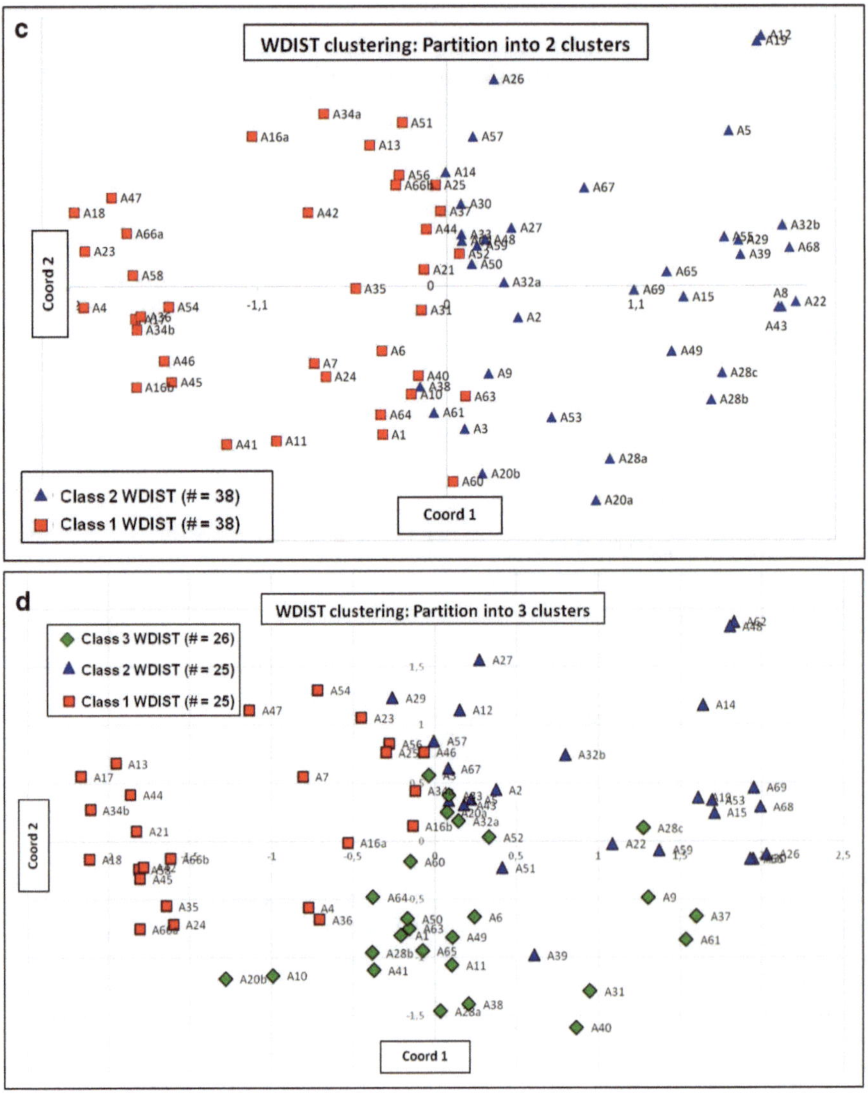

Fig. 5 (continued) (**c**) Visualization of the clustering result based on the measure WDIST. This projection comes from the multidimensional scaling of the corresponding distance matrix. (**d**) Visualization of the three clusters solution based on the measure WDIST. The projection is the same as in Fig. 5c

Acknowledgement Dedicated to the 100th birthday of Fritz Hintze.[1]

[1]The Egyptologist Fritz Hintze (18 April 1915 to 30 March 1993) was Professor in Egyptology at Humboldt University Berlin, "*his principal interest lay in the study of the Meroitic civilization and*

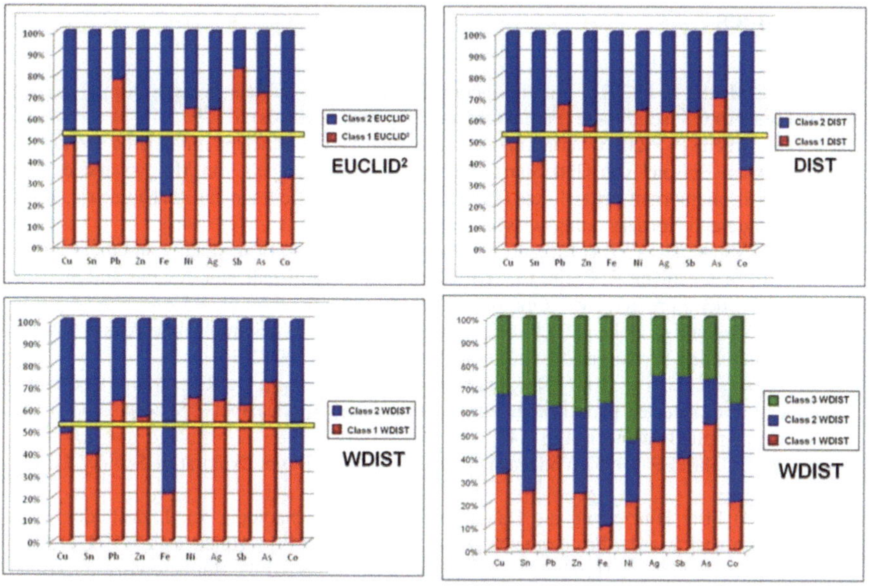

Fig. 6 Visualization of compositional differences of clusters by cluster means

Table 4 Confusion tables of the partitional clustering results based on different measures of distance

Counts	WDIST		
DIST	C1	C2	Total
C1	37	1	38
C2	1	37	38
Total	38	38	76

Counts	WDIST		
EUCLID^2	C1	C2	Total
C1	32	7	39
C2	6	31	37
Total	38	38	76

Counts	DIST		
EUCLID^2	C1	C2	Total
C1	32	7	39
C2	6	31	37
Total	38	38	76

the application of mathematical methods, seriation, and cluster analysis in archaeology" (Dawson and Uphill 1995). The Coptologist Martin Krause characterizes his colleague in a similar manner: "[...] *his later interests on Meroitic studies, the archaeology of the Sudan, and the application of mathematical and statistical analysis to the study of language and archaeological material, are well known.*" (Krause 2010: 69), and the mathematician Peter Ihm emphasized: "[Regarding the use of methods of mathematical statistics and computer science in Egyptian, Meroitic, Nubian philology and archaeology, Fritz Hintze may] *auf seinem wissenschaftlichen Gebiet als einer der Pioniere ansehen werden* (*be viewed as one of the pioneers on his scientific field.*)" (Ihm 2003: 39).

References

Bartel H-G, Mucha H-J (2011) Finding incomparable pairs of subsets by using formal concept. Anal Stat Appl Special Issue (Partial Orders in Applied Sciences):61–79

Bartel H-G, Mucha H-J (2016) Seriation und Clusteranalyse von Objekten mit binären Merkmalen, erläutert am Beispiel von Götterdarstellungen des Löwentempels von Musawwarat es Sufra (Sudan)–In memoriam Fritz Hintze (1915–1993). Berliner Beiträge zur Archäometrie, Kunsttechnologie und Konservierungswissenschaft 24 (in press)

Bartel H-G, Mucha H-J (2014) Measures of incomparability and of inequality and their applications. In: Bruggemann R, Carlsen L, Wittmann J (eds) Multi-indicator systems and modeling in partial order. Springer, New York, pp 47–67

Bartel H-G, Mucha H-J, Dolata J (2003) Über eine Modifikation eines graphentheoretisch basierten partitionierenden Verfahrens der Clusteranalyse. Match 48:209–223

Carlsen L, Bruggemann R (2011) Risk assessment of chemicals in the River Main (Germany) – application of selected partial order ranking tools. Stat Applicazioni – Special Issue (Partial Orders in Applied Sciences):125–140

Dawson WR, Uphill EP (1995) Who was who in Egyptology? Egypt Exploration Society, London, p. 203

Ihm P (2003) Fritz Hintze und die Datenanalyse in der Archäologie. In: Endesfelder E (ed) Von Berlin nach Meroe: Erinnerungen an den Ägyptologen Fritz Hintze (1915–1993) (Asien- und Afrika-Studien der Humboldt-Universität zu Berlin, Band 3). Harrassowitz, Wiesbaden, pp 38–53

Krause M (2010) Coptic texts from western Thebes – recovery and publication from the late nineteenth century to the present. In: Gabra G, Takla HN (eds) Christianity and monasticism in Upper Egypt, Volume 2: Nag Hammadi – Esna. The American University in Cairo Press, Cairo, pp 63–78

Mucha H-J (2002) An intelligent clustering technique based on dual scaling. In: Nishisato S, Baba Y, Bozdogan H, Kanefuji K (eds) Measurement and multivariate analysis. Springer, Tokyo, pp 37–46

Mucha H-J (2014) Classification, clustering, and visualisation based on dual scaling. In: Gaul W, Geyer-Schulz A, Baba Y, Okada A (eds) German-Japanese interchange of data analysis results. Springer, Cham, pp 59–69

Mucha H-J, Bartel H-G, Dolata J (2008) Effects of data transformation on cluster analysis of archaeometric data. In: Preisnach C, Burkhardt H, Schmidt-Thieme L, Decker R (eds) Data analysis, machine learning and applications. Springer, Berlin, pp 681–688

Mucha H-J, Bartel H-G, Dolata J (2014) Dual scaling classification and its application to archaeometry. In: Spiliopoulou M, Schmidt-Thieme L, Janning R (eds) Data analysis, machine learning and knowledge discovery. Springer, Cham, pp 105–113

Riederer J (1992) Metallanalyse der Bronzestatuetten. In: Schoske S, Wildung D (eds) Gott und Götter im alten Ägypten. Zabern, Mainz, pp 223–232

Späth H (1980) Cluster analysis algorithms for data reduction and classification of objects. E. Horwood, Chichester, p 15

Späth H (1985) Cluster dissection and analysis: theory, FORTRAN programs, examples. E. Horwood, Chichester, pp 84–88, 132–135

Incomparable: What Now, IV. Incomparabilities: A Modeling Challenge

Rainer Bruggemann, Lars Carlsen, and Paola Annoni

1 Introduction

An important argument pro MCDA—analysis, as for example by ELECTRE III (Roy 1990), PROMETHEE (Brans and Vincke 1985), TOPSIS (Peters and Zelewski 2007), or TODIM (Gomes and Rangel 2009) is that the introduction of additional parameters allows for getting a ranking. A classical example is the popular MCDA method ELECTRE. In ELECTRE, parameters describe indifferences, significances, vetos and weights for each, single indicator. In this way, they allow for a tailor-made design of the problem. The partial order-based alternative that employs the Hasse diagram technique (HDT) (Bruggemann and Patil 2011) misses a priori such individual design of the decision problem. Tools of HDT are examined as to how far they are helpful in modeling the knowledge of stakeholders or decision makers. In this study, we present one of these tools, namely the weight-intervals method. A practical example from the field of environmental chemistry is used to exemplify the proposed method.

R. Bruggemann (✉)
Department of Ecohydrology, Leibniz—Institute of Freshwater Ecology and Inland Fisheries, Müggelseedamm 310, D-12587 Berlin, Germany
e-mail: brg_home@web.de

L. Carlsen
Awareness Center, Linkøpingvej 35, Trekroner, DK-4000 Roskilde, Denmark

P. Annoni
Economic Analysis Unit, European Commission-Directorate General for Regional and Urban Policy, Avenue de Beaulieu, 5, 1160 Brussels, Belgium

M. Fattore, R. Bruggemann (eds.), *Partial Order Concepts in Applied Sciences*,
DOI 10.1007/978-3-319-45421-4_3

2 Material and Methods

2.1 The Case Study Data Set

The case study is based on environmental chemistry data. The main goal is estimating the hazard these chemicals pose to the environment. In general, neither a measured quantity, nor a mathematical simulation model is available to uniquely order chemicals according to their hazard. Instead, a series of indicators are used as proxy and, in combination, may give a picture of the hazard of the chemical (Fig. 1). For example, the hazard exerted by substances which are persistent (P), bioaccumulating (B), and toxic (T) can be described by three indicators, associated with the P, B, and T characteristics (see e.g., Sailaukhanuly et al. 2013). A system of proxy indicators like this is generally called a Multi-Indicator System—MIS.

By inspection of Fig. 1 some basic information can be derived.

- There are subsets of chemicals which can be uniquely ordered (i.e., can be ranked), e.g., LIN < PCN < DIE < HCL < CHL or PCB < HCB < ALD.
- There are chemicals which cannot be given a mutual rank, such as, e.g., LIN and PCP or DDT, ALD.
- The graph in Fig. 1 is a Hasse diagram and—contrary to rankings—it has a vertical *as well as* a horizontal evolution. The vertical evolution helps to identify rankings; the horizontal one helps to identify contradictions expressed by the indicator values that are called incomparabilities. For example, ALD has a larger value in P, but smaller values in B and T compared to DDT.

2.2 MCDA and Partial Order, Some General Remarks

Most MCDA aims at the derivation of a single ranking index. Such a ranking allows for identifying the best or worst object in a given context and even provides with enough flexibility if the object selected fails to fulfill possible additional constraints. However, the additional parameters requested are typically rather subjective and may easily be questioned. Furthermore, a mathematical mapping to get a ranking index is not only hiding all the valuable information of an MIS but also implies:

- Unwanted compensation across indicators (Munda 2008).
- A rickety ranking, especially when the aggregation is done by a weighted sum and when the indicators are describing different important aspects of the same phenomenon (see Annoni et al. 2015 for an example based on poverty measures).

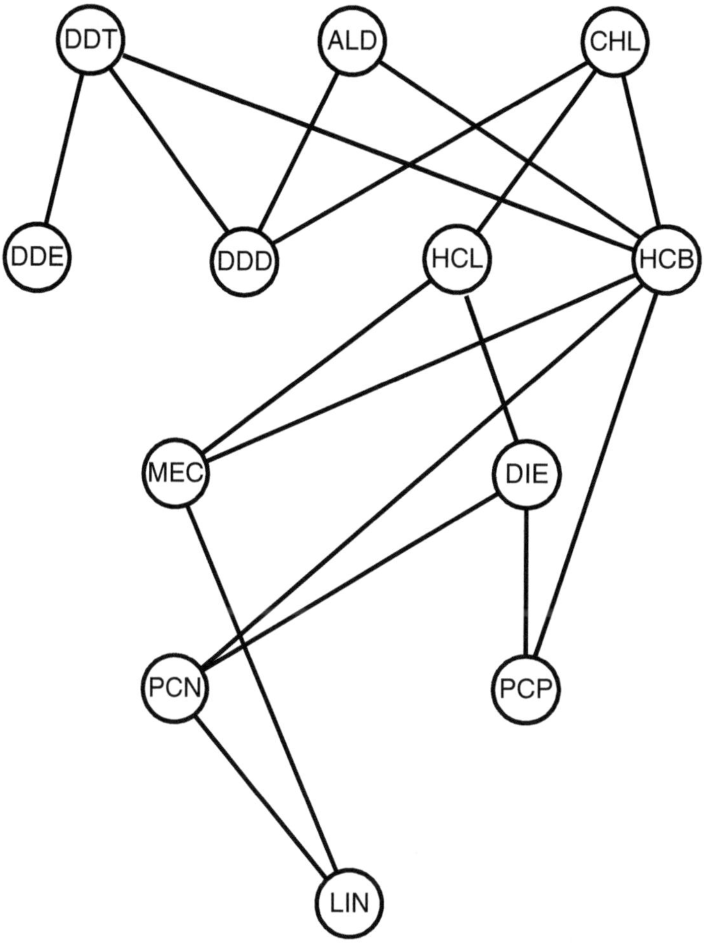

Fig. 1 Simultaneous use of all three indicators, P, B, and T. A partial order results, being visualized by a Hasse diagram (software: PyHasse)

2.3 Basics of Hasse Diagram Technique (HDT)

Some basic elements of partial order tools are given in this section. For a deeper insight, please refer for example to Bruggemann and Patil (2011); Bruggemann and Carlsen (2012); Carlsen and Bruggemann (2014); or Carlsen et al. (2015).

2.3.1 HDT

Let X be the finite set of objects, and IB the set of indicators $q_j, j = 1, \ldots, |\mathrm{IB}|$ (IB: information base), we then define

$$x,\, y \in X : x \leq y :\Longleftrightarrow q_j(x) \leq q_j(y) \text{ for all } q_j \in \mathrm{IB}, \tag{1}$$

where $q_j(x)$ is the (observed/measured/estimated) value of indicator q_j on object x.

Equation (1) represents a partial order, P, which is often denoted as $P = (X, IB)$ to indicate that it consists of the object set X and the set of indicators observed on X, IB. The set X equipped with the partial order structure is called a partially ordered set (poset). Methods based on (1) are often denoted as Hasse diagram techniques (HDT).

When $q_j(x) = q_j(y)$, for all $q_j \in$ IB, x and y are called equivalent objects, denoted as $x \cong y$. With the introduction of the equivalence relation, E, the object set X gets an additional relational structure, namely a partitioning into equivalence classes, Ec. The set of equivalence classes is called a quotient set of X, denoted X/E. Instead of dealing with all elements of X, only one element of each equivalence class is selected, representing all the other equivalent elements in its class. These selected elements are called representative elements, and the set of representative elements is denoted as X_{repr}.

A subset of X (or of X_{repr}), where no pair x, y satisfies (1) is called an antichain. Such pairs, where "x is incomparable with y" are denoted as $x||y$. When analyzing (1), which is a rather strict requirement, a series of points call for attention:

1. In a visual representation of a partial order, e.g., a Hasse diagram, only the strict order $x < y$ is seen. Equivalence classes are kept as background knowledge and may be brought into play in cases the application demands for a complete view on the object set.
2. Partial order will depend on even minor numerical differences among the indicator values. This effect is called "numerical noise." A possible way to lower the effect of noise is provided by ELECTRE with the introduction of indifference levels. In HDT, other concepts are developed (Bruggemann and Carlsen 2016; Wieland and Bruggemann 2013; Carlsen and Bruggemann 2016).
3. Any additional indicator may lead to the increase of incomparabilities because each indicator implies a weak order, and the simultaneous consideration of all indicators is a concatenation of the weak orders. Algebraically, the concatenation is a successive "and"—operator for each order relation $x <_{q_1} y$ and $x <_{q_2} y$ with q_1 and q_2 any two single indicators. As more indicators are considered, the less probable is that the "and" operator for all indicators leads to a comparability. Therefore, a large number of indicators typically lead to a complete antichain, which can be regarded as "information noise." This aspect will not be discussed further in this chapter (see Carlsen (2008) or Annoni et al. (2012) for a sensitivity analysis on the indicators).
4. Assume that the MIS consists of k indicators and that $x(q_j) > y(q_j)$ for $j = 1$, $\ldots k-1$ indicators, whereas for the kth indicator $x(q_k) < y(q_k)$. If k is large, then one may think of x as "generally" larger than y. However, (1) signals the presence of a conflict making the analyst aware that there is one indicator in conflict with the remaining $k-1$. This problem is referred to as the "plurality

problem." It should be noted that there are two other typical problems, i.e., (a) the noise problem, where minor numerical differences influence the structure of the partially ordered set and (b) the "information noise-problem," where too many indicators lead to a high degree of incomparabilities (*see above*).

Partial order methodology pinpoints the issue and further action can be taken for instance to elucidate the actual importance of the *k*th indicator. Partial order methodology gives then the opportunity to judge if the observed incomparability is important or may be neglected (Bruggemann and Voigt 2012; Bartel and Mucha 2014).

2.3.2 Incomparabilities

An important quantity within HDT is the number of incomparabilities, U. The quantity U measures the deviation from a linear order (see Sect. 2.3.3).

$$U = |\{x, y \in X, \ \text{with}\ x \ || \ y\}| \tag{2}$$

When U is calculated from the representative elements, we write U(repr) and (2) is to be applied on X_{repr}. If representative elements represent other elements of X, because they are taken from nontrivial equivalence classes, then we can write $x(e_x)$ to indicate that x represents e_x elements, including x itself. Let x,y be a pair of representative elements for which x y. Then, there are $e_x \times e_y$ incomparable relations on the basis of the object set, obtained from a single incomparable pair. Thus, U(obj), the number of incomparabilities on the basis of the object set is calculated as according to (3) here below, whereas *U(repr)* is a simple count of incomparabilities visualized in a Hasse diagram (4):

$$U\,(\text{obj}) = \sum_{x||y} e_x \cdot e_y \quad x, y \in X_{repr} \tag{3}$$

$$U\,(\text{repr}) = \sum_{x||y} 1 \quad x, y \in X_{repr} \tag{4}$$

When the focus is on data pattern then U(repr) is more important than U(obj). However, when the balance between ranking and incomparability is of interest, then the quantity U(obj) is more informative.

2.3.3 Enrichment

Let $(X, \leq_1)$ be a poset, then a map $(X, \leq_1) \rightarrow (X, \leq_2)$ is called an enrichment (written as $\dashrightarrow_{\text{enrichment}}$), when all order relations of $(X, \leq_1)$ are preserved in $(X, \leq_2)$, and the number of incomparabilities in $(X, \leq_2)$ is equal or lower than in $(X, \leq_1)$ (5). Further, the number of comparabilities may be higher. The main objective of this chapter is to show that enrichment within the framework of HDT is possible and gives HDT the needed flexibility for modeling.

$$(X, \leq_1) \dashrightarrow_{\text{enrichment}} (X, \leq_2) \Longleftrightarrow U(\text{repr})_1 \geq U(\text{repr})_2 \tag{5}$$

The concept of composite indicators is used here to illustrate the enrichment procedure (Bruggemann et al. 2013). Another concept to enrich partial orders is the concept of fuzzy partial order (see for instance Kerber and Bruggemann 2015; Bruggemann et al. 2011).

2.4 *Software*

The calculations described in this chapter are performed using the software PyHasse (Bruggemann et al. 2014). The module HDweightsMC of PyHasse allows including weight intervals (Bruggemann et al. 2013). Recently, an Internet-based version has been launched (see www.pyhasse.org).

3 Incorporation of Knowledge About Weights in the Partial Order Concept

3.1 *Theoretical Basis*

The basic idea is that the weights of the individual indicators are generally unknown. Weight intervals can instead be defined and may be more easily accepted. The introduction of weight intervals instead of point values can be seen as a sort of a relaxation of the decision process.

A ranking index I can then be calculated from a weighted sum:

$$I(x) = \sum w_j \cdot q_j(x), \tag{6}$$

where w_j is a random value in the interval IV(w_j), $j = 1, \ldots k$. The weights w_j are assumed to be [0,1]-normalized:

$$\sum w_j = 1 \tag{7}$$

Let now w be the k-dimensional vector of weights, where all the w_j are selected from IV(w_j). Any point in the space of weights is associated to a vector I with components corresponding to x $\in X$. The vector I is used to derive a linear or weak order (X, $\leq_I$). Selecting all possible $w \in \prod \text{IV}(w_j)$ will generate an infinite set of $I(w)$. Nevertheless, the concatenation:

$$\left(X, \leq_{\prod I(w)}\right) = \bigcap_{w \in \prod \mathrm{IV}(wj)} (X, \leq_w) \tag{8}$$

is the final order, which is in general a partial order. The partial order obtained from (8) is in general an enriched order in comparison to that obtained by (1). To avoid the cumbersome notation of (8), the resulting order is simply denoted as $(X, \leq_w)$.

As it is not possible to perform an infinite number of operations as required by (8), a series of Monte Carlo (MC) runs is carried out. The ideal number of MC runs is the one that assures stable results (convergence). In general, slight changes in w will not change the partial order, as detailed in Bruggemann et al. (2008). This means that generally a reasonable number of MC runs is necessary to reach convergence. Up to now, it is not possible to estimate the specific number of Monte Carlo runs necessary to obtain convergence. The only technique to decide whether or not the number of Monte Carlo runs is sufficient is to test the stability of the results with increasing number of MC runs.

3.2 A Characterizing Quantity

For comparison purposes the volume, $V(k)$, of the k-dimensional weight space is introduced (9). Large uncertainties in the selection of weights may lead to large intervals for the weights $\mathrm{IV}(w_j)$.

$$V(k) := \prod_{j=1}^{k} \left(\max\left(w_j\right) - \min\left(w_j\right) \right) \tag{9}$$

The quantities $\max(w_j)$ and $\min(w_j)$ are the realized maximal and minimal values of the weights for the jth indicator within MC simulations. If $V(k) = 1$, then any weight is considered as possible and U(repr) gets its maximum U(repr, max). On the contrary, if $V(k) = 0$ then either the dimension of the weight space is reduced or the weights are exactly determined and the result is a composite indicator, that is a weak or linear order. For each MIS, a plot can be set up with its U(repr) on the y-axis and $V(k)$ on the x-axis. The curve is starting in (0,0). Connecting the points ($V(k)$, U(repr)) indicates the effect of increasing uncertainty in weights $V(k)$ on posets described by U(repr). Therein, U(repr) is calculated directly from the corresponding poset (4). In contrast to the direct calculation, an estimation of U(repr), denoted as U(pred, repr) is suggested (10):

$$U\,(\text{pred, repr}) = U\,(\text{repr, max}) \cdot (V(k))^p \tag{10}$$

In (10) p is an unknown real number. For a first screening, we set:

$$p = 1/k \tag{11}$$

A power law such as in (10) describes, what we want to call a "normal" behavior: Increasing the uncertainty in selection of weights, increases $V(k)$ and consequently the number of incomparabilities is increasing. Therefore, deviations of U(pred,repr) from U(repr) may indicate peculiarities with respect to the actual weight combination.

3.3 Stability with Respect to Weight Intervals

For a certain weight interval, the composite indicators obtained from the Monte Carlo simulation can be on the one side very different, which causes a high value of U(repr) on the other side they can be similar to each other, i.e., exhibiting only few conflicts. In this latter case, the number of incomparabilities would be small. Comparing the results applying different weight intervals, a partial order with a low U(repr) can be denoted as stable with respect to variations of the weights. However, as U(repr) increases different weight samples taken from the intervals lead to composite indicators whose partial order comprises a high number of incomparabilities. With respect to the weights, such partial orders are considered as unstable.

4 Example

4.1 Inputs

We select the following input scenarios well aware that the selection bears some arbitrariness.

Table 1 shows the input scenarios with the weight intervals IV(w_j) employed for each scenario (note that in principle any interval taken from [0,1] can be selected for any of the indicators).[1]

Table 1 Weight intervals: Different cases

Case	Pers	BioA	Tox
1	[0, 0.25]	[0.25, 0.65]	[0.65, 1]
2	[0.25, 0.65]	[0, 0.25	[0.65, 1]
3	[0.25, 0.65]	[0.65, 1]	[0, 0.25]

MC runs: 1000

[1]In the present example, the single weight intervals are not overlapping but this is not a general requirement.

4.2 Results

In Fig. 2 are the Hasse diagrams shown, corresponding to the three runs.

Equation (10) can now be applied. The parameters are $k = 3$ and $U(\max) = 31$. Data for $V(3)$ are shown in Table 2. The weight volume is not strongly varying; therefore, $V(3) = 0.02$ is selected for all three cases.

$$U\,(\text{pred}, \text{repr}) = 31 \cdot (V(3))^{1/3} = 31 \cdot (0.02)^{1/3} = 8.415$$

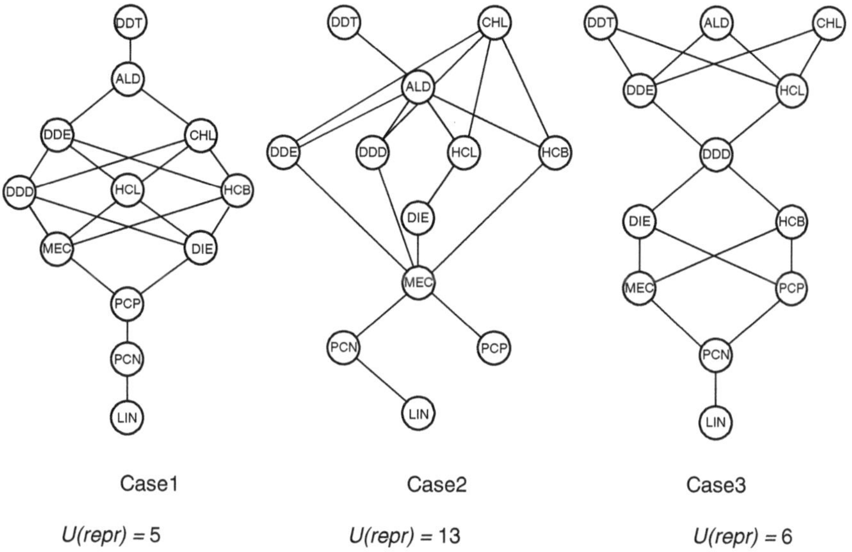

Fig. 2 Hasse diagrams of all three runs with 1000 MC runs

Table 2 Results of applying module HDweightsMC of PyHasse

	Case 1	Case 2	Case 3
	Persistence	BioA	Tox
min P	0.0	0.175	0.176
mean P	0.086	0.32	0.323.
max P	0.261	0.481	0.471
min B	0.178	0.001	0.437
mean B	0.319	0.0.087	0.592
max B	0.491	0.21	0.786
minT	0.447	0.43	0.0
meanT	0.54	0.594	0.085
maxT	0.772	0.783	0.206
Volume of the weight space $V(3)$	0.0205	0.02265	0.02112
Number of incomparabilities	5	13	8

Table 3 Tests of some other weight combinations

	Case 4: large interval for persistence	Case 5: large interval for bioaccumulation	Case 6: large interval for toxicity
Actual IV(w_i)	[0, 0.6]	[0, 0.6]	[0, 0.6]
The other two IV(w_i)	[0.6, 1]	[0.6, 1]	[0.6, 1]
$V(3)$	0.031	0.03	0.03
U(repr)	5	11	7

In case 3, the small weight interval is associated with indicator Tox. The associated value for U(repr) is in comparison to the other two cases the nearest to the one obtained by (10), i.e., the U(repr,pred)—value. We then call case 3 as "normal" as it is following an equation, where no specificities of the partially ordered set are included. In case 1 (small weight interval for Indicator Persistence), U(repr) is overestimated by (10) and in case 2 (weight interval for BioA small) the value U(repr) is underestimated. Obviously, the weight combination in case 2 (BioA gets the narrow range for the weights ∈ [0, 0.25]) leads to more incomparabilities than expected, which could be interpreted as some kind of overall instability of the ranking with respect to the weights, although just the weight interval referring to the weights for indicator BioA is selected as narrow.

Can the same conclusions be reached with wider weight intervals? To answer this question and, consequently, to assess the usefulness of (10), different weight intervals are tested (Table 3):

In all three cases, it can be assumed that $V(3) \approx 0.03$. Applying (10) the expected value of U, U(pred, repr), is 9.6. The value of U for case 5 is the closest to that obtained by (10), showing that with the new sets of weight intervals the indicator bioaccumulation represents the normal behavior.

As U in case 5 is the largest among cases 4, 5, and 6, the "normal" behavior does not exclude that a relative high instability with respect to a ranking may prevail. The weight combination of case 4, where the indicator persistence gets a large weight interval, seems to be pretty stable. In case 6, with a large weight interval for Tox, a slightly more unstable ranking is obtained. However, in both cases U values are below the expectation value U(repr) showing that there is, at least with the realized weight intervals, a surplus on stability with respect to the associated weight combinations.

5 Conclusions

5.1 *What Do We Gain?*

According to the subjectivity in finding exact weight values, we gain more generality by only requesting intervals for the weights. Thus, instead of fixed numbers for selecting weights, weight intervals are the input parameters.

In general, it will probably be easier to agree about some weight intervals instead of fixed weights. We cannot expect to solve all controversies which may arise from the selection of weights. However, the proposed approach is a way to model experiences of stakeholders and decision makers within the framework of partial order theory.

As U(repr) is at its minimum when the weights for the persistence are small, we can conclude that the indicator P (persistence) induces most incomparabilities followed by indicators toxicity and then bioaccumulation. The predominance of persistence is in agreement with a sensitivity study that showed that this indicator was the most important one (Sailaukhanuly et al. 2013). Using weight intervals, we are still left with an (enriched) partial order, i.e., not a linear or weak order can be obtained. The advantage is that there is no need to exactly determine the weight values.

5.2 A Possible Framework for a Systematic Analysis of Parameter-Dependent Posets

The leading quantity is U, mostly U(repr), gives an insight into the degree of enrichment of the poset. Any manipulation, which enriches the partial order can be controlled by U(repr). The number of parameters which govern the enrichment can be very different. In the case of fuzzy posets, it is just one parameter, the "defuzzification" parameter α. In the case of weight intervals, the parameters are the lower and upper limits of the weight intervals, in total $2 \times k$ parameters. In other enrichment procedures, the number of parameters could be k, as for example in the indicator-wise cluster analysis (publication in preparation).

As mentioned above, the realized weight intervals may differ from those originally selected by the analyst. Still, $2 \times k$ parameters remain to be controlled. By the introduction of the weight volume $V(k)$, an aggregated quantity is found and a useful relation between U(pred, repr) and $V(k)$ can be established (10). Indeed, such power laws seem to be valid in other methods too. Hence, it is a good approximation to describe U(pred, repr) in fuzzy partial order by a similar power law, and a power law can be established in the indicator-wise clustering too.

5.3 Future Work

The starting point of the analysis is the question as to how far decisions can be modeled in the framework of HDT. The crucial concept is the partial order incomparability. The analysis leads to the conclusion that parameter-dependent posets can be described by the following steps:

1. Can the decision problem be modeled such that a data matrix and weights for the indicators are suitable? Then:
2. Find a mapping from the set of parameters (specific for the method) to an aggregated quantity.
3. Analyze U as a function of this aggregated quantity. Establish the power law from a theoretically and empirical point of view and
4. Find expressions for the exponent of the power law.

   In the example of weight intervals, the parameters are:

   (a) Lower and upper limits for weights are mapped to the quantity $V(k)$
   (b) $U = U\max \times V(k)^{s}$.
   (c) s seems to be pretty well described by $(1/k)$.

References

Annoni P, Bruggemann R, Saltelli A (2012) Random and quasi-random designs in variance-based sensitivity analysis for partially ordered sets. Reliab Eng Sys Safety 107:184–189

Annoni P, Bruggemann R, Carlsen L (2015) A multidimensional view on poverty in the European Union by partial order theory. J Appl Stat 42:535–554

Bartel H-G, Mucha H-J (2014) Measures of incomparabilities and of inequality and their application. In: Bruggemann R, Carlsen L, Wittmann J (eds) Multi-indicator systems and modelling in partial order. Springer, New York, pp 47–67

Brans JP, Vincke PH (1985) A preference ranking organisation method (The PROMETHEE Method for Multiple Criteria Decision - Making). Manag Sci 31:647–656

Bruggemann R, Carlsen L (2012) Multi-criteria decision analyses. Viewing MCDA in terms of both process and aggregation methods: some thoughts, motivated by the paper of Huang, Keisler and Linkov. Sci Total Environ 425:293–295

Bruggemann R, Carlsen L (2016) An attempt to Understand Noisy Posets. MATCH Commun. Math. Comput. Chem. 75:485–510

Bruggemann R, Patil GP (2011) Ranking and prioritization for multi-indicator systems - introduction to partial order applications. Springer, New York

Bruggemann R, Voigt K (2012) Antichains in partial order, example: pollution in a German region by Lead, Cadmium Zinc and Sulfur in the herb layer. MATCH Commun Math Comput Chem 67:731–744

Bruggemann R, Voigt K, Restrepo G, Simon U (2008) The concept of stability fields and hot spots in ranking of environmental chemicals. Environ Model Softw 23:1000–1012

Bruggemann R, Kerber A, Restrepo G (2011) Ranking objects using fuzzy orders, with an application to refrigerants. MATCH Commun Math Comput Chem 66:581–603

Bruggemann R, Restrepo G, Voigt K, Annoni P (2013) Weighting intervals and ranking. Exemplified by leaching potential of pesticides. MATCH Commun Math Comput Chem 69:413–432

Bruggemann R, Carlsen L, Voigt K, Wieland R (2014) PyHasse software for partial order analysis: scientific background and description of selected modules. In: Bruggemann R, Carlsen L, Wittmann J (eds) Multi-indicator systems and modelling in partial order. Springer, New York, pp 389–423

Carlsen L (2008) Hierarchical partial order ranking. Environ Pollut 155:247–253

Carlsen L, Bruggemann R (2014) The ‘Failed Nations Index’ offers more than just a simple ranking. Soc Indic Res 115:525–530

Carlsen L, Bruggemann R (2016) On the influence of data noise and uncertainty on ordering of objects, described by a multi-indicator system. A set of pesticides as an exemplary case. J Chemom 30(1):22–29
Carlsen L, Bruggemann R, Kenessova O, Erzhigitov E (2015) Evaluation of analytical performance based on partial order methodology. Talanta 132:285–293
Gomes LFAM, Rangel LAD (2009) An application of the TODIM method to the multicriteria rental evaluation of residential properties. Eur J Oper Res 193:204–211
Kerber A, Bruggemann R (2015) Problem driven evaluation of chemical compounds and its exploration. Match Commun Math Comput Chem 73:577–618
Munda G (2008) Social multi-criteria evaluation for a sustainable economy. Springer, Berlin
Peters ML, Zelewski S (2007) TOPSIS als technik zur effizienzanalyse. WiSt Jnuar 2007:9–15
Roy B (1990) The outranking approach and the foundations of the ELECTRE methods. In: Bana e Costa CA (ed) Readings in multiple criteria decision aid. Springer, Berlin, pp 155–183
Sailaukhanuly Y, Zhakupbekova A, Amutova F, Carlsen L (2013) On the ranking of chemicals based on their PBT characteristics: comparison of different ranking methodologies using selected POPs as an illustrative example. Chemosphere 90:112–117
Wieland R, Bruggemann R (2013) Hasse diagram technique and Monte Carlo simulations. MATCH Commun Math Comput Chem 70:45–59

Partial Ordering and Metrology Analyzing Analytical Performance

Lars Carlsen and Rainer Bruggemann

1 Introduction

Development of analytical performance may be a quite troublesome task as a variety of parameters, such as extraction time and temperature, as well as flow rate and temperature in the gas chromatograph, choice of detector, etc. a priori may influence the result. Obviously, a systematic variation of the influential parameters is the basis. However, as a variety of parameters are involved it is not straightforward to analyze the resulting data.

Obviously, one of the indicators to elucidate the analytical performance is the "distance" from the true value, as for example, available from certified standard material. However, it is also immediately clear that evaluating the analytical performance based on this indicator alone will not give a true and fair view on the single analytical methods. Thus, we want to include further indicators as measures for analytical performance. In the present example, we use the z-score, z (JCGM 200 2008; Lærd 2013), the standard deviation of the measurements, s, and the skewness of the data, skew, respectively.

The z-score of the optimal method is by definition being 0 (1). The z-score denotes the standard score (Lærd 2013) or accuracy (JCGM 200, 2008; Stone and Ellis 2009) of the measurements, which enables a comparison between two scores from different normal distributions

$$z = (m - \mu_a)/s_a. \tag{1}$$

L. Carlsen (✉)
Awareness Center, Linkøpingvej 35, Trekroner, 4000 Roskilde, Denmark
e-mail: LC@AwarenessCenter.dk

R. Bruggemann
Department of Ecohydrology, Leibniz—Institute of Freshwater Ecology and Inland Fisheries, Müggelseedamm 310, 12587 Berlin-Friedrichshagen, Germany

M. Fattore, R. Bruggemann (eds.), *Partial Order Concepts in Applied Sciences*,
DOI 10.1007/978-3-319-45421-4_4

In (1) m is the obtained value, i.e., concentration from the analytical method and μ_a and s_a is the true (or accepted) value (mean) and expected (or wanted) standard deviation of the optimal analytical method, respectively. The latter is in the present study set to 0.1.

A further possible indicator to bring into play to elucidate the analytical performance is the skewness of the distribution of the analyses based on the single methods studied (Taylor 2013). This indicator tells about the shape of the distribution and thus to what extent the analyses are in analytical or statistical control (ASTM 2013).

Each indicator (in the present case 3) bears information of its own and a priori—as often seen—the three indicators may be aggregated into a single scalar for handling ranking purposes. Various approaches for aggregating the indicators have been reported (Figueira et al. 2005; Hajkowicz and Higgins 2008). However, such aggregation of the indicators may unequivocally lead to compensation effects where high values in one indicator may compensate low values in another (Munda 2008).

Taking all three indicators into account simultaneously, we face a multi-indicator system (MIS) that advantageously can be analyzed applying partial order methodology also known as Hasse diagram technique (HDT) (Bruggemann and Carlsen 2006; Bruggemann and Patil 2011) as the visual representation of a partial order often is performed by the so-called Hasse diagram.

2 Methods

This chapter describes how selected partial order tools may be applied in the evaluation of analytical performance and thus constituting an advantageous tool in selecting an optimal analytical method, metx, $x = 1,2,3\ \ldots$, taking several indicators simultaneously into account as an alternative to conventional methods to study MIS (Bruggemann and Carlsen 2012).

2.1 *The Basic Equation of Hasse Diagram Technique*

In partial ordering, the only mathematical relation among the objects is "$\leq$" (Bruggemann and Carlsen 2006a; Bruggemann and Patil 2011; Bruggemann and Carlsen 2006b). The "$\leq$"-relation is the basis for a comparison of objects which are characterized by the three indicators z, s, and skewness. Here, the analytical approaches are the objects of our study. A method metx is comparable to a method mety if and only if the relation metx $\leq$ mety holds. The crucial question is, under which conditions it can be claimed that metx $\leq$ mety. Here, a system is described by z, s, and skew, our indicators, which we call for the sake of simplicity r_i ($i = 1,2,3$). This series of indicators r_j, constitutes our MIS. Thus, metx characterized by the a set of indicators r_j(metx), $j = 1, \ldots, m$ can be compared to mety, characterized by the indicators $r_j(y)$, when

$$r_i\,(\text{metx}) \leq r_i\,(\text{mety}) \;\; \text{for all}\; i = 1, \ldots, m \tag{2}$$

Equation (2) is a very hard and strict requirement for establishing a comparison. It demands that all indicators of metx should be better (or at least equal) than those of mety. Further, let X be the group of methods studied, i.e., $X = \{\text{met1,met2, } \ldots \text{ metn}\}$, metx will be ordered lower (better) than mety, i.e., metx < mety, if at least one of the indicator values for metx is lower than the corresponding indicator value for mety and no indicator for metx is higher than the corresponding indicator value for mety. On the other hand, if $r_j(\text{metx}) < r_j(\text{mety})$ for some indicator j and $r_i(\text{metx}) > r_i(\text{mety})$ for some other indicator i, metx and mety will be called incomparable (notation: metx || mety) expressing the mathematical contradiction due to conflicting indicator values. A set of mutual incomparable objects is called an antichain. When all indicator values for metx are equal to the corresponding indicator values for mety, i.e., $r_j(\text{metx}) = r_j(\text{mety})$ for all j, the two methods will have identical rank and will be considered as equivalent, i.e., metx mety. The analysis of (2) results in a graph, the Hasse diagram. Hasse diagrams are unique visualizations of the order relations due to (2).

2.2 *The Hasse Diagram*

The equation (2) is the basic for the Hasse diagram technique (HDT) (Bruggemann and Carlsen 2006a; Bruggemann and Patil 2011; Bruggemann and Carlsen 2006b). The Hasse diagram is a visual representation of the partial order. In the Hasse diagram, comparable objects are connected by a sequence of lines (Bruggemann and Carlsen 2006b; Bruggemann and Münzer 1993; Bruggemann and Voigt 1995, 2008). For construction of the Hasse diagram, a uniform orientation of the indicators should be secured, i.e., high indicator values correspond to "bad" objects and low values to "good" objects. Thus, in the present case the higher an analytical approach is placed in the Hasse diagram the farther away from the reference it is.

A Hasse diagram is characterized by its structure that comprises levels, chains, and antichains (cf. Fig. 1, Sect. 2.3).

- Chain: A subset $X' \subseteq X$, where all objects fulfill (2) is called a chain. A chain has a length, which is $|X'| - 1$. For objects within a chain, e.g., from the bottom to the top of the chain, all indicators are simultaneously non-decreasing.
- Antichain: A subset $X' \subseteq X$, where no object fulfill (2), i.e., all objects in X' are mutually incomparable, is called an antichain. Thus, for any two objects within an antichain there is a conflict in indicator values.
- Level: The horizontal arrangement of objects within a Hasse diagram. Levels per definition constitute an antichain, whereas the reverse not necessarily is true. By convention in cases where an object possess flexibility with regard to level placement, the object is placed at the highest possible level.

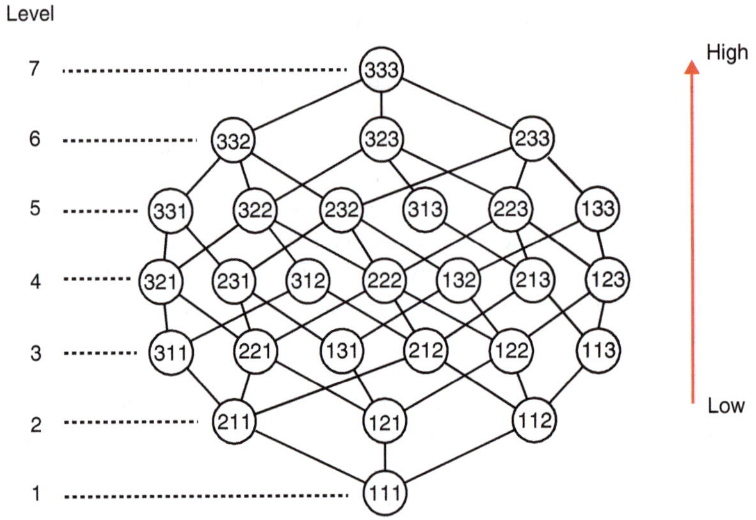

Fig. 1 Visualization (Hasse diagram) of a tool for a 3 × 3 system (event space)

2.3 A Simple Tool

Based on the above it appears rather simple to build a tool that, e.g., may find its application for preliminary studies on analytical performance. Here a system comprising three indicators, each of them possible adopting the values 1 (low), 2 (medium), or 3 (high), respectively, is described. Naming the single objects in this 3 × 3 system according to the values of their three indicators, the object 213 will correspond to an object that is evaluated medium for indicator 1, low for indicator 2 and high for indicator 3, respectively. Hence, this system will in total contain 27 possible objects, 111, 112, 113, 121, 122, 123, ..., 331, 332, 333. With reference to the above relation 1, these 27 possible objects can now be located in a Hasse diagram (Fig. 1). The diagram constitutes a template that subsequently can be applied in specific evaluations independent of the actual number of objects. The only requirement is: three indicators and each indicator takes the discrete values 1, 2, or 3. Sometimes, this kind of representation of a template is called an event space (Bruggemann and Voigt 2008; Carlsen and Bruggemann 2008).

The Hasse diagram (Fig. 1) is organized in a series of levels, in the present case in total 7 comprising 1, 3, 6, 7, 6, 3, and 1 objects, respectively.

Obviously, the idea behind the here shown system with three indicators and three possible evaluations (1, 2, 3) can immediately be transferred to systems with more/less indicators and/or more/less possible evaluations. This will unambiguously lead to additional/fewer elements in the different systems. Thus, a system based on three possible evaluations for four indicators will lead to a system with $3^4 = 81$ elements, whereas a system with four possible evaluations for three indicators contains $4^3 = 64$ elements. An application of the event space concept will be described in Sect. 3.1.

2.4 The More Elaborate Analyses

In the present study, a series of partial ordering tools has been applied for more elaborate analyses of analytical performance including average ranks (Bruggemann and Annoni 2014; Morton et al. 2009; Wienand 2006; De Loof et al. 2006; Lerche et al. 2003, Bruggemann et al. 2004; Bruggemann and Carlsen 2011), peculiarities of data profiles (Bruggemann and Carlsen 2014; Carlsen et al. 2015), scanning analysis (Bruggemann and Carlsen 2014), tripartite graphs (Bruggemann and Voigt 2011), and fuzzy partial order (Bruggemann and Carlsen 2015; Kosko 1992; Van de Walle et al. 1995; Bruggemann et al. 2011) and weight intervals (Bruggemann et al. 2013). For detail information on the single tool, the cited literature should be consulted as a detailed description is outside the scope of this chapter.

2.5 Data

In the present study, the absolute values of the z-score, $|z|$, and the skewness, $|\text{skew}|$, together with the standard deviation, s, are applied as indicators. As already mentioned, any possible triple of values of a certain analytical approach is consequently called r. Thus, each analytical approach is characterized by a data profile, i.e., an ordered set of indicator values (rn_1, rn_2, rn_3) corresponds to ($|z|$,s,$|\text{skew}|$), the indicators rn_i are normalized to a [0,1]-scale by

$$rn_i(x) = (r_i(x) - r_i\,\text{min}) \,/\, (r_i\,\text{max} - r_i\,\text{min}) \tag{3}$$

where $r_i$max and $r_i$min being the maximum, minimum value with respect to the objects. To facilitate the writing, we drop the differentiation r_i and rn_i and thus in the following use r_i for the normalized indicators.

To illustrate the study by partial order methodology, we apply a hypothetical dataset (Carlsen et al. 2015) of 20 different analytical approaches, and it is assumed that each of them has been used for "analyzing" an identical sample 5 times. The "results" of the analyses were randomly generated. Based on these five "measurements," three indicators, i.e., z-score, standard deviation, and skewness were estimated by conventional statistics (R Core Team 2015) and by (1) and are subsequently used for input in the partial order analyses (vide infra) to insight in the various analytical methods. The z-score and the skewness for the reference, REF, were per definition both set to 0 and the standard deviation were fixed to 0.1. In Table 1 the z-score, the standard deviation and the skewness derived from the various analytical approaches are given as well as the [0,1] standardized values and discretized indicator values in the analyses are given.

It is immediate noted (Table 1) that the various analytical approaches may lead to too low or too high z-scores as well as positively or negatively skewed data. However, the absolute values of the z-score and the skewness are applied as indicators for an absolute "distance" from the reference, without judging if positive or negative is good or bad, respectively.

Table 1 Summary statistics of the analytical approaches (methods) together with the corresponding [0,1] normalized and discretized indicators

	Original			[0,1] normalized			Discretized		
Method	*z*-score	s	skew	|*z*|	s	|skew|	|*z*|	*s*	|skew|
met1	−2.1	0.24	0.9	0.724	0.519	1.000	3	2	3
met2	0.9	0.25	−0.46	0.310	0.556	0.511	1	2	2
met3	1.4	0.28	0.28	0.483	0.667	0.311	2	3	1
met4	−0.3	0.22	0.01	0.103	0.444	0.011	1	2	1
met5	2.5	0.21	0.16	0.862	0.407	0.178	3	2	1
met6	−0.1	0.31	0.15	0.034	0.778	0.167	1	3	1
met7	1.2	0.37	−0.42	0.414	1.000	0.467	2	3	2
met8	−0.7	0.1	0.64	0.241	0.000	0.711	1	1	3
met9	−1.6	0.18	0.1	0.552	0.296	0.111	2	1	1
met10	1.2	0.24	0.7	0.414	0.519	0.778	2	2	3
met11	−0.3	0.31	0.21	0.103	0.778	0.233	1	3	1
met12	−0.4	0.2	−0.17	0.138	0.370	0.189	1	2	1
met13	0.4	0.3	−0.54	0.138	0.741	0.600	1	3	2
met14	2.6	0.19	−0.04	0.897	0.333	0.044	3	2	1
met15	1.4	0.19	0.34	0.483	0.333	0.378	2	2	2
met16	1.9	0.18	−0.89	0.655	0.296	0.989	2	1	3
met17	2.9	0.18	0.1	1.000	0.296	0.111	3	1	1
met18	0.9	0.22	0.41	0.310	0.444	0.456	1	2	2
met19	−0.1	0.37	0.07	0.034	1.000	0.078	1	3	1
met20	1.8	0.36	−0.76	0.621	0.963	0.844	2	3	3
REF	0	0.1	0	0.000	0.000	0.000	1	1	1

2.6 *Software*

All partial order analyses were carried out using the PyHasse software (Bruggemann et al. 2014). PyHasse is programmed using the interpreter language Python (version 2.6) (Ernesti and Kaiser 2008; Hetland 2005; Langtangen 2008; Weigend 2006; Python 2015) Today, the software package contains more than 100 modules and is available upon request from the developer, Dr. R.Bruggemann (brg_home@web.de). A simplified version of PyHasse operating on a web-based browser is under continuous development. Some few tools are now available (see www.pyhasse.org).

Statistical analyses in connection with the data generation were carried out applying the freely available software R (R Core Team 2015).

3 Results and Discussions

An often heard criticism of partial order ranking approaches is that the method is lacking of a unique or absolute ranking, i.e., the presence of incomparabilities. Thus, in a Hasse diagram incomparabilities are causing that the diagram (cf. Fig. 1) has not

Table 2 Average rank of the element at the single levels (from bottom to top) as calculated by LPOMext (Bruggemann and Carlsen 2011)

Level	Number of elements	Rank function	Average rank
1	1	1	1
2	3	2	3
3	6	3	8
4	7	4	14
5	6	5	20
6	3	6	25
7	1	7	27

only a vertical (ranking) but also a horizontal geometrical configuration. However, these incomparisons appear to be a source of important information as previously described (Bruggemann and Carlsen 2014, 2015; Carlsen et al. 2015).

In the following, we will first visualize the simple approach for analyzing analytical performance followed by the more elaborate analyses. Subsequently, the two methods will be compared.

3.1 The Simple Approach

As mentioned above, it is clear that the partially ordered objects represented by the event space concept (Fig. 1) cannot be linearly ordered/ranked. However, based on the level structure of the diagram (Fig. 1) it is possible to calculate an average order (rank) for the single objects (Bruggemann and Carlsen 2011). It should here be noted that objects being in the same level by default appear to have the same value of the rank function. These ranks, here calculated by the LPOMext method (Bruggemann and Carlsen 2011), can be used to get an approximate ranking of the 21 objects (Table 2).

Based on the discretized values (Table 1) the 21 objects, i.e., 20 methods and REF, can be placed in the Hasse template (Fig. 1). Obviously, some of the methods by this procedure turn out as equivalent. Thus, as an example both met2 and met18 are denoted as 122. In Table 3, the location of the 21 analytical approaches (methods) in the Hasse template is visualized (Fig. 2). It should be noted that only one representative for each equivalent class is shown, e.g., met2 represents both met2 and met18 (Fig. 2).

It is striking (Fig. 2) that in the realized cases, i.e., those case where the 21 methods are located in the event space, the patterns comprising a “1” for the indicator r_1 are found for 11 times, whereas in the not realized cases this appears only 2 times. Furthermore, a bad quality in both s and skew, i.e., $r_2 = 3$ and $r_3 = 3$, appears only 2 times in the realized cases but 4 times, excluding the pattern (3,3,3) in the non-realized cases. Finally, highly unbalanced cases, i.e., 1 or 3 in r_2 or r_3, appears 4 times in the realized cases and only 2 times in the non-realized cases. This indicates that there exists some potential for an unbalanced quality for the simulated methods studied.

Table 3 Location of the 21 analytical methods within the event space (Fig. 1)

Hasse template	Location of the methods
323	met1
122	met2, met18
231	met3
121	met4, met12
321	met5, met14
131	met6, met11, met19
232	met7
113	met8
211	met9
223	met10
132	met13
222	met15
213	met16
311	met17
233	met20
111	REF
221	
331	
112	
212	
312	
322	
332	
313	
123	
133	
333	

It is worthwhile to remember that this analysis is strongly dependent to the actual cases studied. However, the analysis may lead to valuable information to actual problems associated with the single methods.

3.2 *In-Depth Analyses*

We will now turn to the more elaborate, in depth analyses of the analytical performance of the 20 simulated methods compared to the reference, REF. Thus, for the remaining part of the study we will use the original [0,1] normalized indicators (Table 1). This leads as expected to a somewhat more complicated—at least less symmetric compared to the event space (Fig. 1). In Fig. 3, the Hasse diagram constructed based on the [0,1] normalized indicator values (Table 1) is shown.

Once again the level structure implies a preliminary rough ordering (Table 4). We will return to the average rank corresponding to the Hasse diagram (Fig. 3) below.

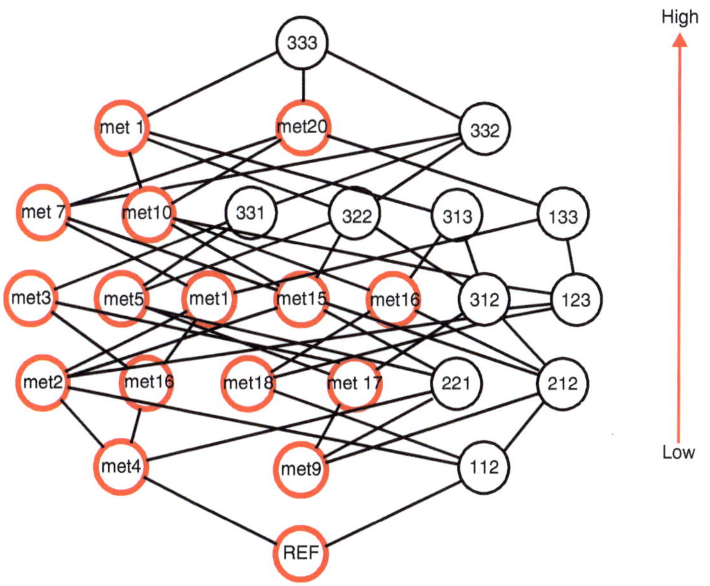

Fig. 2 Location of the 21 analytical approaches in the event space (Fig. 1)

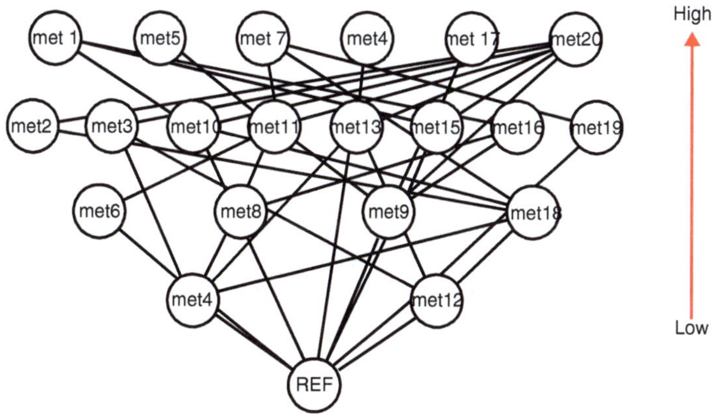

Fig. 3 Hasse diagram based on the [0,1] normalized indicator values (Table 1)

Table 4 Level structure (cf. Fig. 3)

Level	Level composition
5	met1 met5 met7 met14 met17 met20
4	met2 met3 met10 met11 met13 met15 met16 met19
3	met6 met8 met9 met18
2	met4 met12
1	REF

Table 5 Average ranking of the 21 analytical approaches based on the original dataset (Rkav Exact) and level structure of the event space (Fig. 2) (Rkav Simple)

Analytical method	Rkav Exact	Rkav Simple
met1	19.0	25
met20	19.8	25
met7	18.1	20
met10	14.0	20
met3	12.6	14
met5	13.6	14
met13	12.6	14
met14	11.5	14
met15	9.7	14
met16	13.4	14
met2	14.3	8
met6	6.7	8
met8	6.4	8
met11	12.5	8
met17	13.6	8
met18	8.8	8
met19	9.5	8
met4	4.2	3
met9	5.3	3
met12	4.4	3
REF	1.0	1

3.2.1 Average Ranks

As indicated above (Sect. 3.1) a typical request by decision makers is a linear ordering of the studied objects (here analytical methods). It is, however, clear, that taking several indicators into account simultaneous this is not immediately possible. Nevertheless, partial order methodology offers in this connection an attractive alternative, i.e., average ranking, e.g., based on the concept of linear extensions (Morton et al. 2009; Wienand 2006; De Loof et al. 2006; Lerche et al. 2003) or approximate methods (Bruggemann et al. 2004, Bruggemann and Carlsen 2011; Bruggemann and Annoni 2014).

The average rank of a metx is called Rkav(metx). The resulting ranking is in general a weak linear order as several of the methods may have identical Rkav-values, i.e., there are tied ranks. In Table 5, the average ranks are calculated based on the exact method (Morton et al. 2009; Wienand 2006; De Loof et al. 2006). The ranking due to Rkav-values can be compared to the ranking obtained based on the simple approach (cf. Sect. 3.1).

Not unexpectly we do not see a 100 % correlation between the two datasets. However, a correlation coefficient of 0.882 indicates a fairly good correlation and as a first approximation satisfying information concerning the mutual ordering (average ranking) of the analytical approaches, and as such also a first indication as to the distance of a given method from the reference.

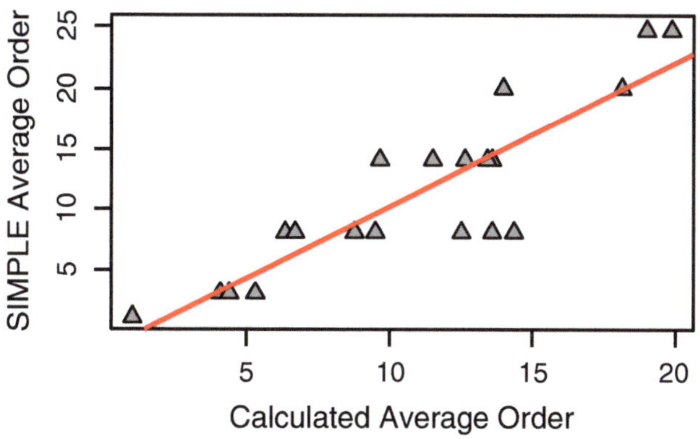

Fig. 4 Correlation between the average order calculated based on the original dataset and the average order based on the here presented simple tool (correlation coefficient: 0.882)

An important factor obviously is the high number of ties observed for the simple tool as all methods placed at the same level automatically is assigned the same average order. This is clearly noted in Fig. 4 where the 6 levels of the event space (Fig. 2) accommodating the 21 methods studied immediately can be recognized.

Remaining to be discussed is how well the average ranks obtained through the application of the simple tool presented in Sect. 3.1 mimic those calculated by the PyHasse software (Morton et al. 2009; Wienand 2006; De Loof et al. 2006). The correlation is depicted in Fig. 4.

3.2.2 A Closer Look at Incomparable Methods

The fact that even minute differences in indicator values may lead to incomparabilities potentially constitutes an obvious drawback of partial ordering. Hence, it may be appropriate to look only at numerically relevant discrepancies. For this purpose, a Δ-values is introduced as a cut-off, i.e., indicator-(metx, mety)-connections with absolute numerical differences below Δ are filtered out as differences below this value are considered as insignificant (Bruggemann and Carlsen 2014). Obviously, increasing Δ will result in a decreasing number of incomparabilities as an increasing number of analytical approaches consequently will be comparable. This is nicely illustrated in Table 6 summarizing the results for two values of Δ (out of a systematic variation of Δ from 0.1 to 1), i.e., a relatively low value ($\Delta = 0.35$) and a larger value ($\Delta = 0.6$), respectively. The Table should be read as follows using the second row (0.35, $|z|$, |skew|, 16) as an example: Here for 16 cases, the indicators $|z|$ and |skew| lead to incomparabilities, as the numerical values are $\geq$0.35.

A graphical representation of incomparabilities can be obtained through the so-called tripartite graphs (Bruggemann and Voigt 2011) (Fig. 5).

As an illustrative example how incomparabilities between single analytical approaches can be visualized two groups are selected from the top level of the Hasse

Table 6 Incomparabilities as function of the *Δ*-value

Δ	Indicator pair	Number of incomparabilities
0.35	|*z*|, *s*	29
0.35	|*z*|, |skew|	16
0.35	*s*, |skew|	18
Sum of realizations		63
0.6	|*z*|, *s*	3
0.6	|*z*|, |skew|	2
0.6	*s*, |skew|	2
Sum of realizations		7

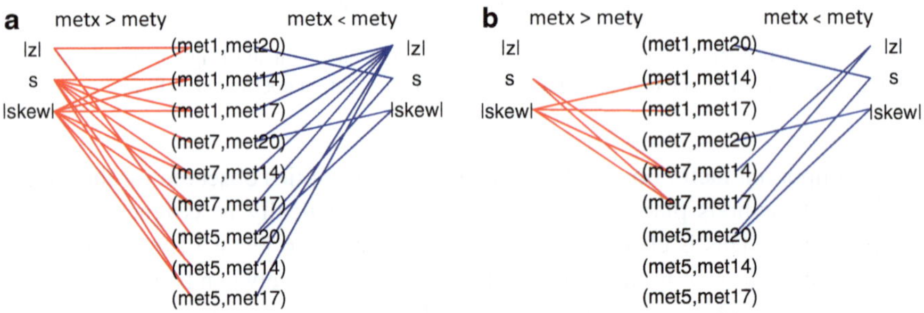

Fig. 5 (**a**) Tripartite graph of two groups of methods from the top level of the Hasse diagram (Fig. 3). (**b**) Tripartite graph, where all indicator—(metx, mety)—connections with absolute numerical differences below 0.35 are filtered out

diagram (Fig. 3). For each of these two groups, three methods are for demonstrative purposes arbitrarily selected. Thus, group 1: met1, met5, and met7; and group 2: met14, met17, met20, respectively. It should be emphasized that any other two groups, in which the methods apparently are not connected, could be selected for such a study. It is obvious (Fig. 3) that the methods of the first group are not connected with those of the second group. Thus, technically the two groups are "separated subsets" (Bruggemann and Patil 2011) and their separation can be explained by the concept of a tripartite graph (Bruggemann and Voigt 2011) (Fig. 5a).

In Fig. 5a, three analytical approaches on the one side (met1, met5, met7) and three approaches on the other side (met7, met14, met20) are displayed with the possible 3 × 3 pairs shown in the middle of the figure. Obviously, incomparability can be caused by different indicator pairs. Thus, (met1, met17) is incomparable due to conflicts in data with respect to (|*z*|,*s*) or (|*z*|,|skew|). To understand the role of indicators, it is of interest to trace back the role of |s| and |skew|. It can be seen (Fig. 5a) that |skew| (on the left side) is connected with seven pairs of methods. Thus, in seven cases met1, met7, met5 have larger values in |skew| than met17, met20, or met14. On the other hand, |*z*| or |*s*| have larger values for met17, met20, or met14. Furthermore, one can see that almost everywhere where |skew| is responsible for a larger value of the objects of the first subset also s is responsible. An exception is (met7, met20) where only s is responsible for the incomparability.

The analysis illustrated in Fig. 5a does not take into account that a relation metx > mety with respect to some indicators and metx < mety with respect to some others may numerically be irrelevant (cf. the above discussion). Hence, it appears appropriate to carry out the tripartite type analysis by filtering out values below a certain cut-off as insignificant. Hence, in Fig. 5b only those indicator pairs are taken into regard whose absolute numerical differences are larger than 0.35 (cf. Table 6). Typically, when a statistical analysis is performed, possible numerical noise would be disclosed and the limiting value, Δ, should be selected accordingly.

Due to the reduced number of incomparabilities through this filtering out procedure a far clearer diagram, which is easier to be interpreted is obtained. From the tripartite graph (Fig. 5b), it is immediately disclosed that now some incomparabilities can be resolved into comparabilities. Thus, the pair (met7, met20) is only connected with |skew| (on the right side of the tripartite graph). It can then be deduced that the incomparability met7 || met20 originally seen is caused by values less 0.35 in s and/or in $|z|$.

Similarly, the pair (met5, met17) is not connected at all with any of the indicators and is as such resolved as equivalent. Thus, the originally seen incomparability between those two analytical approaches was apparently caused by only slight and eventually irrelevant data differences. To round up, it appears clear that if we apply a cut-off of 0.35 only two pairs of methods apparently remains with a "relevant" (or significant) incomparability, i.e., (met7, met17) and (met7, met14).

3.2.3 Peculiar Methods

So far the Hasse diagram, shown in Fig. 3 reveals only the set of all comparisons among the 21 methods. In order to classify an additional object, it appears necessary to have some rules to perform such a classification. The most simple rule is where m single indicators have values in the closed interval [0,1] and to consider a strong proximity of the method to points of an m-dimensional hypercube $h(m)$. Here, $m = 3$ hence we have to analyze the $h(3)$ (cf. Fig. 6) as a classifier (Bruggemann and Carlsen 2014; Carlsen et al. 2015). The extreme points of $h(3)$ are of interest, such as (0,0,0) as the best one (here REF) or (1,0,0) as a point where the z-score is the worst possible, but skewness and standard deviations are as good as the values of the reference. Obviously in general, there are 2^m extreme points. In the present case with three indicators a classification will be performed by 2^3, i.e., by 8 extreme points (corresponding to the corners of a cube, Fig. 6).

We define methods that are found close to the corners (apart from (0,0,0) and (1,1,1) as peculiar, i.e., as deviating from the "main stream" (Bruggemann and Carlsen 2014), where the main stream of data is considered as a cloud of points (i.e., cloud of points representing the methods) in a 3D-space which is

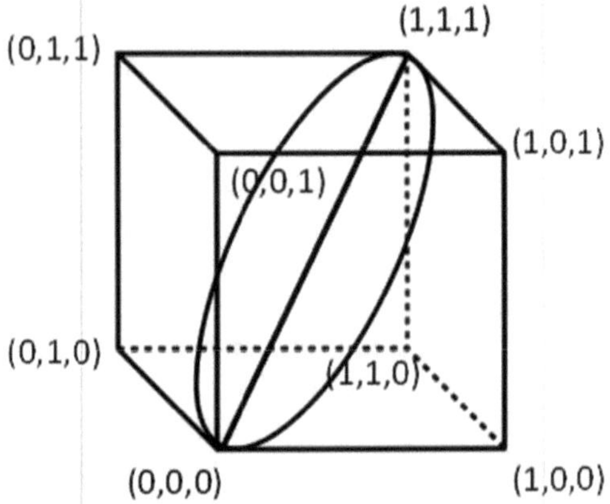

Fig. 6 The extreme points of the cube $h(3)$

oriented in the way that there are methods near the line from (0,0,0) to (1,1,1) (cf. Fig. 6). In more mathematical terms, a "near-enough-factor," f, was introduced (Bruggemann and Carlsen 2014).

In the three-dimensional case, the maximum squared Euclidian distance between (0,0,0) and (1,1,1) in $h(3)$ is $D_{max} = 3$ meaning that the distance between two given points is found on a scale from 0 to 3. Hence, we regard analytical approaches as peculiar if the distance, d, to one of the corners in $h(3)$, $d < f \cdot 3$. These peculiar methods have indicator patterns imbalanced relative to methods with more balanced patterns, i.e., being located around the (0,0,0) to (1,1,1) line. Thus, if f is selected as 0.05 only methods with highly imbalanced indicator profiles will be noted as peculiar, i.e., being located very close to one of the six corners of interest. Thus, $f = 0.05$ can be interpreted as 5 % of the maximal distance, i.e., $d \leq 0.15$ to one of the peculiar corners and will consequently be denoted as peculiar due to the significant imbalance in the indicator profile. In other words, it can be said this corresponds to a 95 % level, meaning that laboratories not being found as extreme on a 95 % level are found located around the (0,0,0)-(1,1,1) straight line.

Referring to the actual corner immediately point to the indicator that causes the peculiarity.

As an increase of one indicator value does not necessarily imply an increase of the value of another indicator, the cloud of points will encompass points which are considered as mutually incomparable. According to the ranking aim, the "main stream" are points which are near the line from (0,0,0) to (1,1,1). We denote points near the other extreme points, i.e., the other corners of $h(3)$ cube as peculiar. In Table 7, the results of the analysis of the 21 methods is summarized using $f = 0.05$. It is seen that 6 out of the 21 methods, i.e., 28.6 %, are classified as peculiar at the 95 % level.

From Table 7 it is immediately seen that in three cases the pattern [0,1,0] is realized, the pattern [0,0,1] only once, and the pattern [1,0,0] appears 2 times.

Table 7 Extreme or peculiar analytical approaches at a 95 % level

Method	Pattern	Interpretation
met6	[0, 1, 0]	This method is close in $\|z\|$ and skew but far away in s
met8	[0, 0, 1]	This method is close in $\|z\|$ and s but far away in $\|skew\|$
met11	[0, 1, 0]	This method is close in $\|z\|$ and skew but far away in s
met14	[1, 0, 0]	This method is close in s and skew but far away in $\|z\|$
met17	[1, 0, 0]	This method is close in s and skew but far away in $\|z\|$
met19	[0, 1, 0]	This method is close in $\|z\|$ and $\|skew\|$ but far away in s

Further it is noted that in the present study no simultaneous appearance of two 1's in the peculiar patterns is observed. The simultaneous appearance of 0 and 1 indicates that some methods are imbalanced, here "good" values in two, but a "bad" value in another aspect. However, obviously situations where more than one indicator value causes the imbalance are possible.

3.2.4 Fuzzy Partial Orders

In the above discussion, it has implicitly been assumed that the indicator values are exact values, i.e., not defective. This may in many cases be an illusion. However, if the indicator values are defective, i.e., approximate rather than exact, it will be appropriate to include these uncertainties in the analyses. This can advantageously be done by application of a fuzzy approach (Bruggemann and Carlsen 2015; Kosko 1992; Van de Walle et al. 1995; Bruggemann et al. 2011; Kruse and Moewes 2015). In the fuzzy approach, the calculations result in a series of the so-called α-cuts that roughly can be translated to points where changes in the mutual relation between the objects appear. The α-cuts formally ranges from 1 to 0, where an α-cut = 1 means that the original data are used and are regarded as exact values, whereas for α-cut = 0 all possible values are accepted, i.e., no discrepancy between the single object is seen as all objects obviously are equivalent. In cases where the distribution of a given values around a mean, e.g., a normal distribution is known it is possible to link the α-cut to the actual uncertainty of the value. In general, such information is not available and especially when several indicator values are involved as in the present study, it is unfortunately not possible to make a direct link between the α-cuts and the actual uncertainties. However, it is still valid that the lower the α-cut the larger the uncertainty on the data, despite a concrete association cannot be made and thus, some indication concerning the sensitivity of the data may be obtained.

Analyzing the present MIS (Table 1) leads to 51 α-cut including α-cut = 1 and 0 and obviously the partial ordering for α-cut = 1 corresponds to the Hasse diagram shown in Fig. 3, whereas an α-cut = 0 corresponds to a situation where all 21 laboratories are equivalent. It can also be mentioned that for an α-cut = 0.576 the 20 methods, met1 ... met20 are equivalent, however, non-equivalent to REF.

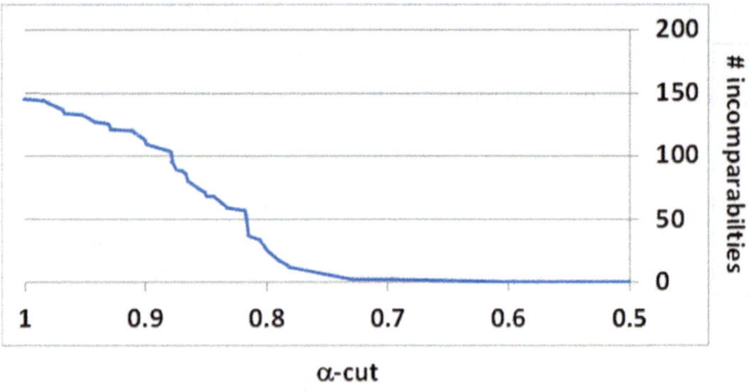

Fig. 7 Number of incomparabilities as function of the α-cut

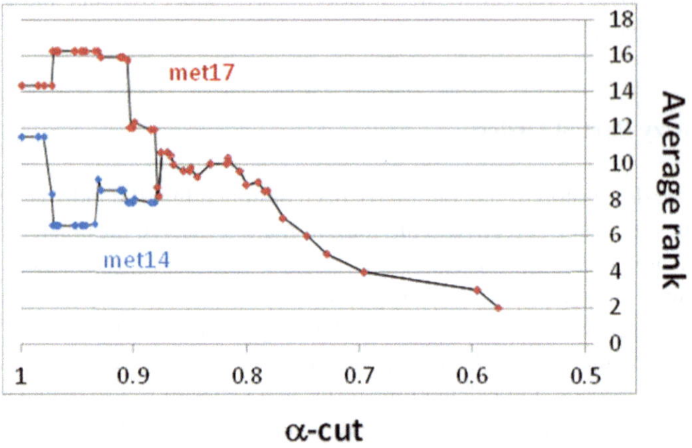

Fig. 8 Variation in the average ranks for met14 and met17 as function of α-cut

Simultaneous the number of incomparabilities between the different analytical approaches will be limited (Fig. 7).

In between the two "extreme," i.e., α-cut = 1 and 0, variations of the average ranks of the single methods will be observed as a function of the actual α-cut. Thus, by following the average rank as function of the α-cuts will give an indication how the mutual ranking between the various objects changes with increasing uncertainty of the data. To illustrate this analysis, we look at two analytical approaches, met14 and met 17, both found in the top level of the Hasse diagram (cf. Fig. 3). In Fig. 8, the variation in the average ranks of the two methods as function of the α-cuts is visualized.

From Fig. 8 it can be deducted that small uncertainties apparently do not cause changes in the average ranks for the two methods. However, further increase in the uncertainty causes a decrease in the rank of met 14, i.e., decreasing the distance to

REF, whereas the opposite is noted for met17. Further increase in the uncertainty eventually causes the two methods to be equivalent.

3.2.5 Weight Intervals

As mentioned above, a typical request is a linear order. One of the simplest way to handle such a request is by introducing the weighted sum, i.e., each indicator is assigned a weight and subsequently an aggregated indicator is generated through the weighted sum.

$$I_{\mathrm{agg}} = \sum w_i r_i \tag{4}$$

where w_i is the weight assigned to the indicator r_i.

Beside the aforementioned problems of compensation (Munda 2008), it may typically lead to problems getting to consensus in determination of the weights. However, it may be a significantly easier process to reach consensus to define weight intervals, where probably the weights could belong to (Bruggemann et al. 2013). Obviously, introducing weight intervals will not lead to a total (linear) order but it will reduce the number of incomparisons and thus enrich the Hasse diagram and therefore strengthen a subsequent ordering based on averaged ranks.

A complete analysis of the effects of introducing weight intervals is outside the scope of the present study. However, a simple example appears appropriate in order to illustrate the advantages of this tool.

In our previous study analyzing the dataset (Table 1), we found that the relative importance of the three indicators were $|z| = 0.397$, $s = 0.444$, and $|\mathrm{skew}| = 0.159$, respectively. Hence, to illustrate the use of weight intervals we will take our onset in these data assuming that the weights for the three indicators would be in the intervals $0.3 < |z| < 0.5$, $0.35 < s < 0.55$, and $0.05 < |\mathrm{skew}| < 0.25$, respectively.

In the original Hasse diagram (Fig. 3), we found 65 comparabilities and 145 incomparabilities. In the weight-based Hasse diagram (Fig. 9) we observe, as predicted a significantly increased number of comparabilities (152) and a simultaneous decreased number of incomparabilities (58). The present example is based on 1000 Monte Carlo simulations.

It is immediately seen that the two Hasse diagrams (Figs. 3 and 9) are distinctly different. The difference is further substantiated with looking at the average ranks of the single elements. In Table 8 the average ranks, as calculated using the LPOMext method (Bruggemann and Carlsen 2011) for the weight-based poset is compared to those based on the original poset.

Obviously, it is not possible based on the figures given in Table 8 to justify that the average ranks based on the weight-based poset is a better ranking. However, as the number of incomparisons is decreasing it seems as a logical assumption. However, it is mandatory to agree on the weight intervals, which still may lead to heated debates.

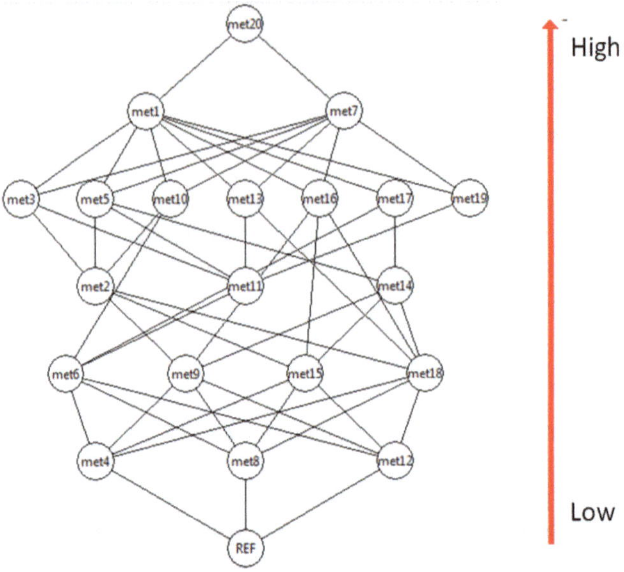

Fig. 9 Hasse diagram assuming weight intervals $0.3 < |z| < 0.5$, $0.35 < s < 0.55$ and $0.05 < |\text{skew}| < 0.25$, respectively, for the three indicators

For comparison, it can be mentioned that if the above-mentioned relative importance for the three indicators is adopted as true weights we will obviously end up with a Hasse diagram consisting of one single chain, i.e., a total order. Hence, the number of comparisons and incomparisons will be 210 and 0, respectively. The differences in the (average) ranks are clearly seen (Table 8).

4 Conclusions and Outlook

When a ranking of methods, based on data concerning the performance of the single analytical methods, is wanted, most decision makers will just aggregate the (normalized) indicator values, and thus find a ranking index and from that they can conclude which methods apparently are the optimal ones. Such ranking index may be obtained by several aggregation methods. However, the scientific field develops highly dynamically and new multi-criteria-decision tools are published in appropriate scientific journals. An overview containing more than 500 pages can be found in (Figueira et al. 2005). There are also established methods, such as PROMETHEE (Brans et al. 1986) or ELECTRE (Roy 1972, 1990; Roy and Bouyssou 1986; Colorni et al. 2001), or the newer TODIM (Gomes and Rangel 2009), which can be applied without any reservation. Partial order concepts are in many of these MCDA methods implicitly used and Roy (Roy 1980) stressed the role, the graphs derived by partial order can have. Nevertheless, it must be questioned, whether or not the partial order variant, the Hasse diagram technique (HDT, with

Table 8 Average ranks of the 21 method based on the weight-based and original posets. For comparison, the ranks based on exact weights ($|z| = 0.397$, $s = 0.444$ and $|\text{skew}| = 0.159$) are given

	Weight-based poset	Original poset	Exact-weight poset
Object	Rkav	Rkav	Rkav
met1	19.667	19.498	19.0
met2	11.012	14.783	10.0
met3	16.0	13.333	15.0
met4	3.0	3.278	3.0
met5	16.54	14.333	18.0
met6	6.071	6.167	6.0
met7	19.4	18.858	20.0
met8	3.0	5.417	2.0
met9	6.565	4.512	5.0
met10	15.467	14.314	14.0
met11	8.629	12.217	9.0
met12	3.0	3.665	4.0
met13	14.683	13.333	12.0
met14	11.621	11.5	13.0
met15	6.565	8.25	8.0
met16	13.983	13.8	17.0
met17	16.133	14.333	16.0
met18	6.11	8.202	7.0
met19	14.167	8.833	11.0
met20	21.0	20.366	21.0
REF	1.0	1.0	1.0

its defining (2)) is similarly established as one of the tools in the field of MCDA. When the aim of MCDA-methods is a unique decision, then the answer is certainly a "no." The partial order concept as it used here, does (in general) not lead to a unique decision, even a unique linear order, the ranking per se, is rarely found by HDT. Indeed, if within an application partial order together with (2) leads to a linear order, it would be a disappointing result, because typically the application of HDT expects a graph, which has not only a linear (and boring) dimension toward increasing ranks, but has also a horizontal evolution, caused by incomparabilities. Incomparabilities are in general caused because HDT is not aggregating the indicators, but let them separated, in order to study their role in the ranking process. This chapter is mainly devoted to the study of the horizontal evolution of the directed, acyclic graph the partial order is providing.

The vertical analysis is not forgotten, but plays a lesser role in this chapter. We initiate the study with the vertical analysis. It is shown that the concept of an event space gives the decision maker an easy understandable measure at hand for a first screening of the ranks these objects, in our cases the analytical methods, have. The average ranks derived from the event space, which is a simple combinatorial exercise) coincides pretty well with the ranks, obtained by more sophisticated approaches. Obviously, the fact that the average rank based on the event space has many ties can be seen a disadvantage.

The horizontal analysis is now equipped with a series of tools, which take care of

1. Numerical effects: Two objects may be incomparable, but the cause can be considered as numerical noise
2. Classification aspects: A ranking assumes that (with [0,1] normalized data) a cloud of points start in the neighborhood of (0,0,... 0) and ends somewhere at (1,1,1,... 1). Peculiar points of an object set are those which are near (0,1)-patterns which are neither (0,... 0) nor (1,... ,1). This is a data pattern-oriented approach and is helpful, if the decision maker would like to see a little bit behind the pure ranking result.
3. Projection possibilities: An analysis is always facilitated if simple graphical schemes like scatter plots are available. By means of the tripartite graph such projections into two dimensions are sometimes possible. However, even if such projection fails, valuable insights are obtained, as how the single indicators contributes to conflicts.

So it can be shown that simple tools obviously have a more versatile nature as it may be found valuable in a variety of cases where a collection of objects is to be ordered based on a series of indicators, i.e., an MIS.

We conclude this chapter by expressing our hope that these partial order tools, even when not usable as a MCDA-method of its own right, are useful instruments to analyze the inherent processes, when an MIS-based ranking is to be done.

Future work will try to improve the tools presented here, but also try to add new tools. Hence, in many studies a huge number of indicators are considered as useful in ranking aims. In the light of partial order, there is a high probability to obtain just an antichain, which is hardly to be interpreted. A meaningful strategy may be to order the indicators hierarchically and to perform the analysis with HDT at different levels of this hierarchy.

References

ASTM (2013) Standard practice for verification and use of control charts in spectrochemical analysis, ASTM E1329-10. ASTM International, West Conshohocken, PA

Brans JP, Vincke PH, Mareschal B (1986) How to select and how to rank projects: the PROMETHEE method. Eur J Oper Res 24:228–238

Bruggemann R, Annoni P (2014) Average heights in partially ordered sets. MATCH Commun Math Comput Chem 71:117–142

Bruggemann R, Carlsen L (eds) (2006a) Partial order in environmental sciences and chemistry. Springer, Berlin

Bruggemann R, Carlsen L (2006b) Introduction to partial order theory exemplified by the evaluation of sampling sites. In: Bruggemann R, Carlsen L (eds) Partial order in environmental sciences and chemistry. Springer, Berlin, pp 61–110

Bruggemann R, Carlsen L (2011) An improved estimation of averaged ranks of partially orders. MATCH Commun Math Comput Chem 65:383–414

Bruggemann R, Carlsen L (2012) Multi-criteria decision analyses. Viewing MCDA in terms of both process and aggregation methods: some thoughts, motivated by the paper of Huang, Keisler and Linkov. Sci Total Environ 425:293–295

Bruggemann R, Carlsen L (2014) Incomparable - what now? MATCH Commun Math Comput Chem 71:694–716

Bruggemann R, Carlsen L (2015) Incomparable - what now, III. Incomparabilities, elucidated by a simple version of ELECTRE III and a fuzzy, partial order approach. MATCH Commun Math Comput Chem 73:277–302

Bruggemann R, Münzer B (1993) A graph-theoretical tool for priority setting of chemicals. Chemosphere 27:1729–1736

Bruggemann R, Patil GP (2011) Ranking and prioritization for multi-indicator systems – introduction. Springer, New York

Bruggemann R, Voigt K (1995) An evaluation of online databases by methods of lattice theory. Chemosphere 31:3585–3594

Bruggemann R, Voigt K (2008) Basic principles of Hasse diagram technique in chemistry. Comb Chem High Throughput Screen 11:756–769

Bruggemann R, Voigt K (2011) A new tool to analyze partially ordered sets - application: ranking of polychlorinated biphenyls and alkanes/alkenes in River main, Germany. MATCH Commun Math Comput Chem 66:231–251

Bruggemann R, Sørensen PB, Lerche D, Carlsen L (2004) Estimation of averaged ranks by a local partial order model. J Chem Inf Comput Sci 44:618–625

Bruggemann R, Kerber A, Restrepo G (2011) Ranking objects using fuzzy orders, with an application to refrigerants. MATCH Commun Math Comput Chem 66:581–603

Bruggemann R, Restrepo G, Voigt K, Annoni P (2013) Weighting intervals and ranking. Exemplified by leaching potential of pesticides. MATCH Commun Math Comput Chem 69:413–432

Bruggemann R, Carlsen L, Voigt K, Wieland R (2014) PyHasse software for partial order analysis. In: Bruggemann R, Carlsen L, Wittmann J (eds) Multi-indicator systems and modelling in partial order. Springer, New York, pp 389–423

Carlsen L, Bruggemann R (2008) Accumulating partial order ranking. Environ Model Softw 23:986–993

Carlsen L, Bruggemann R, Kenessova O, Erzhigitov E (2015) Evaluation of analytical performance based on partial order methodology. Talanta 132:285–293

Colorni A, Paruccini M, Roy B (2001) A-MCD-A, Aide Multi Critere a la Decision, multiple criteria decision aiding. JRC European Commission, Ispra

De Loof K, De Meyer H, De Baets B (2006) Exploiting the lattice of ideals representation of a poset. Fundam Inform 71:309–321

Ernesti J, Kaiser P (2008) Python - Das umfassende Handbuch. Galileo Press, Bonn

Figueira J, Greco S, Ehrgott M (2005) Multiple criteria decision analysis, state of the art surveys. Springer, Boston

Gomes LFAM, Rangel LAD (2009) An application of the TODIM method to the multicriteria rental evaluation of residential properties. Eur J Oper Res 193:204–211

Hajkowicz S, Higgins A (2008) A comparison of multiple criteria analysis techniques for water resource management. Eur J Oper Res 184(2008):255–265

Hetland ML (2005) Beginning Python - from Novice to professional. Apress, Berkeley

JCGM 200 (2008) International vocabulary of metrology — basic and general concepts and associated terms (VIM). http://www.bipm.org/utils/common/documents/jcgm/JCGM_200_2008.pdf. Accessed Jan 2015

Kosko B (1992) Neural networks and fuzzy systems - a dynamical systems approach to machine learning. Pentice Hall, London

Kruse R, Moewes C (2015) Fuzzy systems, Fuzzy set theory., http://fuzzy.cs.uni-magdeburg.de/ci/fs/fs_ch02_fst.pdf. Accessed Jan 2015

Lærd (2013) Standard score. https://statistics.laerd.com/statistical-guides/standard-score.php. Accessed Jan 2015

Langtangen HP (2008) Python scripting for computational science. Springer, Berlin

Lerche D, Sørensen PB, Bruggemann B (2003) Improved estimation of the ranking probabilities in partial orders using random linear extensions by approximation of the mutual ranking probability. J Chem Inf Comput Sci 43:1471–1480
Morton J, Pachter L, Shiu A, Sturmfels B, Wienand O (2009) Convex rank tests and semigraphoids. SIAM J Discrete Math 23:1117–1134
Munda G (2008) Social multi-criteria evaluation for a sustainable economy. Springer, Heidelberg
Python (2015) Python. https://www.python.org/. Assessed Jan 2015
R Core Team (2015) R: a language and environment for statistical computing. R Foundation for Statistical Computing, Vienna, Austria, http://www.R-project.org/. Accessed Jan 2015. ISBN 3-900051-07-0
Roy B (1972) Electre III: Un Algorithme de Classements fonde sur une representation floue des Preferences En Presence de Criteres Multiples. Cahiers du Centre d'Etudes de Recherche Operationelle 20:32–43
Roy B (1980) Selektieren, Sortieren und Ordnen mit Hilfe von Prävalenzrelationen: Neue Ansätze auf dem Gebiet der Entscheidungshilfe für Multikriteria-Probleme. zfbf-Schmalenbachs Zeitschrift für betriebswirtschaftliche Forschung 32:465–497
Roy B (1990) The outranking approach and the foundations of the ELECTRE methods. In: Bana e Costa CD (ed) Readings in multiple criteria decision aid. Springer, Berlin, pp 155–183
Roy B, Bouyssou D (1986) Comparison of two decision-aid models applied to a nuclear power plant siting example. Eur J Oper Res 25:200–215
Stone DC, Ellis J (2009) Aim of statistics in analytical chemistry. Department of Chemistry, University of Toronto. http://www.chem.utoronto.ca/coursenotes/analsci/stats/AimStats.html. Accessed Jan 2015
Taylor C (2013) What is Skewness. http://statistics.about.com/od/Descriptive-Statistics/a/What-Is-Skewness.htm. Accessed Jan 2015
Van de Walle B, De Baets B, Kersebaum KC (1995) Fuzzy multi-criteria analysis of cutting techniques in a nuclear dismantling project. Fuzzy Sets Syst 74:115–126
Weigend M (2006) Objektorientierte Programmierung mit Python. Mitp, Bonn
Wienand O (2006) lcell. http://bio.math.berkeley.edu/ranktests/lcell/. Accessed Jan 2015

Functionals and Synthetic Indicators Over Finite Posets

Marco Fattore

1 Introduction

In this paper, we address the problem of building real functionals on frequency distributions whose domain is a finite poset. The topic may seem very technical, but it is deeply related to relevant open problems in socio-economic statistics. Consider, for example, the measurement of inequality over multidimensional systems of ordinal indicators. This issue, fundamental in a "beyond GDP" perspective, is usually addressed by composing inequality measures over attributes into a single index, through weighted averages. The problem with this approach is that the domain of the joint frequency distribution (i.e., the collection of k-tuples of attribute scores) is considered as an unstructured set, while it is in fact a partially ordered set, whose order relation is the product order of the linear orders corresponding to ordinal attributes. Inequality is conceptually linked to the way achievements are distributed on a "low-high" axis so that the order structure, even if partial, should enter into the computations. Starting from this observation, in this paper we propose a general theory of functionals over finite posets, to provide a sound formal basis for the construction of synthetic indicators in socio-economic sciences. The basic idea of the paper can be introduced as follows. In the classical theory of statistical indices, one is primarily interested in identifying a set of properties assuring the index to capture the concept of interest and to behave in a way that is logically consistent with it. For example, in monetary inequality measurement one requires indices to increase if income is transferred from the "poor" to the "richer." Statistical indices are then seen as functions of the frequency distribution and their properties are assessed in terms of their behavior under different distribution shapes or transformations. The order structure of the domain of the statistical variables,

M. Fattore (✉)
Department of Statistics and Quantitative Methods, University of Milano-Bicocca, Milan, Italy
e-mail: marco.fattore@unimib.it

M. Fattore, R. Bruggemann (eds.), *Partial Order Concepts in Applied Sciences*,
DOI 10.1007/978-3-319-45421-4_5

usually $\mathbb{R}$ or $\mathbb{N}$ with their natural total order relation, does not come into play explicitly, being trivial. But when the domain is multidimensional and different partial order relations can be imposed on it, the role of the order structure becomes essential and one must control for the way indices behave when the frequency distribution is kept fixed, *but the partial order relation changes*. To pursue this approach, it is natural to look at indices as scalar functions defined on the semilattice of posets sharing the same ground set. Properties of functionals may then be connected to algebraic properties of this "poset of posets," leading to a simple and coherent axiomatic theory. The paper is organized as follows. Section 2 provides some basic definitions and sets the notation. Section 3 introduces some fundamental results in the theory of means, namely the Nagumo–Kolmogorov theorem on quasi-arithmetic means and their semigroup representation. Section 4 develops the theory of functionals on finite posets. Section 5 discusses some simple examples. Section 6 concludes.

2 Notation and Basic Definitions

A *partially ordered set* (or a *poset*) $\pi = (X, \leq)$ is a set X (called the *ground set*) equipped with a partial order relation $\leq$, i.e., with a binary relation satisfying the properties of *reflexivity*, *antisymmetry*, and *transitivity* (Davey and Priestley, 2002; Neggers and Kim, 1998; Schröder, 2002):

1. $x \leq x$ for all $x \in X$ (reflexivity);
2. if $x \leq y$ and $y \leq x$, then $x = y$, $x, y \in X$ (antisymmetry);
3. if $x \leq y$ and $y \leq z$, then $x \leq z$, $x, y, z \in X$ (transitivity).

If $x \leq y$ or $y \leq x$, then x and y are called *comparable*, otherwise they are said *incomparable* (written $x \,||\, y$). A partial order π where any two elements[1] are comparable is called a *linear order* or a *complete order*. A subset of mutually comparable elements of a poset is called a *chain*. On the contrary, a subset of mutually incomparable elements of a poset is called an *antichain*. Given $x, y \in \pi$, y is said to *cover* x (written $x \prec y$) if $x \leq y$ and there is no other element $z \in \pi$ such that $x \leq z \leq y$. An element $x \in \pi$ such that $x \leq y$ implies $x = y$ is called *maximal*; if for each $y \in \pi$ it is $y \leq x$, then x is called (the) *maximum* or (the) *greatest element* of π. An element $x \in \pi$ such that $y \leq x$ implies $x = y$ is instead called *minimal*; if for each $y \in \pi$ it is $x \leq y$, then x is called (the) *minimum* or (the) *least element* of π. Given $x \in \pi$, the *down-set* of x (written $\downarrow x$) is the set of all the elements $y \in \pi$ such that $y \leq x$. Dually, the *up-set* of x (written $\uparrow x$) is the set of all the elements $y \in \pi$ such that $x \leq y$. Let π be a poset. If for every choice of $x, y \in \pi$ the intersection of the down-sets of x and y has a maximum (written $\wedge(x, y)$ and called the *meet*

[1] For the sake of simplicity, in the following elements of X partially ordered by $\leq$ will be referred directly as elements of π.

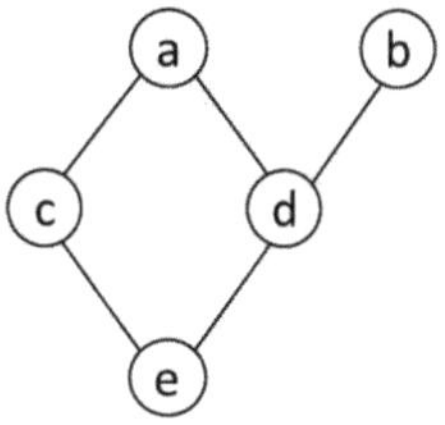

Fig. 1 Hasse diagram of a poset π_0^* on five elements

of x and y), then π is called a *meet-semilattice*. Given two partially ordered sets $\pi = (X, \leq_\pi)$ and $\tau = (X, \leq_\tau)$ on the same set X, we say that τ is an extension of π, if $x \leq_\pi y$ in π implies $x \leq_\tau y$ in τ. In other terms, τ is an extension of π if it may be obtained from the latter turning some incomparabilities into comparabilities. An extension of π which is also a complete order is called a *linear extension*. The set of linear extensions of a poset π is denoted by $\Omega(\pi)$.

We now introduce another mathematical structure which plays a fundamental role in the following discussion. Let π_0 be a poset on a set X and let $\Pi(\pi_0)$ be the family of all of the posets over X that are extensions of π_0. $\Pi(\pi_0)$ can be turned into a poset (Brualdi et al., 1994) (here called *extension poset*) defining the partial order $\sqsubseteq$ as

$$\pi_1 \sqsubseteq \pi_2 \Leftrightarrow \pi_2 \text{ isanextensionof} \pi_1 (\pi_1, \pi_2 \in \Pi(\pi_0)). \quad (1)$$

Endowed with partial order $\sqsubseteq$, the extension poset $\Pi(\pi_0)$ is a *meet-semilattice* (Brualdi et al., 1994), with meet $\wedge$ given by intersection of order relations; its minimum is π_0 and its maximal elements are the linear extensions of π_0.

Example. Figure 1 depicts the Hasse diagram of a poset π_0^* with five elements. The extension poset $\Pi(\pi_0^*)$ "generated" by π_0^* is depicted in Fig. 2. As it can be seen, the linear extensions of π_0^* correspond to the maximal elements of $\Pi(\pi_0^*)$, while its minimum is π_0^* itself.

3 Quasi-Arithmetic Means

In this section, we introduce the concept of *aggregation system* and focus on the so-called *quasi-arithmetic means*, a kind of functionals[2] which are characterized by a set of properties crucial for indicators construction. After providing the formal definitions, we give a particular and useful representation of this class of means, drawing on the concept of semigroup. This semigroup representation is important,

[2]By the term "functional," in this section we mean a function mapping a k-dimensional vector into a real number.

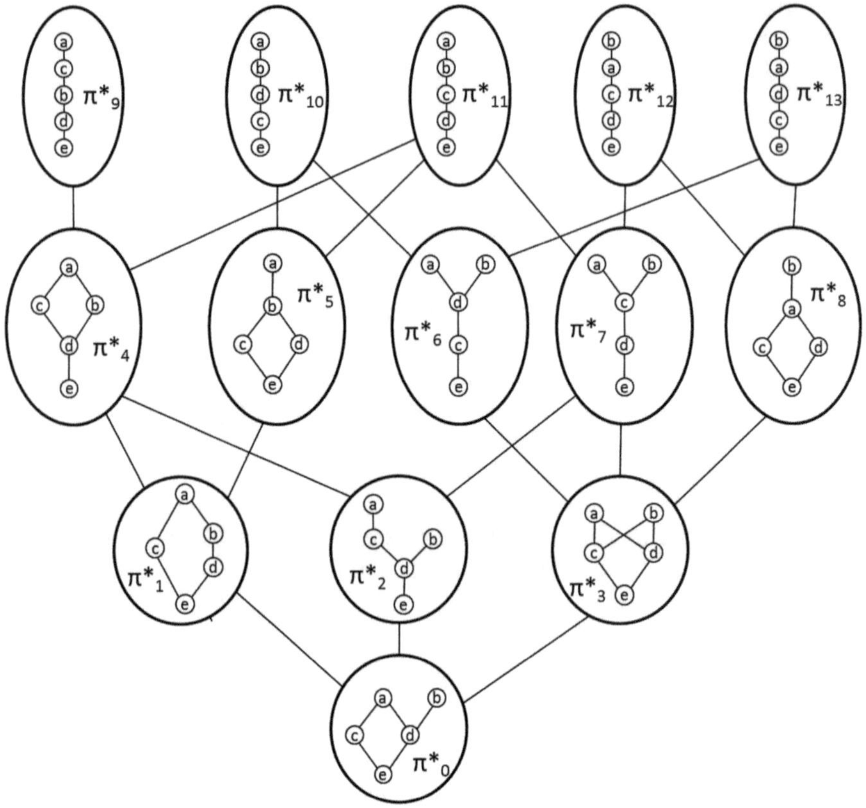

Fig. 2 Extension poset $\Pi(\pi_0^*)$ of the poset π_0^* depicted in Fig. 1 (to improve readability, elements of $\Pi(\pi_0^*)$ have been inserted into ellipses)

in view of the axiomatic theory, since it characterizes the behavior of aggregation functionals, when a sequence of nested partial aggregations is performed on their arguments.

3.1 Aggregation Systems

Let $\boldsymbol{x} = (x_1, \ldots, x_k)$ be a vector of k real numbers in $[0, 1]$, $\boldsymbol{w} = (w_1, \ldots, w_k)$ a vector of k non-negative weights summing to 1, and $g(\cdot)$ a continuous and strictly monotone real function, from $[0, 1]$ to $\mathbb{R}$, called the *generating* function. The *weighted quasi-arithmetic mean* (Beliakov et al., 2007; Grabisch et al., 2009) $M_{g,k}(\cdot)$ is defined as:

$$M_{g,k}(\boldsymbol{x};\boldsymbol{w}) = g^{-1}\left(\sum_{i=1}^{k} w_i g(x_i)\right) \tag{2}$$

where $g^{-1}(\cdot)$ is the inverse of function $g(\cdot)$ which exists and is well-defined from $\mathbb{R}$ to [0, 1], since $g(\cdot)$ is strictly monotone. Many well-known means belong to the class of quasi-arithmetic means, namely the *weighted power means* $M_{[r],k}(\cdot)$:

$$M_{[r],k}(\boldsymbol{x};\boldsymbol{w}) = \left(\sum_{i=1}^{k} w_i x_i^r\right)^{1/r} \qquad (g(x) = x^r,\ r \neq 0); \tag{3}$$

$$M_{[0],k}(\boldsymbol{x};\boldsymbol{w}) = \prod_{i=1}^{k} x_i^{w_i} \quad (g(x) = \log(x)). \tag{4}$$

When $\boldsymbol{w} = (1/k, \ldots, 1/k)$, the above formulas reduce to "classical" means (called *quasi-arithmetic means*):

$$M_{g,k}(\boldsymbol{x}) = g^{-1}\left(\frac{1}{k}\sum_{i=1}^{k} g(x_i)\right); \tag{5}$$

$$M_{[r],k}(\boldsymbol{x}) = \left(\frac{1}{k}\sum_{i=1}^{k} x_i^r\right)^{1/r}; \tag{6}$$

$$M_{[0],k}(\boldsymbol{x}) = \left(\prod_{i=1}^{k} x_i\right)^{\frac{1}{k}}. \tag{7}$$

What makes quasi-arithmetic means of particular relevance for our purposes is that they contain the class of *continuous*, *strictly monotone*, and *decomposable* functionals. To be formal, let us call *aggregation system* a collection $\mathbb{F} = \{F_1(\cdot), F_2(\cdot), F_3(\cdot)\ldots\}$ of functionals with $1, 2, 3\ldots$ arguments, respectively, with $F_1(x) = x$ by convention. Then continuity, strict monotonicity, and decomposability for $\mathbb{F}$ are defined as follows:

1. **Continuity**. An aggregation system $\mathbb{F}$ is continuous if each of its elements $F_k(\cdot) \in \mathbb{F}$ is continuous in each of its k arguments ($k = 1, 2, \ldots$).
2. **Strict monotonicity**. Let $\boldsymbol{x}$ and $\boldsymbol{y}$ be two k-dimensional real vectors; a functional $F_k(\cdot)$ is strictly monotone if $\boldsymbol{x} < \boldsymbol{y}$ in the product order over $\mathbb{R}^k$ implies $F_k(\boldsymbol{x}) < F_k(\boldsymbol{y})$. An aggregation system $\mathbb{F}$ is strictly monotone if each of its elements is strictly monotone.
3. **Decomposability**. An aggregation system $\mathbb{F}$ is decomposable if and only if for all $m, n = 1, 2, \ldots$ and for all $\boldsymbol{x} \in [0, 1]^m$ and $\boldsymbol{y} \in [0, 1]^n$:

$$F_{m+n}(\boldsymbol{x}, \boldsymbol{y}) = F_{m+n}(\underbrace{F_m(\boldsymbol{x}_m), \ldots, F_m(\boldsymbol{x}_m)}_{m \text{ times}}, \boldsymbol{y}). \tag{8}$$

The last formula requires the value of functional $F_{m+n}(\cdot)$ to be obtained substituting to the first m arguments their aggregated value $F_m(\cdot)$, replicated m times. The statement becomes clearer when specialized to arithmetic means. In this case, it simply states that one can compute the average of $m + n$ numbers substituting to each of the first m the average of the first m numbers themselves.

According to the Nagumo–Kolmogorov theorem (Beliakov et al., 2007), an aggregation system $\mathbb{F} = \{F_k(\cdot)\}$ $(k = 1, 2, 3, \ldots)$ is continuous, strictly monotone, and decomposable if and only if there exists a monotone bijective function $g(\cdot)$: $[0, 1] \rightarrow [0, 1]$ such that for $k > 1$, $F_k(\cdot)$ is a quasi-arithmetic mean $M_{g,k}(\cdot)$. A functional F is *homogeneous* if, for every real number $c \in [0, 1]$, it is $F(c \cdot \boldsymbol{x}) = c \cdot F(\boldsymbol{x})$; it can be proved that the only homogeneous quasi-arithmetic means are the power means $M_{[r],k}(\cdot)$ (see again Beliakov et al. 2007). Finally, notice that quasi-arithmetic means are *symmetric*, i.e., they are invariant under permutations of their arguments. As a consequence, they satisfy the property of *strong decomposability* (Grabisch et al., 2009), i.e., they are invariant under the aggregation of any subset of (and not just of consecutive) arguments.

3.2 Semigroup Representation of Quasi-Arithmetic Means

In this paragraph, we show that quasi-arithmetic means can be computed by the repeated application of a binary associative and commutative operation. This will be useful when connecting the properties of functionals to the structure of the extension poset. The presentation follows quite closely (Pursiainen, 2005).

Let $\mathbb{F}$ be an aggregation system. Assume its elements are symmetric (as defined above) and suppose that, if vector $\boldsymbol{x} = (x_1, \ldots, x_m)$ is partitioned into k subvectors $\boldsymbol{x}_{(1)}, \ldots, \boldsymbol{x}_{(k)}$, of length $n_1, \ldots, n_k$, respectively, it holds:

$$F_m(\boldsymbol{x}) = F_k\left(F_{n_1}(\boldsymbol{x}_{(1)}), \ldots, F_{n_k}(\boldsymbol{x}_{(k)})\right) \tag{9}$$

(i.e., suppose that aggregation can be performed "aggregating partial aggregations"). An aggregation system satisfying (9) will be called *consistentinaggregation*. An important special case of the above formula is the following:

$$F_m(x_1, \ldots, x_m) = F_2\left(F_{m-1}(x_1, \ldots, x_{m-1}), x_m\right), \tag{10}$$

which means that vector $\boldsymbol{x}$ can be aggregated in "two steps," the first of which aggregates $m - 1$ components. Using this formula repeatedly, one can reduce $F_m(\cdot)$ to a nested sequence of applications of $F_2(\cdot)$; for example:

$$F_4(x_1, x_2, x_3, x_4) = F_2\left(F_3(x_1, x_2, x_3), x_4\right) = F_2\left(F_2(F_2(x_1, x_2), x_3), x_4\right). \tag{11}$$

One thus sees that $F_2(\cdot,\cdot)$ determines the entire aggregation system $\mathbb{F}$. Thanks to symmetry, $F_2(x_1,x_2) = F(x_2,x_1)$ and also $F_2(F_2(x_1,x_2),x_3) = F_2(x_1,F_2(x_2,x_3))$, i.e., $F_2(\cdot,\cdot)$ is *commutative* and *associative*. Thus $F_2(\cdot,\cdot)$ is a *commutative semigroup operation* and $\mathbb{F}$ is a *commutative semigroup, generated* by $F_2(\cdot,\cdot)$. Denoting $F_2(\cdot,\cdot)$ as $\circ_F$, we can write formula (11) in the following clearer way:

$$F_4(x_1,x_2,x_3,x_4) = (x_1 \circ_F x_2) \circ_F x_3) \circ_F x_4 = x_1 \circ_F x_2 \circ_F x_3 \circ_F x_4, \qquad (12)$$

where the second equality comes from associativity of $\circ_F$.

For our purposes, what is interesting is that weighted quasi-arithmetic means, and power means in particular, are consistent in aggregation and are generated by a suitable choice of $F_2(\cdot,\cdot)$. To show this, some notation must be introduced first. Let x_1 and x_2 be the numbers we want to aggregate, using a weighted quasi-arithmetic mean with weights w_1 and w_2. Put $\boldsymbol{x_1} = (x_1,w_1)$ and $\boldsymbol{x_2} = (x_2,w_2)$; then the following binary operation $\circ_g$ generates $M_{g,k}(\cdot)$:

$$\boldsymbol{x}_1 \circ_g \boldsymbol{x}_2 = \left(g^{-1}\left(\frac{w_1 g(x_1) + w_2 g(x_2)}{w_1 + w_2}\right); w_1 + w_2\right) \qquad (13)$$

Specializing this formula to $g(\cdot) = id(\cdot)$ (identity function) we get the generating operator for the weighted arithmetic mean:

$$\boldsymbol{x}_1 \circ_{id} \boldsymbol{x}_2 = \left(\frac{w_1 x_1 + w_2 x_2}{w_1 + w_2}; w_1 + w_2\right). \qquad (14)$$

Applying recursively this formula to a set of numbers $x_1, \ldots, x_m$, starting with $w_1 = w_2 = 1$, gives the simple arithmetic mean (Pursiainen, 2005).

Remark. Notice that to represent weighted and unweighted quasi-arithmetic means in semigroup terms, it has been necessary to jointly state the update formula for both the values and the weights, so as that each step of the recursion carries over all of the information needed for the next nested application of the semigroup operation.

4 Building Functionals Over Finite Posets

Let us consider a finite poset π_0 and a distribution $\boldsymbol{p}$ of relative frequencies defined on it. Let $\Pi(\pi_0)$ be the extension poset generated by π_0, let $\pi \in \Pi(\pi_0)$ be an extension of π_0, and let $F(\pi,\boldsymbol{p})$ be a functional evaluated on the distribution $\boldsymbol{p}$ over π. From a statistical point of view, when π is a linear order λ, $F(\lambda,\boldsymbol{p})$ can be interpreted as a unidimensional index, in that $F(\cdot)$ applies on a unidimensional poset, which can be seen simply as an ordinal attribute. Properties of $F(\lambda,\boldsymbol{p})$, as the distribution $\boldsymbol{p}$ changes, determine the nature of the functional. We do not address these aspects here, since they pertain to the proper field of the axiomatic theory of univariate statistical indexes. Instead, we focus on the properties of $F(\pi, \boldsymbol{p})$ as the poset π changes in $\Pi(\pi_0)$, while the distribution $\boldsymbol{p}$ is kept fixed. More specifically, we ask ourselves whether $F(\cdot, \boldsymbol{p})$ can be assigned to elements of $\Pi(\pi_0)$ "freely," or whether the algebraic structure of the poset of posets imposes consistency constraints on such assignments. This question has two main motivations. The first is of a formal and technical nature. In order to develop a satisfactory theory of functionals, we need to link their properties to those of the "context" they act upon. Given that here we are dealing with the poset of posets, the only properties that can be considered pertain to extensions and, as it will be seen below, intersections of posets. The second motivation is of an applied nature. In the social sciences, posets may arise from evaluation and comparison processes and may reflect different systems of social values, against which social facts are assessed through some indicators (see, for example, Fattore 2016). In this respect, poset structure is an input to the evaluation process and the behavior of a statistical index as such a structure changes must be taken into account, to assess its effectiveness as a measurement tool. To work out the consistency constraints to be imposed on functionals, some preliminary definitions and technical results must be discussed. They are of a poset theoretical nature and pertain to the properties and the structure of particular subfamilies of elements of the extension poset. The role of the frequency distribution is left aside, until it comes back into play when defining specific statistical indicators, as we do in Sect. 5, in connection to social polarization.

4.1 Non-overlapping Generating Families of Posets

We begin introducing useful ways to represent posets as the intersection of other posets.

Definition. A collection of posets $\pi_1, \ldots, \pi_k \in \Pi(\pi_0)$ will be called a *generating family* for π_0, if $\pi_0 = \wedge(\pi_1, \ldots, \pi_k)$, i.e., if $\pi_0 = \pi_1 \cap \ldots \cap \pi_k$.

Examples of generating families for poset π_0^* of Fig. 1 are reported in Fig. 3.

A generating family $\{\pi_1, \ldots, \pi_k\}$ for π_0 is called *non-overlapping* if the sets $\Omega(\pi_i)$ of linear extensions of its elements are disjoint, i.e., if

$$\bigcap_{i=1}^{k} \Omega(\pi_i) = \emptyset \tag{15}$$

(for the sake of clarity, we stress that in the above formula, we are not taking the intersection of different linear extensions, but of different families of linear extensions, i.e., we are imposing to these families to have no linear extension in common). A generating family is called *complete* if the union of the sets of linear extensions of its elements equals the set of linear extensions of π_0, i.e., if:

$$\bigcup_{i=1}^{k} \Omega(\pi_i) = \Omega(\pi_0). \tag{16}$$

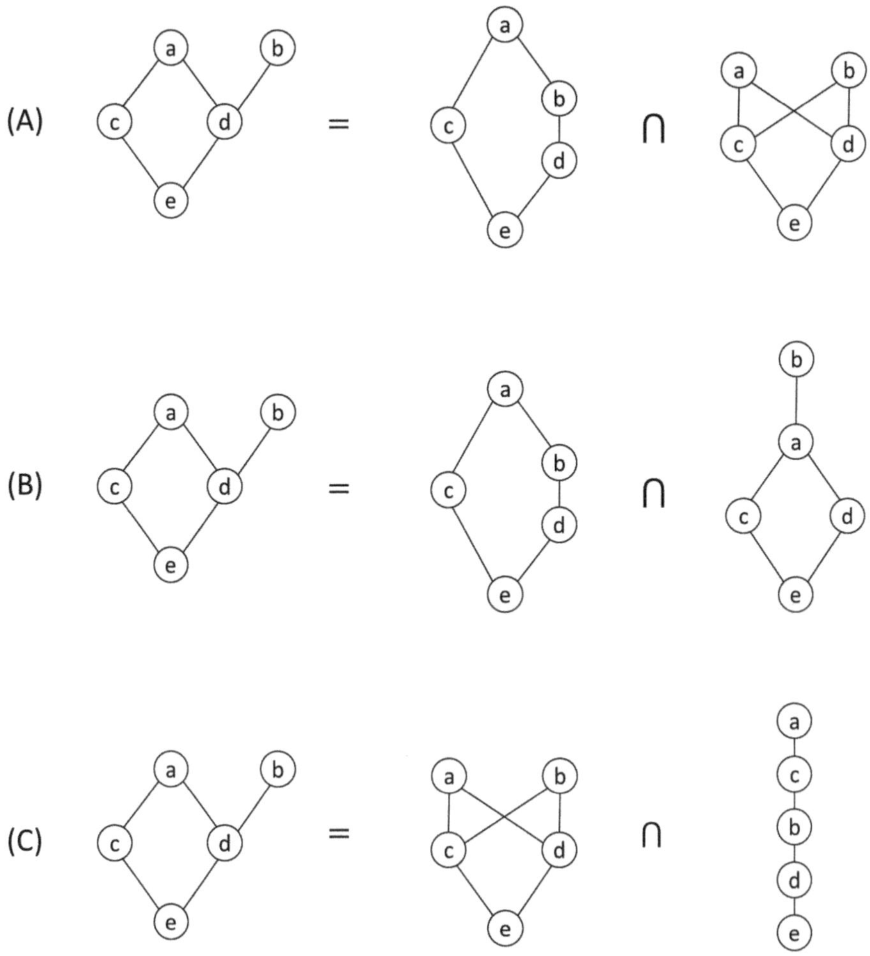

Fig. 3 Examples of generating families for poset π_0^* of Fig. 1

Definition. A generating family which satisfies both condition (15) and condition (16) is called a *non-overlapping complete* (NOC) generating family for π_0.

Given a poset π_0, at least one NOC generating family exists, namely the set $\Omega(\pi_0)$ of its linear extensions. Other examples of NOC generating families for poset π_0^* of Fig. 1 are reported in Fig. 3 (panels A, B and C). NOC generating families have some simple but relevant properties, summarized in the following proposition:

Proposition. Let $G = \{\pi_1, \ldots, \pi_k\}$ be an NOC generating family for π_0. (A) Sets $\Omega(\pi_1), \ldots, \Omega(\pi_k)$ provide a partition of $\Omega(\pi_0)$. (B) For a fixed index h ($1 \leq h \leq k$), let $G_h = \{\pi_{h_1}, \ldots, \pi_{h_s}\}$ be an NOC generating family for π_h; then $G \backslash \{\pi_h\} \cup G_h$ is an NOC generating family for π_0.

Proof. (A) This is just a restatement of (15) and (16). (B) That $G \backslash \{\pi_h\} \cup G_h$ is a generating family for π_0 is evident. Since G and G_h are NOC families, the collection $\{\Omega(\pi_i)\}$ $(\pi_i \in G)$ is a partition of Ω and $\{\Omega(\pi_j)\}$ $(\pi_j \in G_h)$ is a partition of $\Omega(\pi_h)$. Consequently, $\bigcup_{\pi \in G \backslash \pi_h \cup G_h} \Omega(\pi)$ is a partition of $\Omega(\pi_0)$. Thus $G \backslash \{\pi_h\} \cup G_h$ is an NOC generating family for π_0.

q.e.d.

In practice, property (B) states that if an element of an NOC generating family for π_0 is substituted by one among its own NOC generating families, the resulting collection of posets is again an NOC generating family for π_0.

4.2 Axiomatic Properties of Functionals on $\Pi(\pi_0)$

In this paragraph, we specify the properties we want functionals on posets to satisfy, in order to provide useful aggregation tools for statistical indicator construction. At the heart of the axiomatic system, there is the consistency between the behavior of functionals and the nested structure of NOC generating families, that will lead to quasi-arithmetic means.

Let $\varphi(\pi_0, \boldsymbol{p})$ be a functional (e.g., a statistical index) over a distribution $\boldsymbol{p}$ of relative frequencies defined on a poset π_0 (since here we are interested in the behavior of $\varphi(\pi_0, \boldsymbol{p})$ as the poset changes, $\boldsymbol{p}$ being fixed, we drop the second argument and write simply $\varphi(\pi_0)$; this should not produce any confusion). Since π_0 may be reconstructed from its generating families, it is natural to assume $\varphi(\pi_0)$ to be expressible as a function of the values of $\varphi(\cdot)$ on the elements of such families. In other words, once $\varphi(\cdot)$ is assigned on a generating family for π_0, its value on π_0 itself should be univocally given. Formulated this way, however, this statement is not really useful. To see why, consider again Fig. 2 and notice that $\{\pi_9^*, \pi_{13}^*\}$, $\{\pi_{10}^* . \pi_{12}^*\}$, and $\{\pi_9^*, \pi_{10}^*, \pi_{12}^* . \pi_{13}^*\}$ are generating families for π_0^*; thus, two functions $F_2(\cdot)$ and $F_4(\cdot)$ (subscripts stand for the number of arguments) should exist such that one could equivalently write:

$$
\begin{aligned}
\varphi(\pi_0^*) &= F_2(\varphi(\pi_9^*), \varphi(\pi_{13}^*)) \\
\varphi(\pi_0^*) &= F_4(\varphi(\pi_9^*), \varphi(\pi_{10}^*), \varphi(\pi_{12}^*), \varphi(\pi_{13}^*)) \\
\varphi(\pi_0^*) &= F_2(\varphi(\pi_{10}^*), \varphi(\pi_{12}^*)).
\end{aligned}
\tag{17}
$$

As a consequence

$$F_2(\varphi(\pi_9^*), \varphi(\pi_{13}^*)) = F_4(\varphi(\pi_9^*), \varphi(\pi_{10}^*), \varphi(\pi_{12}^*), \varphi(\pi_{13}^*)) = F_2(\varphi(\pi_{10}^*), \varphi(\pi_{12}^*)) \tag{18}$$

so that, on the one hand, $F_4(\cdot)$ should be independent of $\varphi(\pi_{10}^*)$ and $\varphi(\pi_{12}^*)$ (first equality) and, on the other hand, it should be independent of $\varphi(\pi_9^*)$ and $\varphi(\pi_{13}^*)$ (second equality). In practice, $F_4(\cdot)$ and $F_2(\cdot)$ should be constant functions, leading to a trivial theory. These problems arise since any subset of $\Pi(\pi_0)$ comprising a generating family is a generating family itself. To overcome them, we still require $\varphi(\pi_0)$ to be a function of the values of $\varphi(\cdot)$ on other posets, but restricting them to elements of NOC generating families for π_0. Formally, given two NOC families $\{\pi_1, \ldots, \pi_k\}$ and $\{\tau_1, \ldots, \tau_m\}$, we require that an aggregation family $\mathbb{F}$ exists such that:

$$\varphi(\pi_0) = F_k(\varphi(\pi_1), \ldots, \varphi(\pi_k)) = F_m(\varphi(\tau_1), \ldots, \varphi(\tau_m)) \tag{19}$$

where $F_k(\cdot), F_m(\cdot) \in \mathbb{F}$. This *invariance property* has two main consequences:

1. $\varphi(\pi_0)$ can be computed as a function of the values of $\varphi(\cdot)$ on the linear extensions of π_0. In fact, $\Omega(\pi_0)$ is an NOC family and thus one can set:

$$\varphi(\pi_0) = F_\omega(\varphi(\lambda_1), \ldots, \varphi(\lambda_\omega)) \tag{20}$$

   where $\lambda_1, \ldots, \lambda_\omega$ are the linear extensions of π_0.
2. $F_k(\cdot)$ behaves in a consistent way, when an element of an NOC family is replaced by one of its NOC families. To see this, suppose $\{\pi_1, \pi_2\}$ is an NOC family for π_0 and let $\{\tau_1, \ldots, \tau_m\}$ be an NOC family for π_1. Since then $\{\tau_1, \ldots, \tau_m, \pi_2\}$ is an NOC family for π_0, the "invariance" principle requires that:

$$\varphi(\pi_0) = F_2(\varphi(\pi_1), \varphi(\pi_2)) = F_{m+1}(\varphi(\tau_1), \ldots, \varphi(\tau_m), \varphi(\pi_2)). \tag{21}$$

   Similarly, it must be

$$\varphi(\pi_1) = F_m(\varphi(\tau_1), \ldots, \varphi(\tau_m)) \tag{22}$$

   so that

$$\varphi(\pi_0) = F_2(F_m(\varphi(\tau_1), \ldots, \varphi(\tau_m)), \varphi(\pi_2)). \tag{23}$$

   The above equalities lead to

$$F_2(F_m(\varphi(\tau_1), \ldots, \varphi(\tau_s)), \varphi(\pi_2)) = F_{m+1}(\varphi(\tau_1), \ldots, \varphi(\tau_s), \varphi(\pi_2)) \tag{24}$$

   which is essentially a "consistency in aggregation" requirement on nested NOC generating families.

The first property links functionals over general finite posets to functionals over the simplest kind of posets, i.e., linear orders [so justifying the use of the concept of linear extensions in many theoretical and applied studies (see, for example, Bruggemann and Annoni 2014 and Fattore 2016)]. The second property resembles formula (10), which expresses the semigroup nature of consistent-in-aggregation functionals.

In addition to the above properties, we require the aggregation system $\mathbb{F}$ to be continuous and monotone (see Sect. 3). The first is a technical, but natural, requirement; the second is a natural condition too, in view of application to statistical indicator construction: suppose that the value of the statistical indicator of interest grows on each linear extension of π_0 (for example, since the underlying frequency distribution changes in a specific way), then we want it to increase on π_0 as well. According to the discussion of Sect. 3, the most natural candidate aggregation system for frequency distributions on finite posets is thus the class of quasi-arithmetic means. Since in applications to statistical indicators it is also customary to impose homogeneity to aggregation functions, we finally restrict the class of functionals to power means.

According to the semigroup representation of quasi-arithmetic means (and hence of power means), it is easy to state how to compute $F(\pi_0)$, given an NOC generating family. To be simple, let $\{\pi_1, \pi_2\}$ be an NOC family and let $\Omega_1 = \Omega(\pi_1)$ and $\Omega_2 = \Omega(\pi_2)$ be the corresponding sets of linear extensions. Since $\{\pi_1, \pi_2\}$ is an NOC family, it is $\Omega_1 \cap \Omega_2 = \emptyset$ and $\Omega = \Omega(\pi_0) = \Omega_1 \cup \Omega_2$. Then $\varphi(\pi_0)$ is simply the power mean of $\varphi(\pi_1)$ and $\varphi(\pi_2)$, with weights given by the relative cardinality of Ω_1 and Ω_2:

$$\varphi(\pi_0) = M_{[2];w}\left(\varphi(\pi_1), \varphi(\pi_2); \left(\frac{|\Omega_1|}{|\Omega|}, \frac{|\Omega_2|}{|\Omega|}\right)\right). \tag{25}$$

This formula can be directly extended to NOC families with more than two elements. In particular, for the NOC family composed of the linear extensions of π_0, it reduces to a non-weighted power mean of the values of the functional $\varphi(\cdot)$ on the linear extensions themselves.

Final remark. The above discussion may be informally summarized as follows: to be consistent with the meet-semilattice structure of the extension poset, functionals over $\Pi(\pi_0)$ are required to be invariant and consistent in aggregation with respect to NOC families. As a consequence, once functionals are assigned to maximal elements of $\Pi(\pi_0)$ (i.e., to the linear extensions of π_0), they can be extended to any poset in $\Pi(\pi_0)$, through a power mean $M_{[k],w}(\cdot)$. The axiomatic theory cannot impose any constraint on the properties of functionals on linear extensions (in view of statistical indicators construction, these "intrinsic" properties depend upon the nature of the univariate index corresponding to the functional and upon the way it "reacts" to changes in the frequency distribution). It can only give prescriptions on the "structural" properties of the aggregation system $\mathbb{F}$, for the extension process to be consistent with the algebraic structure of $\Pi(\pi_0)$.[3]

5 Application to Synthetic Index Construction: An Example

For exemplificative purposes, in this section, we apply the extension procedure to the construction of bi-polarization[4] indices (Wolfson, 1994) on $\Pi(\pi_0^*)$ (see Fig. 2) for different frequency distributions. We must first introduce a univariate bi-polarization index, to "start" the process. Let $\boldsymbol{p} = (p_{a_1}, \ldots, p_{a_m})$ be a distribution of relative frequencies on a linear order $a_1 < a_2 < \cdots < a_m$. The following functional (Berry and Mielke, 1992):

$$L(\boldsymbol{p}_i) = \frac{2}{m-1} \sum_{i=1}^{m} \sum_{j=1}^{m} p_{a_i} p_{a_j} |r(a_i) - r(a_j)| \tag{26}$$

(where $r(x)$ stands for the rank[5] of x) is called the Leti index (Leti, 1983) and can be interpreted as a normalized measure of bi-polarization on ordinal attributes. We consider six different frequency distributions on π_0^*, to analyze the behavior of the functionals in different cases. As often in statistics, here we aggregate through arithmetic means. Frequency distributions are reported in Table 1. We denote by $L(\pi_j^*, \boldsymbol{p}_i)$, where $i = 1, \ldots, 6$, $j = 0, \ldots, 13$, the Leti functional "extended" to poset π_j^* and computed on the frequency distribution $\boldsymbol{p}_i$, i.e., the functional on poset π_j^* derived by $L(\boldsymbol{p}_i)$. Formally:

$$L(\pi_j^*, \boldsymbol{p}_i) = \frac{1}{|\Omega(\pi_j^*)|} \sum_{\lambda \in \Omega(\pi_j^*)} L(\lambda, \boldsymbol{p}_i) \tag{27}$$

where $L(\lambda, \boldsymbol{p}_i)$ is the unidimensional Leti index defined in (26), specialized to the linear order corresponding to linear extension λ. Table 2 reports the values of the extended Leti functionals in the various cases. Reading the table by rows, one can see how inequality changes as poset structure changes, while the frequency

[3]This means, the other way around, that the axiomatic theory presented above imposes constraints on the way scalar values can be assigned to elements of $\Pi(\pi_0)$.

[4]The term "bi-polarization" originally refers to economic variables and attributes that tend to be concentrated on low and high values. Bi-polarization corresponds to the "vanishing of the middle class" phenomenon, typical of the so-called developed countries, in the last decades.

[5]The rank of an element x in a linear order ℓ equals 1 plus the number of elements below x in ℓ. In the linear order $a_1 < a_2 < \cdots < a_m$, the rank of a_i equals i and expression (26) coincides with the original formulation, given in Berry and Mielke (1992).

Table 1 Frequency distributions over the poset π_0^* depicted in Fig. 1

	a	b	c	d	e
$\boldsymbol{p}_1$	0.2	0.2	0.2	0.2	0.2
$\boldsymbol{p}_2$	0.3	0.0	0.4	0.0	0.3
$\boldsymbol{p}_3$	0.5	0.0	0.0	0.0	0.5
$\boldsymbol{p}_4$	0.0	0.3	0.0	0.4	0.3
$\boldsymbol{p}_5$	0.0	0.5	0.0	0.0	0.5
$\boldsymbol{p}_6$	0.0	0.3	0.5	0.2	0.0

Table 2 Extended Leti functionals over $\Pi(\pi_0^*)$ for the six distributions of Table 1

	π_0^*	π_1^*	π_2^*	π_3^*	π_4^*	π_5^*	π_6^*
$\boldsymbol{p}_1$	0.800	0.800	0.800	0.800	0.800	0.800	0.800
$\boldsymbol{p}_2$	0.756	0.840	0.770	0.735	0.840	0.840	0.735
$\boldsymbol{p}_3$	0.900	1.000	0.917	0.875	1.000	1.000	0.875
$\boldsymbol{p}_4$	0.672	0.560	0.630	0.735	0.525	0.630	0.735
$\boldsymbol{p}_5$	0.800	0.666	0.750	0.875	0.625	0.750	0.875
$\boldsymbol{p}_6$	0.498	0.413	0.453	0.520	0.390	0.415	0.565
	π_7^*	π_8^*	π_9^*	π_{10}^*	π_{11}^*	π_{12}^*	π_{13}^*
$\boldsymbol{p}_1$	0.800	0.800	0.800	0.800	0.800	0.800	0.800
$\boldsymbol{p}_2$	0.735	0.630	0.840	0.840	0.840	0.630	0.630
$\boldsymbol{p}_3$	0.875	0.750	1.000	1.000	1.000	0.750	0.750
$\boldsymbol{p}_4$	0.735	0.840	0.420	0.630	0.630	0.840	0.840
$\boldsymbol{p}_5$	0.875	1.000	0.500	0.750	0.750	1.000	1.000
$\boldsymbol{p}_6$	0.475	0.625	0.410	0.460	0.370	0.580	0.670

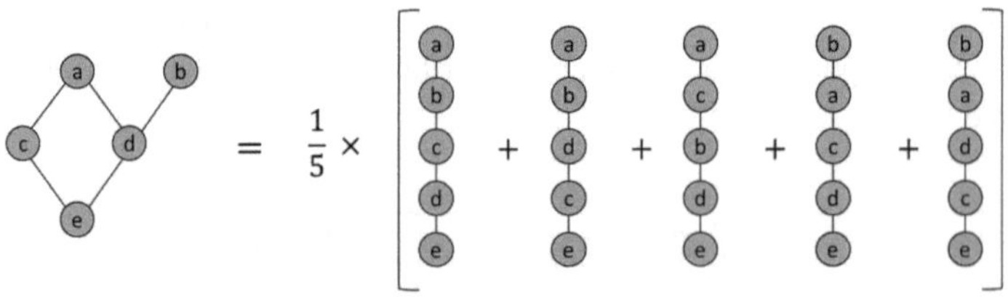

Fig. 4 Evaluation of the extended Leti index, for a generic frequency distribution, based on the NOC family of the linear extensions

distribution is kept fixed; reading it by columns, one sees how the index changes when the frequency distribution changes on the same poset. Values associated with the various posets satisfy the consistency-in-aggregation constraint. To help grasping this, Figs. 4 and 5 report "symbolically" the computation of the extended Leti index on π_0^*, for different NOC families (shaded Hasse diagrams stand for the evaluation of the Leti index on the corresponding poset, i.e., they represent scalars).

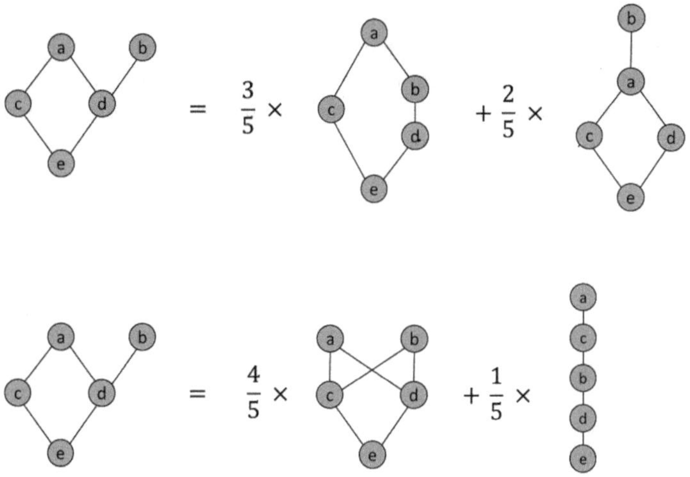

Fig. 5 Evaluation of the extended Leti index, for a generic frequency distribution, based on the NOC families reported in Fig. 3

6 Conclusion

In this paper, we have developed an axiomatic theory of continuous, monotone, and consistent-in-aggregation functionals on frequency distributions defined on finite posets, as a step towards a general theory for the construction of synthetic indicators over multidimensional systems of ordinal attributes. In concrete applications to socio-economic measurement, the formal setting outlined before must be integrated with the axiomatics of specific classes of indicators, since what specifies a particular functional as an inequality, a polarization, or a variability index (to make a few examples) lays outside the general theory here proposed. This, in fact, may be seen as a "trick" to extend to partially ordered sets univariate indicators, i.e., indicators defined on linear orders. This "device" has been exemplified in the case of the Leti bi-polarization index, but it may be applied to any kind of indicator and may be seen as a general procedure to produce well-behaved "multidimensional versions" of univariate indices. The advantage of this approach over classical attempts to get multidimensional generalizations of univariate measures is that the axiomatic setting is neat and general and does not involve any attempt to generalize "core" axioms, which are "context specific." A prototypical example of this situation concerns again the measurement of bi-polarization. While it is clear how to (partially) order univariate distributions in terms of bi-polarization (Allison-Foster criterion), it is not clear how to do this neatly in the multidimensional case. Correspondingly, while an axiom of "monotonicity" can be easily and unambiguously stated for univariate polarization indices, the same is not true for the corresponding multidimensional versions. These problems are circumvented when one imposes the index to behave consistently under changes in the order structure of the domain of the frequency distributions. The resulting multidimensional index is coherent as a bi-polarization index since (1) on linear orders it behaves as expected and (2) on multidimensional posets it behaves consistently with changes of the partial order structure. As usual when dealing with finite posets, computations may be difficult due to the huge number of linear extensions. The problem is well known and can be addressed using sampling procedures (Bubley and Dyer, 1999) to estimate the values of statistical indicators, rather than to compute them exactly. However, the primary interest of the paper is of a theoretical kind, so we have not considered this issue here.

References

Beliakov G, Pradera A, Calvo T (2007) Aggregation functions: a guide for practitioners. Studies in fuzziness and soft computing, vol 221. Springer, Berlin

Berry JK, Mielke Jr PW (1992) Assessment of variation in ordinal data. Percept Mot Skills 74:63–66

Brualdi RA, Hyung CJ, Trotter Jr WT (1994) On the poset of all posets on n elements. Discret Appl Math 50:111–123

Bruggemann R, Annoni P (2014) Average heights in partially ordered sets. MATCH Commun Math Comput Chem 71:101–126

Bubley R, Dyer M (1999) Faster random generation of linear extensions. Discret Math 201:81–88

Davey BA, Priestley BH (2002) Introduction to lattices and order. Cambridge University Press, Cambridge

Fattore M (2016) Partially ordered sets and the measurement of multidimensional ordinal deprivation. Soc Indic Res 128(2):835–858. doi: 10.1007/s11205-015-1059

Grabisch M, Marichal J-L, Mesiar R, Pap E (2009) Aggregation functions. Cambridge University Press, Cambridge

Leti G (1983) Statistica descrittiva. Il Mulino, Bologna

Neggers J, Kim SH (1998) Basic posets. World Scientific, Singapore

Pursiainen H (2005) Consistent aggregation methods and index number theory. Ph.D. Dissertation

Schröder B (2002) Ordered set. An introduction. Birkäuser, Boston

Wolfson CM (1994) When inequalities diverge. Am Econ Rev 84(2):353–358

Evaluation, Considered as Problem Orientable Mathematics Over Lattices

Adalbert Kerber

1 Evaluations Using Parameters

Toxicological and Environmental Chemistry as well as their applications are usually accompanied by an evaluation of objects (for example, chemical compounds like refrigerants) with respect to relevant properties (good or bad ones, for example, ozone depletion potential, etc., expressed in terms of parameter values) in order to evaluate their behavior and to select best possible of the objects considered. This is a problem "of the real world," and so we have to go beyond the classical binary thinking of mathematics. An object is mostly no longer either good or bad, it is usually good or bad "in a certain sense." Moreover the qualities of two of the objects quite often are incomparable. And usually there are several objects that are "best possible" (which means that there are no better ones among the evaluated objects), as we shall see.

Hence, the result of an evaluation of a set of objects with respect to a set of attributes is quite often a table of elements of a partial order or, in mathematical terms, of elements in a lattice L. This is described in some detail and underlined by examples. At first glance it may sound as if this were a disadvantage, making things more complicated. But it will be shown that it has an interesting advantage, since we can do a problem oriented approach by choosing a suitable logic, depending if we want to be more or less strict in our interpretation of the given evaluation or not. Here are, to begin with, two examples which illustrate this.

A. Kerber (✉)
Department of Mathematics, University of Bayreuth, Universitätsstraße, 95440 Bayreuth, Germany
e-mail: kerber@uni-bayreuth.de

M. Fattore, R. Bruggemann (eds.), *Partial Order Concepts in Applied Sciences*,
DOI 10.1007/978-3-319-45421-4_6

1.1 An Evaluation of Refrigerants Under Ecological Aspects, $L = [0, 1]^3$

In Bruggemann et al. (2011) and Kerber and Bruggemann (2015) we considered 18 of the 40 refrigerants evaluated by G. Restrepo in his dissertation (Restrepo, 2008). He evaluated these objects with respect to the ecologically relevant parameters ozone depletion potential (*ODP*), general warming potential (*GWP*), and atmospheric lifetime (*ALT*). After normalizing these parameters so that the values of the normalized parameters *nODP*, *nGWP*, and *nALT* are contained in the interval [0, 1], the following matrix of parameter values shows up:

Refr.	*nODP*	*nGWP*	*nALT*	*Chem.Formula*
RE_1	0.19608	0.31622	0.01406	CCl_3F
RE_2	0.16078	0.72432	0.03125	CCl_2F_2
RE_6	0.02353	0.04818	0.00290	$C_2H_3Cl_2F$
RE_7	0.01275	0.15338	0.00559	$C_2H_3ClF_2$
RE_8	0.00008	0.96689	0.08437	CHF_3
RE_{16}	0	0.00135	0.00001	C_3H_8
RE_{21}	0	0.00007	0.03750	CO_2
RE_{22}	1	0.08784	0.00343	$CBrClF_2$
RE_{23}	0	0.67568	1	C_4F_8
RE_{29}	0	1	0.05156	C_2HF_5O
RE_{32}	0.00392	0.00108	0.00040	CH_3Cl
RE_{33}	0.17647	0.40541	0.02656	$C_2Cl_3F_3$
RE_{35}	0.16667	0.66216	0.09375	$C_2Cl_2F_4$
RE_{36}	0	0.00007	0.00003	CF_3I
RE_{37}	0	0.00007	0	C_2H_6O
RE_{38}	0	0	0.00007	NH_3
RE_{39}	0	0.04432	0.00178	$C_2H_3F_3O$
RE_{40}	0	0.04709	0.00125	$C_3H_3F_5O$

Here is a first, trivial but important

Remark 1.1. If we consider this table as a matrix with three columns of parameter values, it is a matrix with entries in the interval of real numbers [0, 1]. This set [0, 1], consisting of all the real numbers between 0 and 1, is a *total order*, a set where any two different elements can be compared, one of them is the smaller number, the other one the bigger. But if we consider the above table as a matrix with a single column, containing *triples* of real numbers, it is a matrix over the lattice $L = [0, 1]^3$, which is *a partial and not a total order, and therefore incomparabilities show up.* And it is this embedding into L which is the desired evaluation of the refrigerants.

The "new" matrix looks like that:

Refr.	*triples* $(nODP, nGWP, nALT)$	*Chem.Formula*
RE_1	$(0.19608, 0.31622, 0.01406)$	CCl_3F
RE_2	$(0.16078, 0.72432, 0.03125)$	CCl_2F_2
RE_6	$(0.02353, 0.04818, 0.00290)$	$C_2H_3Cl_2F$
RE_7	$(0.01275, 0.15338, 0.00559)$	$C_2H_3ClF_2$
RE_8	$(0.00008, 0.96689, 0.08437)$	CHF_3
RE_{16}	$(0, 0.00135, 0.00001)$	C_3H_8
RE_{21}	$(0, 0.00007, 0.03750)$	CO_2
RE_{22}	$(1, 0.08784, 0.00343)$	$CBrClF_2$
RE_{23}	$(0, 0.67568, 1)$	C_4F_8
RE_{29}	$(0, 1, 0.05156)$	C_2HF_5O
RE_{32}	$(0.00392, 0.00108, 0.00040)$	CH_3Cl
RE_{33}	$(0.17647, 0.40541, 0.02656)$	$C_2Cl_3F_3$
RE_{35}	$(0.16667, 0.66216, 0.09375)$	$C_2Cl_2F_4$
RE_{36}	$(0, 0.00007, 0.00003)$	CF_3I
RE_{37}	$(0, 0.00007, 0)$	C_2H_6O
RE_{38}	$(0, 0, 0.00007)$	NH_3
RE_{39}	$(0, 0.04432, 0.00178)$	$C_2H_3F_3O$
RE_{40}	$(0, 0.04709, 0.00125)$	$C_3H_3F_5O$

As smaller parameter values are the better ones, for example, the triples that evaluate the ecological properties of RE_{37} and RE_{38} are **incomparable**, none of them is the best of these two, none of them is the worst. The resulting *evaluation* of the eighteen refrigerants according to their ecological qualities, expressed in terms of these three parameters, is just the following *canonical* order on their triples:

$$(\alpha, \beta, \gamma) \leq (\alpha', \beta', \gamma') \iff \alpha \leq \alpha', \beta \leq \beta', \gamma \leq \gamma'.$$

Remark 1.2. We prefer the word evaluation, as the word ranking suggests that there is a single object of highest rank, a best refrigerant, then one second best, a single third best, and so on, which is definitely not the case! Rankings are usually obtained by weighting the parameters until a ranking is obtained. This is certainly a kind of manipulation!

Using Bruggemann's PyHasse, cf. Bruggemann et al. (2014), we can easily visualize this evaluation. Here is the Hasse diagram that we obtain:

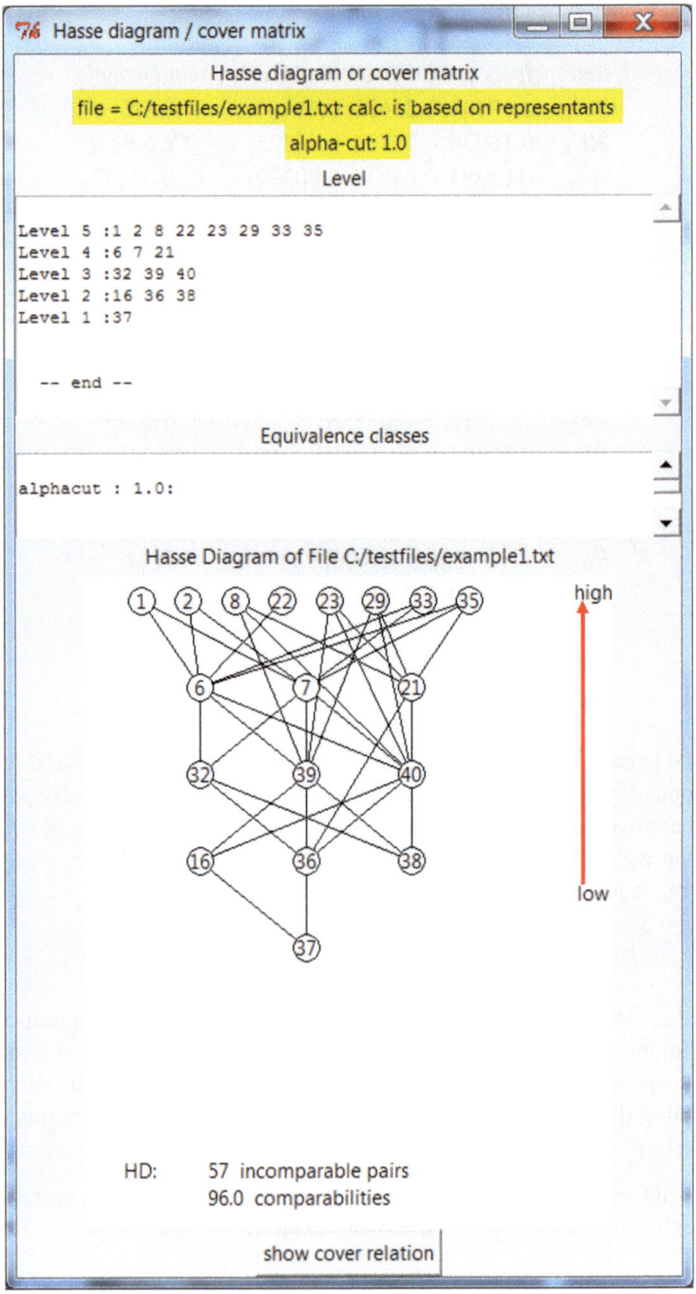
Hasse diagram / cover matrix
Hasse diagram or cover matrix
file = C:/testfiles/example1.txt: calc. is based on representants
alpha-cut: 1.0
Level
Level 5 :1 2 8 22 23 29 33 35
Level 4 :6 7 21
Level 3 :32 39 40
Level 2 :16 36 38
Level 1 :37
-- end --
Equivalence classes
alphacut : 1.0:
Hasse Diagram of File C:/testfiles/example1.txt
1
2
8
22
23
29
33
35
6
7
21
32
39
40
16
36
38
37
high
low
HD:
57 incomparable pairs
96.0 comparabilities
show cover relation

As smaller values of the parameters are the better ones, it shows that the refrigerants RE_{37} and RE_{38} turn out to be *best possible,* i.e., there are no refrigerants better than RE_{37} or better than RE_{38}, while the eight refrigerants on the highest level are *worst*, i.e., there are no refrigerants worse than RE_1, RE_2, RE_8,... RE_{35}.

Hence, the desired evaluation is exactly what we obtain when we consider the matrix of parameter values as a matrix consisting of a single column with its values in the lattice $L = [0, 1]^3$. And this has in fact further advantages as we shall see in a minute. But before we do this let us consider a second example of an evaluation that also uses values in a lattice, and it shows that evaluations over lattices occur "in daily life":

1.2 A Test of Electronic Devices for CDs and DVDs, $L = \{\ominus\ominus / \ominus\ominus, \ldots\}$

Such evaluations can be found in the computer journal c't, they may look like that:

	$wDVD$	wCD	$rDVD$	rCD	$eDVD/CD$	$nDVD/CD$
a	$\oplus\oplus/\oplus$	$\ominus/\ominus$	$\ominus\ominus/\bigcirc$	$\oplus/\ominus$	$\ominus\ominus/\ominus$	$\oplus/\bigcirc$
b	$\oplus/\ominus$	$\oplus/\ominus$	$\ominus/\ominus$	$\bigcirc/\ominus$	$\ominus\ominus/\oplus$	$\ominus/\bigcirc$
c	$\bigcirc/\bigcirc$	$\oplus\oplus/\oplus\oplus$	$\bigcirc/\oplus$	$\oplus/\bigcirc$	$\oplus\oplus/\ominus\ominus$	$\ominus/\ominus$
d	$\oplus/\oplus$	$\bigcirc/\ominus$	$\bigcirc/\bigcirc$	$\ominus/\oplus$	$\oplus/\oplus$	$\oplus/\bigcirc$

The entries are pairs of test results because there are two norms taken into account, $+R$ and $-R$. In contrast to the evaluation of refrigerants, considered as a matrix with three columns, the resulting pairs of the present test results are elements of a partial order, a lattice L, where incomparable entries occur, $L = \{\ominus\ominus / \ominus\ominus, \ldots, \oplus\oplus / \oplus\oplus\}$. This lattice is a cartesian square of a total order (e.g., replace $\ominus\ominus$ by -2, $\ominus$ by -1, $\bigcirc$ by 0, $\oplus$ by 1, and $\oplus\oplus$ by 2):

$$L \equiv \{-2, -1, 0, 1, 2\}^2.$$

Its Hasse diagram looks as follows and it shows many incomparabilities:

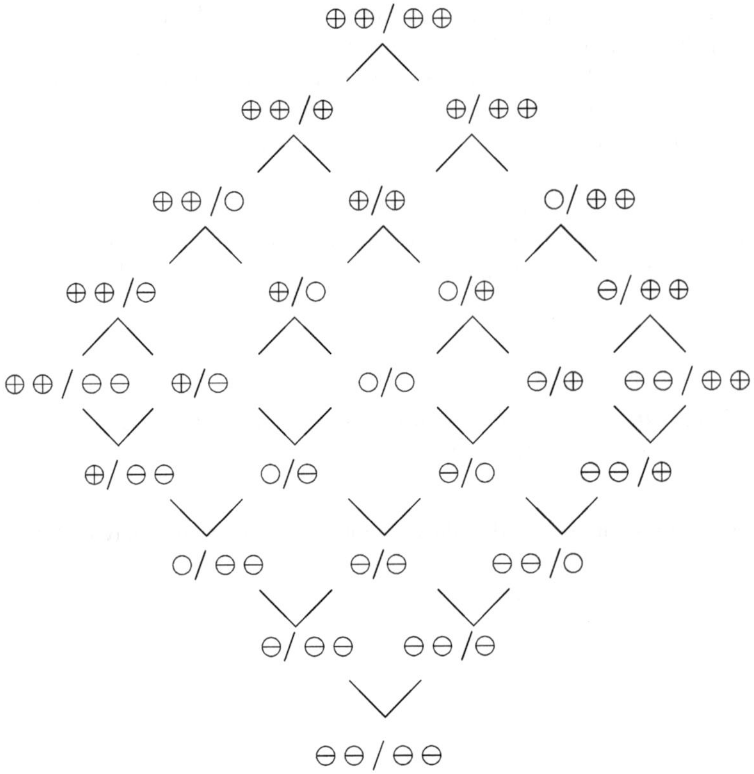

There are various possibilities to use these test results:

- Using the same approach as in the case of the refrigerants, we can consider this table as a one-column table, where the lattice is L^6, obtaining that the four devices are pairwise incomparable which does not help very much.
- Another approach suggests to separate the two norms $\pm R$, obtaining two tables:

$+R$	*wDVD*	*wCD*	*rDVD*	*rCD*	*eDVD/CD*	*nDVD/CD*
a	⊕⊕	⊖	⊖⊖	⊕	⊖⊖	⊕
b	⊕	⊕	⊖	○	⊖⊖	⊖
c	○	⊕⊕	○	⊕	⊕⊕	⊖
d	⊕	○	○	⊖	⊕	⊕

and also here the four devices are pairwise incomparable, while with respect to $-R$ we get

$-R$	$wDVD$	wCD	$rDVD$	rCD	$eDVD/CD$	$nDVD/CD$
a	⊕	⊖	○	⊖	⊖	○
b	⊖	⊖	⊖	⊖	⊕	○
c	○	⊕⊕	⊕	○	⊖⊖	⊖
d	⊕	⊖	○	⊕	⊕	○

> Comparing the test results for a and d, for example, we see that with respect to $-R$ device d is better than a and also better than b, while d and c are incomparable, which also holds for b and c, a and c, as well as a and b.

In many cases evaluations use parameters, say $p_1,\ldots,p_n$, and they have their values in sets $P_1,\ldots P_n$, like $[0,1]$ or $\{-2,-1,0,1,2\}$, for example, which are total orders (and therefore they can be reordered so that the smaller values are the better ones). Hence the resulting matrix can be considered as a one-column matrix with its values contained in the cartesian product

$$L = P_1 \times \cdots \times P_n$$

of the parameter sets, which is in fact a lattice, a set with an infimum:

$$(p_1(x),\ldots,p_n(x)) \wedge (p_1(y),\ldots,p_n(y)) = (p_1(x)\wedge p_1(y),\ldots,p_n(x)\wedge p_n(y)),$$

where the infima $p_i(x)\wedge p_i(y)$ are defined as the minima of the two parameter values:

$$p_i(x)\wedge p_i(y) := \min\{p_i(x),p_i(y)\}$$

and, correspondingly, the suprema

$$(p_1(x),\ldots,p_n(x)) \vee (p_1(y),\ldots,p_n(y)) = (p_1(x)\vee p_1(y),\ldots,p_n(x)\vee p_n(y)),$$

where we put

$$p_i(x)\vee p_i(y) := \max\{p_i(x),p_i(y)\}.$$

For this reason we feel justified to use the following approach.

2 A General Approach to Evaluation Over *L*

Definition 2.1. An evaluation of a finite set O of objects o_i, $1 \le i \le m$, with respect to a finite set A of attributes a_k, $1 \le k \le n$, over a lattice L, can be considered as a mapping

$$\mathcal{E}\colon O \times A \to L, (o_i, a_k) \mapsto \mathcal{E}(o_i, a_k)$$

that yields an $m \times n$-matrix of values $\mathcal{E}(o_i, a_k) \in L$:

$$(\mathcal{E}(o_i, a_k))_{1\le i\le m, 1\le k\le n}.$$

The value $\mathcal{E}(o_i, a_k) \in L$ is the "truth-value" associated with the statement "object o_i has attribute a_k".

In the first example such a truth value is a triple of parameter values like $(0, 0.00007, 0)$, in the second example it is a value like $\oplus \oplus / \oplus$, say. This general approach to multi-parameter evaluations has advantages over the standard situation, where the values are contained in $\{0, 1\}$ or $[0, 1]$: It allows problem orientation and an exploration, as we shall briefly sketch now (for more details see Pollandt 1997) and *this should be useful in many applications*.

The first step is an interpretation of such an evaluation as an L-subset which allows a problem oriented approach. For this purpose we use the interpretation of subsets as mappings:

2.1 Evaluations as L-subsets

We introduce the basic notions of the theory of L-subsets, in order to consider evaluations as such subsets, which is helpful, see, e.g., Bruggemann et al. (2011), De Baets and De Meyer (2003), Kerber and Bruggemann (2015) and Naessens et al. (1999).

- Recall the notion of *crisp* (or *Boolean*) sets, i.e., of the "usual" sets X, for which an object x is either contained in X or not. The subsets of such crisp sets can be identified with mappings S from X to $\{0, 1\}$, and therefore the set of all crisp subsets of X can be identified with

$$\{0, 1\}^X = \{S \mid S : X \to \{0, 1\}\},$$

  as we can associate such a mapping S with the subset of elements x that are mapped onto 1, by S, $S(x) = 1$. Such a mapping (or subset) may be called a $\{0, 1\}$-*subset* of X.
- More generally, we can identify the *L-subsets* of X with the elements $\mathcal{S}$ of the set of mappings

$$L^X = \{\mathcal{S} \mid \mathcal{S} : X \to L\},$$

  where we may interpret the value $\mathcal{S}(x)$ as the *truth value* of x being an element of the L-subset $\mathcal{S}$.

2.2 Set Theories for L-subsets

The advantage is that we can choose between several set theories (and corresponding logics) for L-subsets:

- Our main example of an L-subset is, of course, the notion of evaluation given above,

$$\mathcal{E}: O \times A \to L, (o, a) \mapsto \mathcal{E}(o, a),$$

  an L-subset of the cartesian product $O \times A$ of the set O of objects and the set A of attributes, where the value $\mathcal{E}(o, a)$ evaluates the truth of the statement "object o has attribute a".
- We introduce *L-inclusion* $\subseteq_L$ between two L-subsets of X by the equivalence

$$\mathcal{S} \subseteq_L \mathcal{S}' \iff \forall\, x \in X : \mathcal{S}(x) \leq \mathcal{S}'(x).$$

- In order to describe the necessary parts of the set theory of L-subsets, we introduce *t-norms* that are used in order to define an L-intersection of such sets. These are the mappings $\tau : L \times L \to L$ with
  - *symmetry*: $\tau(\lambda, \mu) = \tau(\mu, \lambda)$, for all $\lambda, \mu \in L$,
  - *monotony*: $\mu \leq \nu$ implies $\tau(\lambda, \mu) \leq \tau(\lambda, \nu)$,
  - *associativity*: $\tau(\lambda, \tau(\mu, \nu)) = \tau(\tau(\lambda, \mu), \nu)$,
  - *side-condition*: $\tau(\lambda, 1_L) = \lambda$.
- A t-norm τ yields *τ-intersections* $\mathcal{I}$:

$$\mathcal{I}(x) = (\mathcal{S} \cap_\tau \mathcal{S}')(x) = \tau(\mathcal{S}(x), \mathcal{S}'(x)),$$

  σ-unions $\mathcal{S} \cup_\sigma \mathcal{S}'$ can be introduced analogously via t-conorms σ.
- The most important t-norms are
  - *standard* norm: $s(\lambda, \mu) = \lambda \wedge \mu$,
  - *drastic* norm

$$d(\lambda, \mu) = \begin{cases} \lambda & \mu = 1_L, \\ \mu & \lambda = 1_L, \\ 0_L & \text{otherwise}, \end{cases}$$

  - and if the lattice is $L = [0, 1]^n$, where we can use addition and multiplication of real numbers, there are the *algebraic product ap* and the *bounded difference bd*, defined as follows:

$$ap(\lambda, \mu) = \lambda \cdot \mu, \text{ and } bd(\lambda, \mu) = \text{Max}\{0, \lambda + \mu - 1\}.$$

  - There are various other t-norms, and in particular the following is true:

$$d(\lambda, \mu) \leq \tau(\lambda, \mu) \leq s(\lambda, \mu)$$

which means that we have a choice by picking any suitable t-norm τ, i.e., a t-norm between d and s, *a choice between more or less strict argumentations.*

This means that we have a way of *problem-orientation* at hand since evaluations $\mathcal{E}$ can be considered as L-subsets where the set theory can be chosen by picking a suitable pair (τ, σ), a norm, and a conorm that fit together.

An illustration of the difference between t-norms is the following example, showing models of $\mathcal{SA}$, the linguistic expression of *strong acid*, and $\mathcal{WA}$, the linguistic expression *weak acid*,

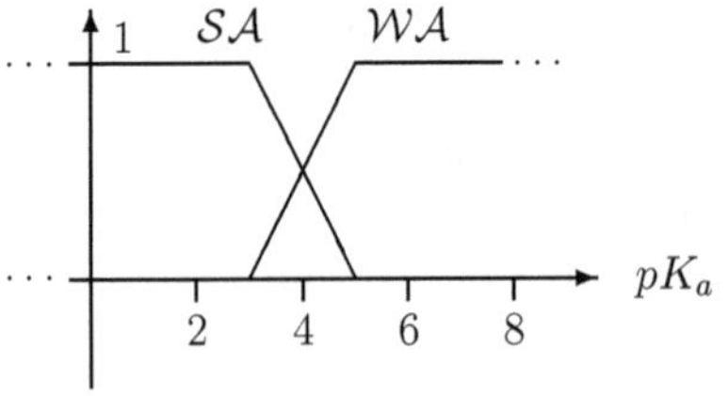

If we choose $\tau = s$, an acid with pK_a-value 4 is both a strong acid and a weak acid. If $\tau = d$, we do not accept that there is any acid being both a strong acid and a weak acid, as

$$(\mathcal{SA} \cap_d \mathcal{WA})(r) = 0$$

(although $\mathcal{SA}(4) = \mathcal{WA}(4) = 0.5$). We are in fact using kind of *semantic notion of truth*, based on τ.

2.3 A Corresponding Logic

In many cases, there is another important mapping, closely related to the chosen norm, see Pollandt (1997):

- $\tilde{\tau}: L \times L \to L$ is a *residuum* of τ, if it satisfies the following equivalence, expressed in terms of the partial order $\leq$ in the given lattice L,

$$\tau(\lambda, \mu) \leq \nu \iff \lambda \leq \tilde{\tau}(\mu, \nu).$$

- In addition, if $\tau(\lambda, \bigvee M) = \bigvee_{\mu \in M} \tau(\lambda, \mu)$ holds, where $\bigvee M$ denotes the supremum of a subset $M \subseteq L$, then we obtain the values of $\tilde{\tau}$ as follows:

$$\tilde{\tau}(\lambda, \mu) = \bigvee \{\nu \mid \tau(\lambda, \nu) \leq \mu\}.$$

In this case τ is called a *residual* t-norm.

- *This yields a logic* corresponding to the lattice L, the chosen t-norm τ and its residuum $\tilde{\tau}$. In particular we obtain the following *"truth value" of $\mathcal{A}$ being an L-subset of $\mathcal{B}$*, i.e., for L-inclusion of $\mathcal{A}, \mathcal{B} \in L^X$,

$$\tilde{\tau}(\mathcal{A} \subseteq_L \mathcal{B}) = \bigwedge_{x \in X} \tilde{\tau}(\mathcal{A}(x), \mathcal{B}(x)).$$

2.4 Examples of Residua for L = [0, 1]

Here are a few examples (in a simple case of lattice) that are easy to check:

$$\tilde{s}(\alpha, \beta) = \begin{cases} 1 \text{ if } \alpha \leq \beta, \\ \beta \text{ otherwise,} \end{cases}$$

$$\tilde{d}(\alpha, \beta) = \begin{cases} \beta \text{ if } \alpha = 1, \\ 1 \text{ otherwise,} \end{cases}$$

$$\tilde{a}(\alpha, \beta) = \begin{cases} \beta/\alpha \text{ if } \alpha \neq 0, \\ 1 \quad\quad \text{otherwise,} \end{cases}$$

$$\tilde{b}(\alpha, \beta) = \text{Min}\{1, 1 - \alpha + \beta\}.$$

3 Exploration of the Evaluation $\mathcal{E}$

In order to *explore* an evaluation we need to describe the implications that we can obtain from $\mathcal{E}$. For this purpose we deduce from the foregoing section the following items:

- An object $o \in O$ *has* attribute $a \in A$ if and only if $\mathcal{E}(o, a) > 0$.
- The row corresponding to the object $o \in O$ is the sequence of values $\mathcal{E}(o, a), a \in A$, i.e., in terms of mappings, it is

$$\mathcal{E}(o, -) \colon A \to L, a \mapsto \mathcal{E}(o, a), \text{ in formal terms } \mathcal{E}(o, -) \in L^A.$$

- Now we use that we can assume that smaller values of the parameters are the better ones, so that $\mathcal{A} \in L^A$ *holds* for the object $o \in O$ if and only if

$$\mathcal{A} \subseteq_L \mathcal{E}(o, -).$$

- This means that the truth value, let us denote it by $\mathcal{A}'(o)$, of "$\mathcal{A}$ holds for o in $\mathcal{E}$" is

$$\mathcal{A}'(o) = \bigwedge_{a \in A} \tilde{\tau}(\mathcal{A}(a), \mathcal{E}(o, a)).$$

- Finally we want to evaluate if, for $\mathcal{A} \in L^A$ and $\mathcal{B} \in L^A$, $\mathcal{A}$ implies $\mathcal{B}$ (in the evaluation $\mathcal{E}$). In formal terms, we want to derive the truth value of the implication $\mathcal{A} \Rightarrow_{\mathcal{E}} \mathcal{B}$ (read "$\mathcal{A}$ implies $\mathcal{B}$, according to the evaluation $\mathcal{E}$"). Thus, we have first to evaluate both $\mathcal{A} \subseteq_L \mathcal{E}(o, -)$ and $\mathcal{B} \subseteq_L \mathcal{E}(o, -)$, and we already know that these are $\mathcal{A}'(o)$ and $\mathcal{B}'(o)$, respectively. Hence, the truth value of "$\mathcal{A}$ implies $\mathcal{B}$ in $\mathcal{E}$," let us denote it by $\mathrm{tv}(\mathcal{A} \Rightarrow_{\mathcal{E}} \mathcal{B})$, is

$$\begin{aligned} \mathrm{tv}(\mathcal{A} \Rightarrow_{\mathcal{E}} \mathcal{B}) &= \bigwedge_{o \in O} \tilde{\tau}(\mathcal{A}'(o), \mathcal{B}'(o)) \\ &= \bigwedge_{o \in O} \tilde{\tau}\Big(\bigwedge_{a \in A} \tilde{\tau}(\mathcal{A}(a), \mathcal{E}(o, a)), \bigwedge_{a \in A} \tilde{\tau}(\mathcal{B}(a), \mathcal{E}(o, a))\Big). \end{aligned}$$

- In addition it is obvious that $\mathcal{A} \Rightarrow_{\mathcal{E}} \mathcal{B}$ *holds* in $\mathcal{E}$ iff $\mathrm{tv}(\mathcal{A} \Rightarrow_{\mathcal{E}} \mathcal{B}) = 1$, i.e., if and only if $\mathcal{A}' \subseteq_L \mathcal{B}'$, which needs that $\mathcal{A}'(o) \leq \mathcal{B}'(o)$, for each $o \in O$.
- We note that the main tool is the "derivation" $\mathcal{A} \mapsto \mathcal{A}'$ and similarly $\mathcal{O} \mapsto \mathcal{O}'$ for L-subsets $\mathcal{O} \in L^O$, so that we also have $\mathcal{A}' \mapsto \mathcal{A}''$ as well as $\mathcal{O}' \mapsto \mathcal{O}''$. Particular pairs of such L-subsets are of the form

$$(\mathcal{A}'', \mathcal{A}'), \text{ for an } \mathcal{A} \in L^A,$$

or, if you prefer,

$$(\mathcal{O}', \mathcal{O}''), \text{ for an } \mathcal{O} \in L^O.$$

They are called *concepts* and form the central notion in FCA, the *formal concept analysis*, an important mathematical model of the notion "concept" (due to philosophy) and introduced by R. Wille and co-workers, cf. Ganter and Wille (1996). $\mathcal{A}'' \in L^A$ (or $\mathcal{O}' \in L^A$) is called the *extent* of this concept, and $\mathcal{A}' \in L^O$ (or $\mathcal{O}'' \in L^O$) the *intent* of it, and the concepts are partially ordered with respect to L-inclusion of the extents:

$$(\mathcal{A}'', \mathcal{A}') \leq (\mathcal{B}'', \mathcal{B}') \iff \mathcal{A}'' \subseteq_L \mathcal{B}''.$$

In this case $(\mathcal{A}'', \mathcal{A}')$ is called a *subconcept* of $(\mathcal{B}'', \mathcal{B}')$, and correspondingly, we call $(\mathcal{B}'', \mathcal{B}')$ a *superconcept* of $(\mathcal{A}'', \mathcal{A}')$. The concepts form a lattice, *the lattice of concepts of the evaluation* $\mathcal{E}$.
- We saw that the extents of concepts are of the form $\mathcal{A}''$, for an $\mathcal{A} \in L^A$, while the intents are $\mathcal{O}''$, for an $\mathcal{O} \in L^O$.

These notions are used in the following theorem that is decisive for our problem, the exploration of a given evaluation. It is due to Duquenne and Guigues, and it is an important result in data bank theory: We define *pseudo-contents* $\mathcal{P} \in L^A$ by

$$\mathcal{P} \neq \mathcal{P}'' \text{ and for each pseudo-content } \mathcal{Q} \subset_L \mathcal{P}: \ \mathcal{Q}'' \subseteq_L \mathcal{P}.$$

The crucial point is, cf. Duquenne (1987):

Theorem 3.1. *From the following set* $\mathbb{P}$ *of implications one can deduce every implication between sets of attributes which holds in* $\mathcal{E}$. *For this reason it is called the Duquenne/Guigues-basis after its inventors (in the binary case):*

$$\mathbb{P} = \{\mathcal{P} \Rightarrow_{\mathcal{E}} (\mathcal{P}'' \setminus \mathcal{P}) \mid \mathcal{P} \textit{ pseudo-content}\}.$$

Thus, we may very well call the calculation of this basis the *exploration* of the evaluation considered. Here is an example:

3.1 The Exploration of a Binary Evaluation of the Refrigerants

We consider again the 18 refrigerants, a subset of the altogether 40 refrigerants evaluated in Restrepo (2008). The intention is to show an evaluation $\mathcal{E}^*$ of the 18 refrigerants and an extended set A^* of attributes, over the lattice $L = \{0, 1\}$ of the refrigerants in order to demonstrate its exploration, using available free software so that the interested reader can do such explorations himself. (Free software for evaluations over general lattices is not yet available, so we better restrict attention to classical contexts over the trivial lattice $L = \{0, 1\}$.)

Attributes will be parameters $nODP^*$, $nGWP^*$, and $nALT^*$, obtained from $nODP$, $nGWP$, and $nALT$ as follows: Using that the third quartiles (as we want to find out the reasons for refrigerants being bad with respect to the ecological properties) of the refrigerants are 0.162 for $nODP$, 0.666 for $nGWP$, and 0.041 for $nALT$, hence, we define the following parameters:

$$nODP^*(RE) = \begin{cases} 1 \text{ if } nODP(RE) > 0.162, \\ 0 \text{ otherwise,} \end{cases}$$

$$nGWP^*(RE) = \begin{cases} 1 \text{ if } nGWP(RE) > 0.666, \\ 0 \text{ otherwise,} \end{cases}$$

$$nALT^*(RE) = \begin{cases} 1 \text{ if } nALT(RE) > 0.041, \\ 0 \text{ otherwise.} \end{cases}$$

In addition we consider as attributes the existence of Cl, F, Br, I in RE, the function ether, i.e., if RE is an ether, and whether or not RE is CO_2 or NH_3. Moreover we use a parameter nC which is defined as follows:

$$nC(RE) = \begin{cases} 1 \text{ if } RE \text{ contains at least two single bonded carbon atoms,} \\ 0 \text{ otherwise.} \end{cases}$$

We obtain the following evaluation $\mathcal{E}^*$ over $L = \{0, 1\}$:

$\mathcal{E}^*$	$nODP^*$	$nGWP^*$	$nALT^*$	nC	Cl	F	Br	I	Ether	CO_2	NH_3
1	1	0	0	0	1	1	0	0	0	0	0
2	0	1	0	0	1	1	0	0	0	0	0
6	0	0	0	1	1	1	0	0	0	0	0
7	0	0	0	1	1	1	0	0	0	0	0
8	0	1	1	0	0	1	0	0	0	0	0
16	0	0	0	1	0	0	0	0	0	0	0
21	0	0	0	0	0	0	0	0	0	1	0
22	1	0	0	0	1	1	1	0	0	0	0
23	0	1	1	1	0	1	0	0	0	0	0
29	0	1	1	1	0	1	0	0	1	0	0
32	0	0	0	0	1	0	0	0	0	0	0
33	1	0	0	1	1	1	0	0	0	0	0
35	1	0	1	1	1	1	0	0	0	0	0
36	0	0	0	0	0	1	0	1	0	0	0
37	0	0	0	1	0	0	0	0	1	0	0
38	0	0	0	0	0	0	0	0	0	0	1
39	0	0	0	1	0	1	0	0	1	0	0
40	0	0	0	1	0	1	0	0	1	0	0

The evaluation is binary, and so $\mathcal{A}'$ is the set of objects that have all the attributes contained in $\mathcal{A}$ while $\mathcal{O}'$ consists of all the attributes that all the objects $o \in \mathcal{O}$ have. For example, the following sets $\mathcal{A}$ consisting of a single attribute yield:

$$\{nODP^*\}' = \{RE_1, RE_{22}, RE_{33}, RE_{35}\},$$
$$\{nGWP^*\}' = \{RE_2, RE_8, RE_{23}, RE_{29}\},$$
$$\{nALT^*\}' = \{RE_8, RE_{23}, RE_{29}, RE_{35}\}.$$

An application of the mapping $-'$ to $\{nODP^*\}'$ yields

$$\{nODP^*\}'' = \{nODP^*, Cl, F\},$$
$$\{nGWP^*\}'' = \{nGWP^*, F\},$$
$$\{nALT^*\}'' = \{nALT^*, F\}.$$

Now we note that

$$\mathcal{P} = \{nODP^*\} \neq \{nODP^*\}'' = \mathcal{P}''.$$

This inequality together with $\emptyset = \emptyset''$ (which implies that the empty set is not a pseudo-content) shows that the 1-element set of attributes $\{nODP^*\}$ is a pseudo-content, and similarly for $\{nGWP^*\}$ and $\{nALT^*\}$. Thus, we obtain the following attribute implications contained in the basis of the attribute implications, the Duquenne/Guigues-basis:

$$\{nODP^*\} \Rightarrow_{\mathcal{E}^*} \{Cl, F\},$$
$$\{nGWP^*\} \Rightarrow_{\mathcal{E}^*} \{F\},$$
$$\{nALT^*\} \Rightarrow_{\mathcal{E}^*} \{F\}.$$

They show that $nODP^*(RE) = 1$, i.e., a high *nODP* implies the existence of *Cl* and *F*, while a high *nGWP* or a high *nALT* needs the presence of *F* but not of *Cl*. *But we should carefully note that this is true for just the 18 refrigerants evaluated by the binary evaluation shown above.* Of course, this is more or less trivial. Our focus here is not the set of implications per se but the fact that they were obtained *automatically and that they contain the whole knowledge about attribute implications contained in the evaluation of the 18 refrigerants that we consider*. The complete Duquenne/Guigues-basis is obtained as follows: Download conexp-1.3 (developed by Yevtushenko 2000) from

http://www.sourceforge.net

and apply it to the evaluation $\mathcal{E}^*$, the binary context given above. Doing so, the complete basis is obtained:

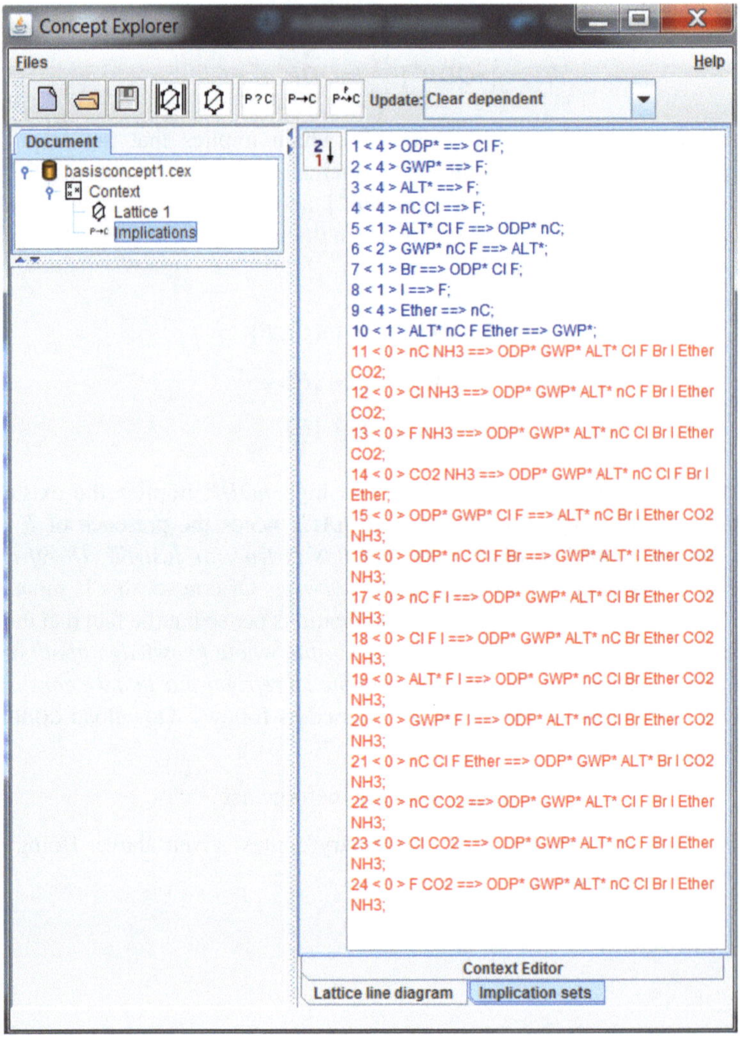

Skipping implications which are true but do not help, e.g., the implication number 22, where the assumption (that the molecule is CO_2 and contains three carbon atoms) is wrong but the implication is true because everything follows from a wrong assumption, we obtain a *reduced Duquenne/Guigues-basis*:

$$\begin{aligned} \{nODP^*\} &\Rightarrow_{\mathcal{E}} \{Cl, F\} \\ \{nGWP^*\} &\Rightarrow_{\mathcal{E}} \{F\} \\ \{nALT^*\} &\Rightarrow_{\mathcal{E}} \{F\} \\ \{nC, Cl\} &\Rightarrow_{\mathcal{E}} \{F\} \end{aligned}$$

$$\begin{aligned}
\{nALT^*, Cl, F\} &\Rightarrow_{\mathcal{E}} \{nODP^*, nC\} \\
\{nGWP^*, nC, F\} &\Rightarrow_{\mathcal{E}} \{nALT^*\} \\
\{Br\} &\Rightarrow_{\mathcal{E}} \{nODP^*, Cl, F\} \\
\{I\} &\Rightarrow_{\mathcal{E}} \{F\} \\
\{ether\} &\Rightarrow_{\mathcal{E}} \{nC\} \\
\{nALT^*, nC, F, ether\} &\Rightarrow_{\mathcal{E}} \{nGWP^*\}
\end{aligned}$$

It can be considered as a set of hypotheses on refrigerants in general, and this set is obtained from the evaluation of 18 refrigerants and its exploration. More details can be found in Kerber and Bruggemann (2015).

References

Bruggemann R, Kerber A, Restrepo G (2011) Ranking objects using fuzzy orders with an application to refrigerants. MATCH Commun Math Comput Chem 66:581–603

Bruggemann R, Carlsen L, Voigt K, Wieland R (2014) PyHasse software for partial order analysis: scientific background and description of selected modules. In: Bruggemann R, Carlsen L, Wittmann J (eds) Multi-indicator systems and modelling in partial order. Springer, New York, pp 389–423

De Baets B, De Meyer H (2003) On the existence and construction of t-transitive closures. Inf Sci 152:167–179

Duquenne V (1987) Contextual implications between attributes and some properties of finite lattices. In: Ganter B, Wille R, Wolff KE (eds) Beiträge zur Begriffsanalyse. BI Wissenschaftsverlag, Mannheim

Ganter B, Wille R (1996) Formal concept analysis (mathematical foundations). Springer, Berlin, pp x+286 [ISBN 3-540-60868-0]

Kerber A, Bruggemann R (2015) Problem driven evaluation of chemical compounds and its exploration. MATCH Commun Math Comput Chem 73:577–618

Naessens H, De Baets B, De Meyer H (1999) Generating Hasse trees of fuzzy preorder closures: approach. In: EUSFLAT–ESTYLF joint conference, p 387, 2403–2411

Pollandt S (1997) Fuzzy-Begriffe. Springer, Berlin

Restrepo G Assessment of the environmental acceptability of refrigerants by discrete mathematics: cluster analysis and Hasse diagram technique. Doctoral Thesis, University of Bayreuth, Germany. Available via the address. http://opus.ub.uni-bayreuth.de/frontdoor.php?source_opus=393&la=de

Yevtushenko SA (2000) System of data analysis "concept explorer". In: Proceedings of the 7th national conference on artificial intelligence, Russia, K II-2000, pp 127–134. Software available from http://sourceforge.net/projects/conexp

A Combined Lexicographic Average Rank Approach for Evaluating Uncertain Multi-indicator Matrices with Risk Metrics

Elvis Hernández-Perdomo, Johnathan Mun, and Claudio M. Rocco

1 Introduction

In many situations, projects are characterized by several criteria or attributes that can be assessed from multiple perspectives (financial, economic, etc.).

Each criterion is quantified via performance values (PV), which can either be numerical or categorical. This information is typically structured in a multi-indicator matrix **Q**. A typical problem faced by a decision maker is to define an aggregate quality (AQ) able to synthesize the global characteristics of each project and then derive the rankings from the best to the worst (base-case ranking).

Ranking techniques can be classified as parametric and nonparametric. A parametric technique requires information about decision-maker preferences (e.g., criterion weights). According to Dorini et al. (2011), some examples of parametric techniques include the ELECTRE methods (Roy 1968) and PROMETHEE—Preference Ranking Organization Methods for Enrichment Evaluations (Brans and Vincke 1985). Nonparametric techniques, such as Partial Order Ranking (Bruggemann et al. 1999) and Copeland Scores (Al-Sharrah 2010), do not require information from the decision maker. In general, all of these techniques are able to produce a ranking of the alternatives from the best to the worst.

Therefore, given a matrix **Q**, the selected procedure generates a ranking, defined as the base-case rank (BCR). As a result of this assessment, for each alternative a specific rank R_i that considers the multiple perspectives defined by the decision

E. Hernández-Perdomo
University of Hull, Hull, UK

J. Mun
U.S. Naval Postgraduate School, Monterey, CA, USA

C.M. Rocco (✉)
Facultad de Ingeniería, Universidad Central de Venezuela, Caracas 1050, Venezuela
e-mail: croccoucv@gmail.com

M. Fattore, R. Bruggemann (eds.), *Partial Order Concepts in Applied Sciences*,
DOI 10.1007/978-3-319-45421-4_7

maker is obtained. The set of R_i corresponds to the global evaluation under the first synthetic attribute, defined in this chapter and named as *base ranking*, capable of characterizing the alternatives in the base case.

However, in real-life situations each performance value could be affected by uncertain factors. Several approaches have been presented for analyzing how the uncertainty in the performance values (the input) affects the ranking of the objects (the output) (Rocco and Tarantola 2014; Corrente et al. 2014; Hyde et al. 2004; Hyde and Maier 2006; Yu et al. 2012). The approaches, based on Monte Carlo simulation, consider each uncertain factor as a random variable with known probability density functions. As a result, the AQ of each alternative and, therefore, its ranking also become random variables, with approximated probability distributions. In such situations, the decision maker could perform probability distribution evaluations. For example, the decision maker could be interested in determining not only what the worst rank of a specific alternative is, but also its probability and volatility (risk evaluation).

In this chapter, the probability of an alternative being ranked as in the BCR is selected as the synthetic attribute *probability* able to characterize the alternatives under uncertainty.

The stochastic nature of the AQ of each alternative could be further assessed in order to reflect the risk evaluation induced by uncertainty. In this case, it is required to compare several random variables synthesized through its percentiles and statistical moments. Several approaches have been proposed to this end, such as a simple comparison of the expected value, the expected utility (Von Neumann and Morgenstern 1947), the use of low order moments (Markowitz 1952), risk measures (Jorion 2007; Mansini et al. 2007; Rockafellar and Uryasev 2000), the Partitioned Multiobjective Risk Method (PMRM) (Asbeck and Haimes 1984; Haimes 2009), and the stochastic dominance theory (Levy 2006), among others.

In order to consider the risk evaluation induced by uncertainty, each alternative is represented by the third synthetic attribute: *compliance*. This new attribute is based on a simultaneous assessment of several risk measures and some moments of each AQ distribution.

At this point, each alternative is assessed from three different angles:

1. Multiple decision-making perspectives that include several aspects such as economic, financial, technical, and social (*base ranking*)
2. Uncertainty propagation on performance values (*probability*)
3. A risk evaluation, based on the generated probability distribution (*compliance*)

These perspectives are then used for defining a new multi-indicator matrix $\mathbf{Q}_1$ correlated to projects and synthesized using a ranking technique. However, in some situations, as in the case study to be analyzed, decision makers need to select projects following their most-preferred criteria successively. For this reason, an aggregation ranking technique that allows compensation is useless.

Therefore, in this chapter the final assessment is derived by using a combined approach based on a *nonparametric aggregation rule* (using the concept of average rank) for attributes 1 and 2; a simple procedure for score assignment for attribute 3;

and a *lexicographic rule*. In addition, a preliminary analysis of the alternatives is performed by using a Hasse diagram (Bruggemann and Patil 2011). To the best of our knowledge, this type of combined assessment has not been reported in the literature.

The remainder of the chapter is organized as follows: Sect. 2 provides an overview on partial order, the lexicographic approach, and risk metrics. Section 3 develops the proposed approach. A case study example follows in Sect. 4. Concluding remarks are provided in Sect. 5.

2 Background on Average Rank, Lexicographic Approach, and Risk Measures

2.1 Average Rank

Let P define a set of n objects (e.g., alternatives) to be analyzed and let the descriptors $q_1, q_2, \ldots, q_m$ define m different attributes or criteria selected to assess the objects in P (e.g., cost, availability, environmental impact). It is important that attributes are defined to reflect, for example, that a low value indicates low rankings (best positions), while a high value indicates high ranking (worst positions) (Restrepo et al. 2008). However, for a given problem or case study, this convention could be reversed.

If only one descriptor is used to rank the objects, then it is possible to define a total order in P. In general, given $x, y \in P$, if $q_i(x) \leq q_i(y)$ $\forall i$, then x and y are said to be comparable. However, if two descriptors are used simultaneously, the following could happen: $q_1(x) \leq q_1(y)$ and $q_2(x) > q_2(y)$. In such a case, x and y are said to be incomparable (denoted by $x||y$). If several objects are mutually incomparable, set P is called a partially ordered set or *poset*. Note that since comparisons are made for each criterion, no normalization is required.

The objects in a poset can be represented by a directed acyclic graph whose vertices are the objects $\in P$, and there is an edge between two objects only if they are comparable and one covers the other, that is, when no other element is in between the two. Such a chart is termed a Hasse diagram (Bruggemann et al. 1995).

A Hasse diagram is, then, a nonparametric ranking technique and can perform ranking decisions from the available information without using any aggregation criterion. However, while it cannot always provide a total order of objects, it does provide an interesting overall picture of the relationships among objects.

A useful approach to produce a ranking is based on the concept of the average rank of each object in the set of linear extensions of a poset (De Loof et al. 2011). Since the algorithms suggested for calculating such average ranks are exponential in nature (De Loof et al. 2011), special approximations have been developed, such as the Local Partial Order Model (LPOM) (Bruggemann et al. 2004), the extended LPOM (LPOMext) (Bruggemann and Carlsen 2011), or the approximation suggested by De Loof et al. (2011).

From the Hasse diagram, several sets can be derived (Bruggemann and Carlsen 2011). If $x \in P$:

1. $U(x)$, the set of objects incomparable with $x : U(x) := \{y \in P : x \,||\, y\}$
2. $O(x)$, the *down* set: $O(x) := \{y \in P : y \le x\}$
3. $S(x)$, the successor set: $S(x) := O(x) - \{x\}$
4. $F(x)$, the *up* set: $F(x) := \{y \in P : x \le y\}$

Then, the following average rank indexes are defined:

(a) $LPOM(x) = (|S(x)| + 1) \times (n + 1) \div (n + 1 - |U(x)|)$

(b) $LPOMext(x) = |O(x)| + \sum_{y \in U(x)} \frac{p_y^<}{p_y^< + p_y^>}$

where n is the number of objects,
$|V|$ defines the cardinality of the set V,

$$p_y^< = |O(x) \cap U(y)|, \quad p_y^> = |F(x) \cap U(y)|, \quad \text{and} \quad y \in U(x)$$

2.2 Lexicographic Approach

A lexicographic approach allows decision makers to introduce decision rules in which they select more objects impacting on their most-preferred criteria. According to Saban and Sethuraman (2014), when two objects have the same impact on the most-preferred criteria, decision makers prefer the one with the highest impact on the second most-preferred criteria, and so forth. This lexicographic representation models problems where decision makers strictly prefer one criterion over another or they are managing noncompensatory aggregation (Yaman et al. 2011; Pulido et al. 2014).

Finally, decision makers can model their strong preferences over the criteria selected mainly because, after further analysis of the problem, they are not indifferent or only weakly sure about their preferences on the criteria taken into consideration. In other words, they will always prefer one criterion to another without considering criterion weights explicitly.

2.3 Risk Metrics and Compliance

Risk metrics are statistical indicators or measurements that allow decision makers to analyze the dispersion (volatility) of certain events or outcomes. Hence, a random variable can be evaluated using statistical moments (e.g., mean, variance, skewness, kurtosis) or risk measurements can be used to analyze extreme values, such as Value at Risk (VaR) and Conditional VaR (Bodie et al. 2009; Fabozzi 2010; Matos 2007; Mun 2015).

In decision problems, risk metrics play an important role in analyzing the volatility or stability of a set of options or a portfolio of alternatives; for example, in financial risk management (Chong 2004), portfolio risk management (Bodie et al. 2009), and enterprise risk management (Scarlat et al. 2012), as well as a variety of other areas (Fabozzi 2010; Szolgayová et al. 2011).

In order to determine how risky an object is and its relationship with other objects, a compliance approach is followed, that is, the definition of a set of rules to guide decision makers (Hopkins 2011). Several approaches have been proposed for assessing the compliance. For example, Barrett and Donald (2003) propose a stochastic dominance analysis to compare probability distributions before establishing a hierarchy; Boucher et al. (2014) rely on risk metrics and forecasting to adjust models by historical performance; and Zanoli et al. (2014) analyze impacts of risk factors on noncompliance in UK farming.

The compliance approach implemented in this chapter is more user-friendly for decision-making because it allows evaluating whether an object performs according to decision makers' preferences over defined risk metrics. The basic idea is to dichotomize the risk continuum (Hopkins 2011). Therefore, the higher the compliance with a defined risk metric, the higher the alignment with the decision makers' preferences. Similar approaches are considered by Scarlat et al. (2012) and Tarantino (2008) relying on key risk indicators.

3 Combined Approach

This section describes a four-step process for performing the analysis of several projects.

Step 1: Determine *Base Ranking*.

Given a multi-indicator matrix **Q**:

- Perform a preliminary Hasse diagram.
- Determine the BCR for each alternative, using the average rank methods.

Step 2: Determine *Probability*.

Given the probability distribution for each PV, perform an uncertainty propagation evaluation using the following Monte Carlo procedure:

(a) Random deviates are generated for each uncertain factor according to their known probability distribution function. These sets of values define a multi-indicator matrix $\mathbf{Q}_{\text{sample}}$.
(b) A ranking of alternatives is obtained using $\mathbf{Q}_{\text{sample}}$ and the average rank method.
(c) Steps (a) and (b) are repeated *N sample* times. The result of the Monte Carlo approach provides an approximated probability distribution function of the aggregated quality and rank position of each alternative.
(d) For each alternative, select the probability of being ranked as in the BCR (step 1).

Step 3: Determine *Compliance*.

Using the probability distribution of the AQ of each alternative (derived in step 2), determine the risk metrics: mean-to-variability (mv) ratio, kurtosis, skew, Value at Risk (VaR), and Conditional VaR. In this chapter, the compliance of each alternative is determined using the following simple procedure, based on a 0/1 condition (Hopkins 2011; Müller and Supatgiat 2007):

- Assign 1 point to the 25 % of alternatives with the highest mean-to-variability (mv) ratio, 0 otherwise
- Assign 1 point to alternatives with negative skewness, 0 otherwise
- Assign 1 point to alternatives with positive kurtosis, 0 otherwise
- Assign 1 point to the 25 % of alternatives with best VaR, 0 otherwise
- Assign 1 point to the 25 % of alternatives with best CVaR, 0 otherwise

The final compliance score is then the normalized sum of the individual scores (in [0,1]). Of course, other approaches could be used for assessing the compliance attribute (e.g., by using a weighted average of the individual compliance factors, as suggested by Müller and Supatgiat 2007).

Step 4: Perform a *Combined Assessment*.

Build a new multi-indicator matrix $\mathbf{Q}_1$ considering the performance of each alternative on the three synthetic attributes derived in the previous steps. Given a lexicographic decision rule, determine the ranking of each alternative.

4 Case Study

The case study analyzed here corresponds to a real situation in portfolio management in Venezuela, where a set of projects are used by a decision maker and analysts to address the planning of a distributed electricity generation facility.

The decision maker and representatives collected the data and estimated the impacts by using their own prospective models, which include financial, economic, technical, and environmental aspects. Projects were evaluated according to their performance on four criteria ($m = 4$) previously selected by the decision unit:

1. Net Present Value (NPV): A classical financial indicator in project management that reflects the time value of money of an investment and its effect on the company's profitability (Ross et al. 2003). A higher value is better.
2. Return-to-Risk (RtR) Ratio: A real options valuation approach that helps decision makers dealing with uncertainty and flexibility in the project management process (Hull 1999; Mun 2015, 2016). It could be interpreted as how much value an active decision maker adds to the project management by each unit of volatility (derived from its underlying asset). A higher value is better.
3. Impact Assessment (IA): A criterion that considers the impact of the project on the environment measured by annual emissions. A lower value is better.

Table 1 Multicriteria information base case (%)

Project	NPV	RtR	*IA	*ELOE	Project	NPV	RtR	*IA	*ELOE
P1	2	**100**	4.55	33.24	P11	58	4.65	21.53	49.2
P2	20	3.17	31.75	57.98	P12	13.33	2.54	−12.37	12.23
P3	20	2.96	30.75	57.95	P13	66.67	6.13	63.34	89.1
P4	53.33	37.21	53.47	56.38	P14	66.67	9.3	50.67	76.6
P5	80	10.99	-9.36	7.98	P15	2.67	62.79	30.49	55.59
P6	40	17.97	38.84	58.24	P16	**100**	26.22	23.4	65.16
P7	40	18.82	38.84	58.16	P17	11.33	13.32	8.22	38.56
P8	14.67	11.42	**100**	71.54	P18	57.6	2.73	72.78	**100**
P9	16	3.38	57.67	55.05	P19	10.4	4.86	15.64	62.23
P10	16.67	14.16	−8.22	14.63	P20	20	3	21.46	61.7

4. Expected Loss of Energy (ELOE): A technical criterion related to assessing the impact of each project on the power electric sector, specifically, the expected amount of energy not supplied. A lower value is better.

Table 1 lists the base **Q** matrix of 20 projects ($n = 20$) under analysis. Each column represents the performance value $PV_{i,j}$ for each project ($i = 1,\ldots,20$) and for each attribute ($j = 1,\ldots,4$).

To facilitate the example, all attributes are transformed in such way that maximization is preferred. In this case, IA and ELOLE are denoted as *IA and *ELOLE, respectively. All of the values were normalized due to confidentiality. For each attribute, the best projects are 16, 1, 8, and 18, respectively (numbers in bold in Table 1 represent the best values for each attribute considered). There is no single dominant project.

Step 1: Determine *Base Ranking*.

- Perform a preliminary analysis based on a Hasse diagram.
  Figure 1 shows the Hasse diagram for the base case (Bruggemann et al. 2014). At the top level, there are seven projects that are incomparable. Note that projects 1 and 15 appear as isolated projects since they do not dominate any of the projects, and they are not dominated by any (note that these projects have the best RtR values). Hasse locates incomparable objects at the top level. Project 12 is located at the lower level.
  From the Hasse diagram some complete rankings can be extracted. For example, $12 < 3 < 10 < 4$ and those projects form a chain. This order is preserved no matter the importance weights used for the set of criteria, under any parametric multicriteria technique.
- Determine the BCR using the average rank approach.
  To overcome the incomparability, a BCR ranking and the corresponding aggregate quality are derived by using the LPOMext method.

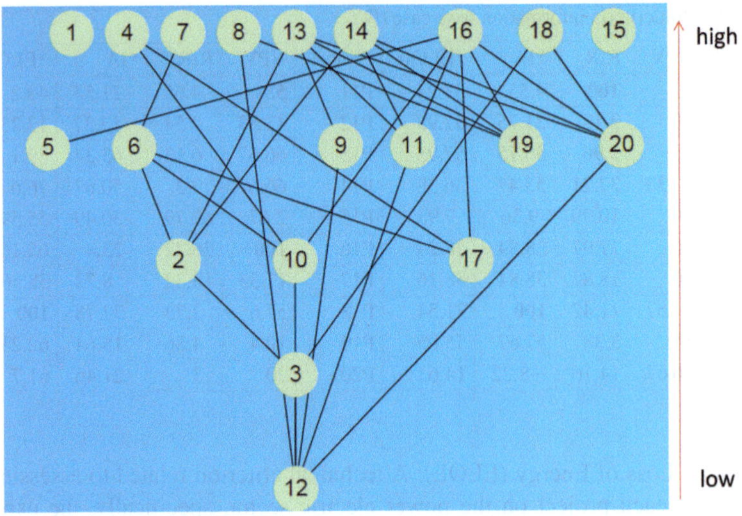

Fig. 1 Hasse diagram for the base case

Table 2 Final results produced by the LPOMext method (the higher the values, the better the ranking positions)

Project	LPOMext	Rank	Project	LPOMext	Rank
P1	10.500	11	P11	6.800	7
P2	7.682	8	P12	1.358	1
P3	4.289	2	P13	18.129	19
P4	15.450	16	P14	17.555	17
P5	8.000	9	P15	10.500	11
P6	14.924	14	P16	**18.217**	20
P7	17.962	18	P17	4.467	3
P8	14.167	13	P18	15.417	15
P9	9.333	10	P19	4.467	3
P10	5.638	5	P20	5.721	6

Table 2 shows the results of the evaluation. From here, Project 16 has the highest AQ (shown in bold) and, therefore, it is ranked as the best in terms of the global hierarchy (a higher position means better).

In the original Hasse diagram, projects 1 and 15 were isolated objects, while the aggregation method ranks both in the 11th position. For this reason, this preliminary assessment is important to analyze object sets by partial order set theory (Bruggemann and Patil 2011). Note that for this example, the LPOMext method does not produce a total order.

Step 2: Determine *Probability*.

To illustrate this step, it is assumed that each PV_{ij} is modeled by uniform distributions with bounds at $\pm 10\,\%$ of the base-case values. Figure 2 (x-axis shows projects, and y-axis shows base-case ranks) illustrates the resulting

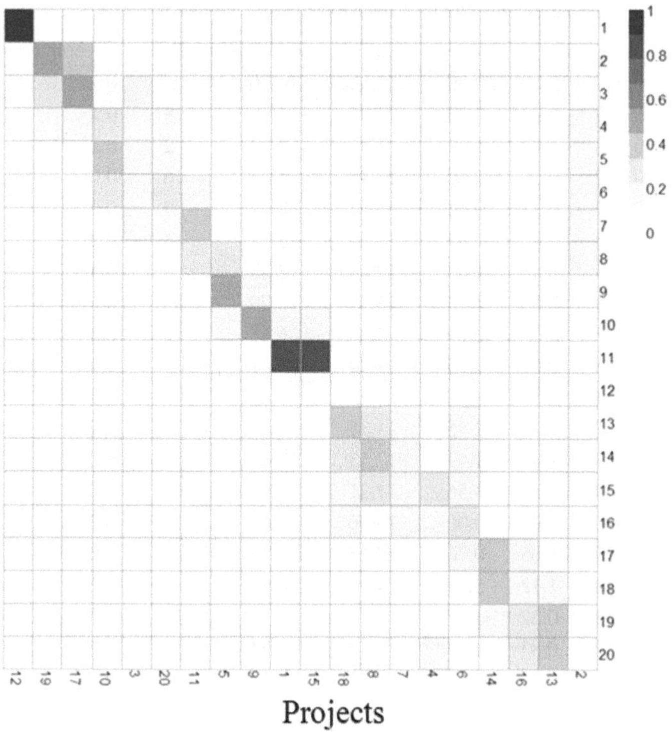

Fig. 2 Probability heat map of the uncertainties' impact on projects' ranking

probability heat map ($N\ sample = 10{,}000$). The more highly probable ranked alternative appears in the upper left, and the darker the intersection, the higher the probability that the ranking should be applied to the alternative.

For example, under uncertainty, project 16 could be ranked from the 17th to the 20th positions. However, the 20th (the best) and 19th (the second best) positions are the most probable. Note that project 12 has a high probability of remaining in the worst position.

It is important to emphasize that if the uncertainty does not impact on the global hierarchy, the diagonal principal of the heat map would be darker, corresponding to high probability values. Otherwise, it can be observed that some projects are highly affected by the uncertainty, and, therefore, they experience a higher dispersion on their rankings.

Step 3: Determine *Compliance*.

The left side of Table 3 shows the risk metrics' figures derived from the probability distribution of AQ, in step 2, for each project. The right side of Table 3 shows the multi-indicator matrix $\mathbf{Q}_1$ using *base ranking*, *compliance*, and *probability* attributes (the compliance column is derived by using the rules described in Sect. 3's step 3).

Table 3 Risk measurements and final decision-making approaches

Project	mv	sk	Kurtosis	VaR(95)	CVaR(95)	BCR	Compliance	Prob
P1	0.000	0.00	0.00	10.50	10.50	11	0.2	0.868
P2	0.003	0.83	0.20	4.94	4.70	8	0.2	0.136
P3	0.003	1.04	1.34	4.31	3.97	2	0.2	0.052
P4	0.012	–0.04	–1.40	15.45	15.45	16	0.6	0.175
P5	0.379	–0.63	–0.64	7.67	7.62	9	0.4	0.454
P6	0.005	–0.46	–0.97	12.80	12.28	14	0.2	0.146
P7	0.005	–0.46	–1.00	12.88	12.58	18	0.4	0.092
P8	0.019	0.00	1.93	13.17	12.29	13	0.6	0.300
P9	0.005	0.33	0.29	7.20	6.48	10	0.2	0.549
P10	0.125	–0.03	6.55	5.44	5.31	5	0.4	0.382
P11	0.014	1.19	3.88	5.75	5.69	7	0.2	0.341
P12	0.377	2.73	12.85	1.35	1.34	1	0.4	1.000
P13	0.333	–1.44	1.49	17.55	17.43	19	1.0	0.368
P14	0.469	–0.89	3.56	17.21	17.05	17	1.0	0.391
P15	0.000	0.00	0.00	10.50	10.50	11	0.2	0.868
P16	0.130	–1.05	0.76	17.25	16.89	20	1.0	0.307
P17	0.078	1.12	5.64	4.27	4.09	3	0.2	0.131
P18	0.005	–0.22	–1.17	11.83	11.70	15	0.2	0.222
P19	0.010	0.83	1.63	3.47	3.31	3	0.2	0.295
P20	0.005	0.07	–0.55	4.31	3.97	6	0.0	0.231

Step 4: Perform a *Combined Assessment*.

Figure 3 shows a graphical representation of the $\mathbf{Q}_1$ matrix. The x-axis shows a 0–1 scale associated with the compliance (solid bar) or probability (checkered bar) levels. The y-axis shows the projects ranked according to the base-case results (best rank at the top).

The lexicographic approach allows decision makers to set a rule by which they select more projects following their most-preferred criteria successively. In terms of investment decision-making, there is a strong preference for those projects with high rankings, level of compliance, and probability of keeping the highest ranking positions (Hernández 2015).

In order to illustrate the approach, the following set of lexicographic decision rules is considered ("$\succ\succ$" means that one criterion is highly preferred in relation to another):

1. Rule "Base Case Rank $\succ\succ$ % Compliance $\succ\succ$ Probability." Under this rule, the results obtained are equal to the base case's ranks for those projects that are not tied (e.g., $14 < 7 < 13 < 16$ highest ranked projects). Note that projects 1 and 15 are tied in the base-case evaluation and also when considering compliance and probability attributes. However, projects 17 and 19 were also tied in the third position in the base-case evaluation. Both have the same % compliance but project 19 has a higher probability. In this case, $17 < 19$.

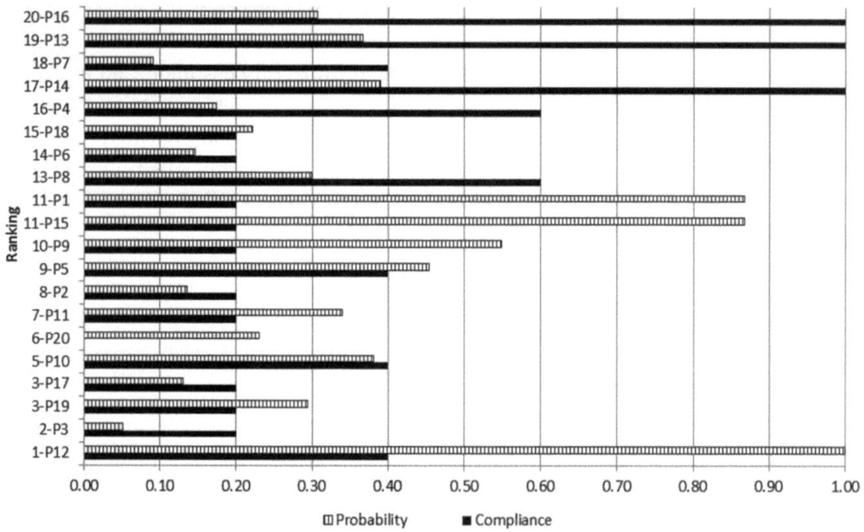

Fig. 3 Graphical representation of the decision-making criteria for the lexicographic approach (x-axis is compliance or probability levels, y-axis is rank, and best base rank is at the top)

2. Rule "% Compliance = 100 ≻≻ Probability ≻≻ Base Case Rank." Under this rule, $16 < 13 < 14$.

According to the different noncompensatory lexicographic rules, different projects can be selected as the best candidates in terms of their rankings, compliance, and probability.

5 Conclusions

In this chapter, a lexicographic multicriteria approach, where the decision makers can select more options from the most preferred criteria in a noncompensatory way, is revised to robustly rank alternatives using a combination of methods derived from the partial order theory, compliance with risk indicators, and probability analysis.

The Hasse diagrams along with the average rank method allow decision makers to solve the ranking problem by focusing on the dominant and incomparability relationships without considering decision-maker preferences. The compliance with risk methods analysis evaluates how stable and robust the aggregate quality of each project is, and the probability analysis does the same with the global hierarchy obtained. These two aspects take into consideration both uncertainty and risk analysis.

The approach is illustrated by assessing a portfolio of projects in the resource planning of a distributed electricity generation facility under different criteria, financial performance, and real options in place; impact assessment; and expected loss of energy.

The decision-making rules using the lexicographic approach provide interesting insights for future research, for example, analyzing the impact of other noncompensatory rules obtained from the ordered weighting operators (Max–Min, Min–Max, among others) (Yager 1988,1996).

References

Al-Sharrah G (2010) Ranking using the Copeland Score: a comparison with the Hasse Diagram. J Chem Inf Model 50:785–791

Asbeck E, Haimes YY (1984) The partitioned multiobjective risk method. Large Scale Syst 6(1):13–38

Barrett GF, Donald SG (2003) Consistent tests for stochastic dominance. Econometrica 71(1):71–104

Bodie Z, Kane A, Marcus AJ (2009) Investments, 8th edn. McGraw-Hill, New York

Boucher CM, Danielsson J, Kouontchou PS, Maillet BB (2014) Risk models-at-risk. J Bank Finance 44:72–92

Brans JP, Vincke PH (1985) A preference ranking organization method: the PROMETHEE method for multiple criteria decision making. Manag Sci 31(6):647–656

Bruggemann R, Carlsen L (2011) An improved estimation of averaged ranks of partial orders. Match Commun Math Comput Chem 65:383–414

Bruggemann R, Patil GP (2011) Ranking and prioritization for multi-indicator systems: introduction to partial order applications. Springer, New York

Bruggemann R, Schwaiger J, Negele RD (1995) Applying Hasse diagram technique for the evaluation of toxicological fish tests. Chemosphere 30:1767–1780

Bruggemann R, Bücherl C, Pudenz S, Steinberg C (1999) Application of the concept of partial order on comparative evaluation of environmental chemicals. Acta Hydrochim Hydrobiol 27:170–178

Bruggemann R, Sorensen P, Lerche D, Carlsen L (2004) Estimation of averaged ranks by a local partial order model. J Chem Inf Comput Sci 44:618–625

Bruggemann R, Carlsen L, Voigt K, Wieland R (2014) Hasse software for partial order analysis: scientific background and description of selected modules. In: Bruggemann R, Carlsen L, Wittmann J (eds) Multi-indicator systems and modelling in partial order. Springer, New York, pp 389–423

Chong YY (2004) Investment risk management. Wiley, Oxford

Corrente S, Figueira JR, Greco S (2014) The SMAA-PROMETHEE method. Eur J Oper Res 239:514–522

De Loof K, De Baets B, De Meyer H (2011) Approximation of average ranks in posets. Match Commun Math Comput Chem 66:219–229

Dorini G, Kapelan Z, Azapagic A (2011) Managing uncertainty in multiple-criteria decision making related to sustainability assessment. Clean Techn Environ Policy 13:133–139

Fabozzi FJ (2010) An introduction to financial markets, business finance, and portfolio management. Wiley, Hoboken

Haimes YY (2009) Risk modeling, assessment, and management, 3rd edn. Wiley, Hoboken

Hernández E (2015) A multicriteria methodology to evaluate engineering projects under risk and uncertainty using real options theory. Doctoral Thesis in Engineering Science, Universidad Central de Venezuela (in Spanish)
Hopkins A (2011) Risk-management and rule-compliance: decision-making in hazardous industries. Saf Sci 49:110–120
Hull JC (1999) Options, futures and other derivatives, 5th edn. Prentice Hall, Upper Saddle River
Hyde KM, Maier HR (2006) Distance-based and stochastic uncertainty analysis for multi-criteria decision analysis in Excel using visual basic for applications. Environ Model Software 21:1695–1710
Hyde KM, Maier HR, Colby CB (2004) Reliability-based approach to multicriteria decision analysis for water resources. J Water Resour Plan Manag 130(6):429–438
Jorion P (2007) Value at risk: the new benchmark for managing financial risk, 3rd edn. McGraw-Hill, New York
Levy H (2006) Stochastic dominance: investment decision making under uncertainty. Springer, New York
Mansini R, Ogryczak W, Speranza MG (2007) Conditional value at risk and related linear programming models for portfolio optimization. Ann Oper Res 152(1):227–256
Markowitz HM (1952) Portfolio selection. J Financ 7(1):77–91
Matos MA (2007) Decision under risk as a multicriteria problem. Eur J Oper Res 181(3):1516–1529
Müller S, Supatgiat C (2007) A quantitative optimization model for dynamic risk-based compliance management. IBM J Res Dev 51(3/4):295–307
Mun J (2015) Modeling risk: applying risk simulation, strategic real options, stochastic forecasting, business analytics, and portfolio optimization, 3rd edn. Thomson-Shore and ROV Press, California
Mun J (2016) Real options analysis: tools and techniques for making strategic decisions, 3rd edn. Thompson-Shore, Dexter
Pulido FJ, Mandow L, de la Cruz JLP (2014) Multiobjective shortest path problems with lexicographic goal-based preferences. Eur J Oper Res 239(1):89–101
Restrepo G, Bruggemann R, Weckert M, Gerstmann S, Frank H (2008) Ranking patterns, an application to refrigerants. Match Commun Math Comput Chem 59:555–584
Rocco CM, Tarantola S (2014) Evaluating ranking robustness in multi-indicator uncertain matrices: an application based on simulation and global sensitivity analysis. In: Bruggemann R, Carlsen L, Wittmann J (eds) Multi-indicator systems and modelling in partial order. Springer, New York, pp 275–292
Rockafellar RT, Uryasev S (2000) Optimization of conditional value-at-risk. J Risk 2:21–41
Ross SA, Westerfield RW, Jaffe JF (2003) Corporate finance, 6th edn. McGraw Hill, New York
Roy B (1968) Classement et choix en presence de points devue multiples (la methode ELECTRE). Rev Inform Rech Oper 6(8):57–75
Saban D, Sethuraman J (2014) A note on object allocation under lexicographic preferences. J Math Econ 50:283–289
Scarlat E, Chirita N, Bradea IA (2012) Indicators and metrics used in the enterprise risk management (ERM). Econ Comput Econ Cybernet Stud Res 46(4):5–18
Szolgayová J, Fuss S, Khabarov N, Obersteiner M (2011) A dynamic CVaR-portfolio approach using real options: an application to energy investments. Eur Trans Electr Power 21(6):1825–1841
Tarantino A (2008) Governance, risk, and compliance handbook: technology, finance, environmental, and international guidance and best practices. Wiley, Hoboken
Von Neumann J, Morgenstern O (1947) Theory of games and economic behavior. Princeton University Press, Princeton
Yager RR (1988) On ordered weighted averaging aggregation operators in multi-criteria decision making. IEEE Trans Syst Man Cybern 18:183–190
Yager RR (1996) Quantifier guided aggregating using OWA operators. Int J Intell Syst 11(1):49–73

Yaman F, Walsh TJ, Littman ML, Desjardins M (2011) Democratic approximation of lexicographic preference models. Artif Intell 175(78):1290–1307

Yu O-Y, Guikema SD, Briaud J-L, Burnett D (2012) Sensitivity analysis for multi-attribute system selection problems in onshore environmentally friendly drilling (EFD). Syst Eng 15(2):153–171

Zanoli R, Gambelli D, Solfanelli F, Padel S (2014) Assessing the risk of non-compliance in UK organic agriculture. Br Food J 116(8):1369–1382

Part II
Partial Order Theory in Socio-economic Sciences

Peculiarities in Multidimensional Regional Poverty

Paola Annoni, Rainer Bruggemann, and Lars Carlsen

1 Introduction and Background Data

Some basic tools of partial order theory are employed to analyze poverty patterns in the European Union (EU) regions. Poverty is intrinsically a multidimensional concept and its measure must take into account a wide variety of aspects that do not always display identical patterns.

The starting point is a multi-indicator system (MIS) comprising three indicators measuring regional poverty in 88 different regions and countries of the European Union (Annoni and Weziak-Bialowolska 2016): Absolute Poverty Index (*API*), Relative Poverty Index (*RPI*), and Earnings and Incomes Index (*EII*), respectively (Annoni et al. 2015). These indexes evaluate poverty in absolute and in relative terms, taking into account monetary and non-monetary aspects. A total of 13 raw indicators are used to compute the three poverty indexes (Fig. 1).

API measures the individual capacity of affording basic needs, as classically defined in the literature (Törmälehto and Sauli 2010). It is based on seven non-monetary indicators of material deprivation including material deprivation rate and intensity, capacity of making ends meet, quality of the housing, and affordability of health and dental care.

P. Annoni (✉)
European Commission Directorate General for Regional and Urban Policy, Economic Analysis Unit, Avenue de Beaulieau 5, Brussels 1160, Belgium
e-mail: Paola.ANNONI@ec.europa.eu

R. Bruggemann
Department of Ecohydrology, Leibniz—Institute of Fresh Water Ecology and Inland Fisheries, Berlin, Germany

L. Carlsen
Awareness Center, Linkøpingvej 35, Trekroner, Roskilde 4000 Denmark

M. Fattore, R. Bruggemann (eds.), *Partial Order Concepts in Applied Sciences*,
DOI 10.1007/978-3-319-45421-4_8

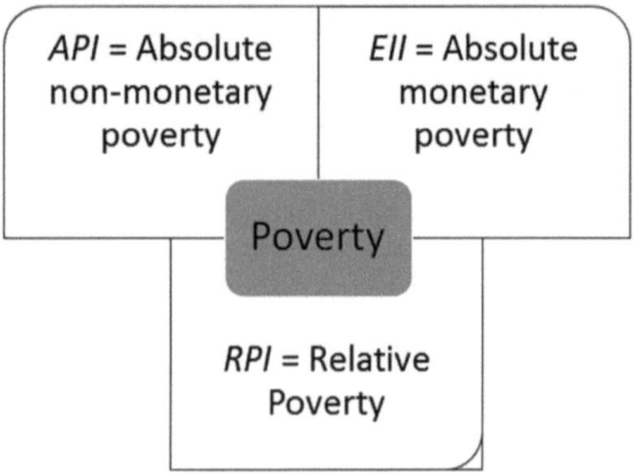

Fig. 1 The three indexes included in the poverty MIS

RPI is computed by aggregating three well-known poverty statistics: poverty incidence P_0, depth P_1, and severity P_2 (Foster et al. 1984). These statistics are based on the national poverty line defined as 60 % of the national country income median. Relative poverty statistics capture the condition of living of an individual compared to the people surrounding him. The general formulation of a relative poverty statistics is defined for each integer, non-negative number α as:

$$P_\alpha(\mathbf{y}, z) = \frac{1}{n}\sum_{i=1}^{q}\left(\frac{z - y_i}{z}\right)^\alpha \quad (1)$$

where $\mathbf{y} = (y_1, y_2, \ldots, y_n)$ is a vector of properly defined income in increasing order, $z > 0$ is a predefined poverty line, $(z - y_i)$ is the income gap of individual i, within a group of n individuals, and q is the number of individuals having income not greater than the poverty line z. When $\alpha = 0$, $P_\alpha = P_0$ is the share of poor people where poverty is defined with respect to the poverty line and is known as at-risk-of-poverty rate. When $\alpha = 1$, $P_\alpha = P_1$ is the normalized income gap measure and indicates the average relative gap between the incomes of poor individuals and the poverty line. Finally, when $\alpha = 2$ $P_\alpha = P_2$ is called severity indicator because it measures the degree of inequality in the distribution of income among poor people.

EII describes different types of income and includes: (1) the median regional income, computed from the individual income distribution within each region, and two other measures derived from the Eurostat database (ec.europa.eu/eurostat/data/database); (2) compensation of employees; and (3) net adjustable household income.

Each poverty index, *API*, *RPI*, and *EII*, is computed as a generalized mean of order 0.5 of the selected indicators, after standardization (Annoni and Weziak-Bialowolska 2016; Decancq and Lugo 2013).

To better clarify the orientation of poverty indexes used here, we remind that:

- Poverty, as conventionally understood, has a negative orientation, so if we say "high poverty" it means that people are generally poor;

- However, for technical reasons linked to the aggregating functional form chosen in the original analysis, the three poverty indexes are all oriented in order to have the higher the region score, the lower the level of poverty.

Region scores on the three poverty indexes are shown in the Table in Appendix, where high scores mean high levels of quality of life, i.e., low levels of poverty. The distribution of the poverty indexes across the regions shows the presence of severe pockets of poverty, as *API* and *RPI* distributions are significantly negatively skewed. On the other hand, the *EII* distribution is characterized by few but very high values showing that there are certain regions with extremely high income levels.

We follow a multi-criteria approach to shed more light into the poverty pattern of the EU regions.

The chapter is organized as follows. Section 2 provides a brief overview of partial order concepts applied in the analyses. Section 3 presents and discusses main results while Sect. 4 concludes the chapter.

2 Partial Order Approach, General Remarks

The concepts of partial order theory allow for an alternative view on MIS because the need to derive a one-dimensional scalar as a ranking index is avoided. Instead, a simultaneous view on the role of the three poverty indexes is provided, which is not easily possible when aggregation methods are applied. The analysis of multidimensional poverty and deprivation can also be done applying a conceptual evaluation framework as the one recently discussed in Fattore (2015).

2.1 *The Hasse Diagram Technique*

The central point in partial order theory is the introduction of a binary relation between any two objects, which fulfills the axioms of partial order (Birkhoff 1984; Neggers and Kim 1998; Trotter 1992).

Let x, $y \in X$, where X is the set of objects of interest, and $q_i \in IB$, IB being the set of indicators observed on the objects in X. The governing relation between the objects derived from IB is given by:

$$x \leq y : q_i(x) \leq q_i(y) \text{ for all } q_i \in IB \tag{2}$$

Relation (2) is the basic of the special variant of partial order theory, which is known as Hasse diagram technique (HDT), see for example Bruggemann and Patil (2011) for an overview of the technique. In partial order terminology, objects which follow (2) are said to be comparable, i.e., they can be ordered. Objects which do not follow (2) are said to be incomparable (notation: $x \| y$). Given object x, the number of objects incomparable to x is called $|U(x)|$.

Usually, a partial order based on (2) is visualized by a Hasse Diagram, HD. Comparable objects are connected by a sequence of lines. A HD provides insight on subsets of objects for which a complete ranking can be found, called chains (see the formal definition below), without the need of aggregation or additional information such as weights. Also, it allows for identifying subsets of incomparable objects whose scores will be influenced by the particular type of aggregating function used.

As a downside, a HD can be very complicated as can be seen from the HD corresponding to the here studied poverty MIS (Fig. 2). There exist software packages, such as PyHasse used here (Bruggemann et al. 2014), which provide support in navigating across a HD.

Some further relevant definitions are:

Chain: A subset $X' \subseteq X$, where all objects fulfill (2) is called a chain. A chain has a length, which is $|X'| - 1$. For objects within a chain, say from the bottom to the top of the chain, all indicator values are simultaneously non-decreasing.

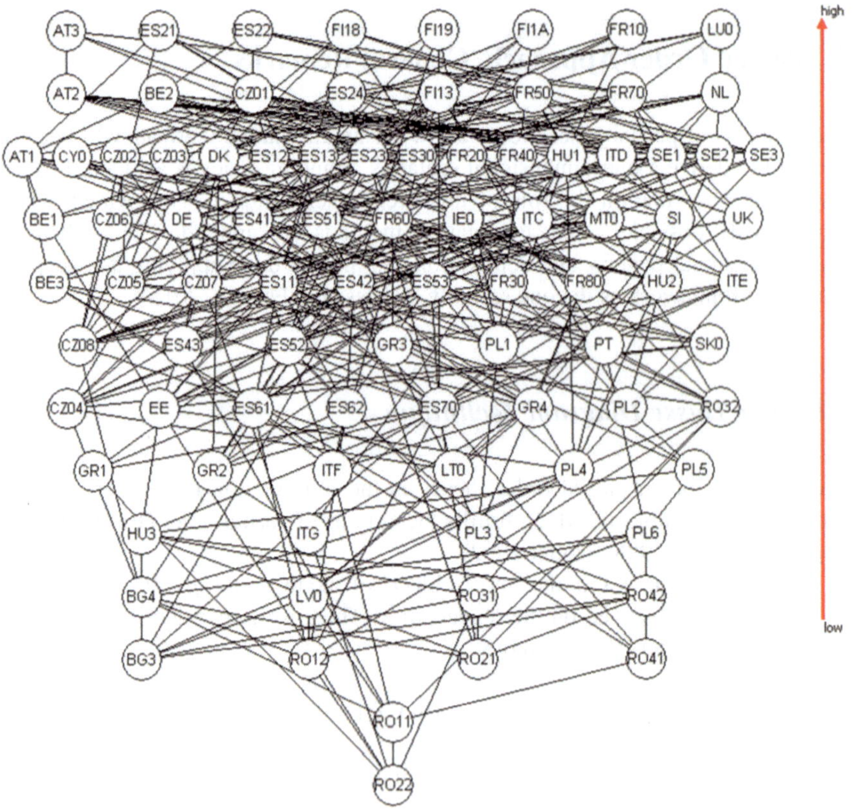

Fig. 2 Hasse Diagram of the 88 regions based on the three poverty measures API, RPI, and EII (PyHasse software)

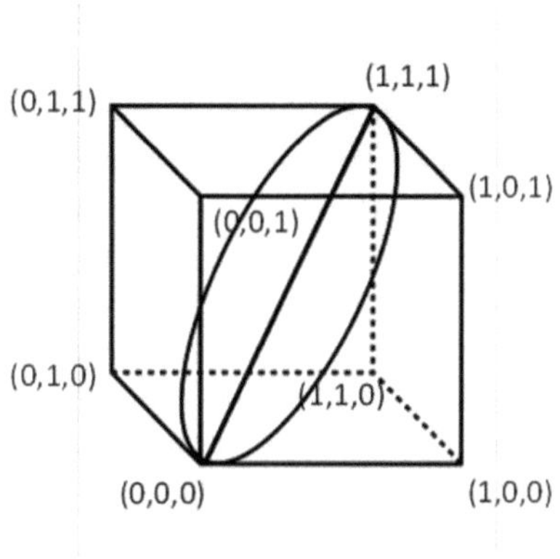

Fig. 3 The extreme points of the cube $h(3)$

Antichain: A subset $X' \subseteq X$, where no object fulfill (2), i.e., all objects in X' are mutually incomparable, is called an antichain. Thus, for any two objects within an antichain there is a conflict in indicator values.

Local HD: An object is selected and all the objects comparable upwards and downwards are visualized. One says: object x is "generating" its local HD.

2.2 *The Peculiarity Rule*

The Hasse diagram, shown in Fig. 2, reveals by its set of connecting lines only the set of all comparisons among the regions. It is possible to further classify the regions in order to identify peculiar ones. The first step is then to define the peculiarity concept.

In the general case with m indicators, a simple rule to define peculiarity is through normalizing the indicator values in the interval [0,1] and considering the m-dimensional hypercube $h(m)$. In our case $m = 3$; hence, we have to analyze the $h(3)$ (Fig. 3) as a classifier (Bruggemann and Carlsen 2014; Carlsen et al. 2015). The three dimensions of the cube represent the three poverty indexes, *API*, *RPI*, and *EII*. The vertexes of the cube $h(3)$ can be then interpreted in terms of extreme poverty levels: for example, the points (0,0,0) represents the worst and (1,1,1) the best condition, respectively, while, e.g., (1,0,0) is the point where *API* is the best possible, but *RPI* and *EII* are the worst. In general, there are 2^m extreme points (vertexes). Often an increase in one indicator will be accompanied by an increase in the other indicators. Thus, regions which are centered around the line between (0,0,0) and (1,1,1) can be denoted as "main stream" regions. Whereas regions close to one of the other vertices, e.g., (1,0,0) or (0,1,1) can be classified as peculiar regions, i.e., regions deviating from the "main stream" (Fig. 3).

A "near-enough-factor"-$f \in [0, 1]$-can be introduced to define the level of proximity of the points to the vertices of $h(3)$ (Bruggemann and Carlsen 2014). In the three-dimensional case, the maximum squared Euclidian distance between (0,0,0) and (1,1,1) in $h(3)$ is $D_{\max} = 3$. This means that in the three-dimensional case the (squared) distance between two given points (regions) is always in the interval [0, 3]. Regions are considered as peculiar if their distance, d, to one of the corners in $h(3)$ is $d < f \cdot 3$. The closer d to 0, the closer the region, characterized by its (normalized) values with respect to *API*, *RPI*, and *EII*, will be to one of the vertices of $h(3)$. As (0,0,0) and (1,1,1) are the extreme cases of the "mainstream," the interest is focused on the set of corners $h(3)' := h(3) - \{(0,0,0), (1,1,1)\}$. Regions apart from (0,0,0) and (1,1,1) are denoted peculiar regions due to their imbalanced pattern relative to the regions belonging to the main steam. If f is set to 0.05 only regions with highly imbalanced poverty measure profiles will be classified as peculiar. The value $f = 0.05$ can be interpreted as 5 % of the maximal distance, i.e., $d \leq 0.15$ to one of the peculiar vertices.

By looking at the actual vertex of proximity, the particular poverty index that causes the peculiarity can be spotted out. Thus, for example a region close to the (1,0,0) vertex has a markedly higher API than regions located in the "main stream," meaning lower levels of absolute poverty.

As an increase of one poverty index does not necessarily imply an increase of another poverty index, the cloud of points will encompass points which are considered as mutually incomparable.

3 Results

3.1 Hasse Diagram

Figure 2 shows the Hasse Diagram—HD—associated to the poverty MIS. As expected, given the relatively high number of objects, not much information can be directly extracted from this HD. However, it can be seen that:

- Poor regions, as measured by MIS are found at the bottom of the diagram; for example, BG3, RO21, and RO22;
- Rich regions are at the top of the diagram; for example, AT3, ES21, and LU0;
- The number of incomparable regions is very high (1754 total number of incomparabilities);
- There is a high number of comparable regions; for example, those in the chain AT3 > AT2 > SE1 > DE or NL > SE1 > UK > ES70 > GR1 > RO22. Two thousand seventy-four comparabilities are found in total.

The last point means that the HD is characterized by many chains of different lengths. As already said, the presence of chains is interesting as each chain is a realization of a unique ranking for a subset of objects. As an example, the chain:

AT3 > AT2 > ES12 > MTO > CZO5 > SKO > CZO4 > LTO > PL3 > RO42 > BG3

is informative of the fact that all the three poverty measures decrease (weakly) monotonously from AT3 with profile (5.8, 6.4, 5.8) to BG3 with profile

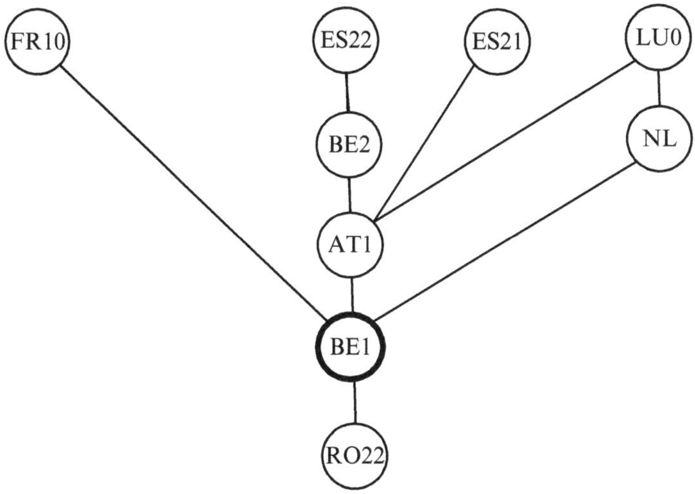

Fig. 4 Local partial order for the region BE1 with profile (4.6, 2.1, 5.9) visualized by the HD

(1.6, 3.5, 2.8) (see Table in Appendix). Many other chains connecting AT3 with BG3 are present (in total 66 chains including 11 regions).

3.2 *Local HD*

The analysis of a complex HD as the one showed in Fig. 2 needs proper navigation tools.

The concept of local HD tool is here applied. Assume to select the object $x = \text{BE1}$, then a "local HD" can be constructed as visualized in Fig. 4. The local HD is the visualization of the partial order of all objects y fulfilling either $y < \text{BE1}$ or $y > \text{BE1}$, so that only regions which are comparable with BE1 are shown. In this case, there are seven regions higher then BE1, showing a lower level of multidimensional poverty; however, only one region is lower than BE1, having worse poverty levels, namely RO22. Many of these regions with a lower degree of poverty than BE1 are mutually incomparable, indicating that the reasons for having a lower degree of poverty are different. For example, it is found FR10 > BE1 as well as ES21 > BE1, but FR10 || ES21. Specifically, FR10 < ES21 with respect to the poverty measure *API*; but FR10 > ES21 with respect to *EII*.

Region BE1 is comparable to 8 other regions only, while the number of incomparable regions with BE1 is rather high (79). Hence, 91 % of the 87 regions are incomparable with BE1. The profile of BE1 (API = 4.6, RPI = 2.1, EII = 5.9) interestingly shows that BE1 has a very low level of *RPI*. Only one region has even lower values, namely RO22 (RPI = 1.84). Thus, almost all regions perform better than BE1 with respect to *RPI*. As a result of this, the data profile of BE1 has a potential to crisscross (a high/low value of one poverty measure implying a low/high value in all the others) with all those regions whose data profile display higher values in *RPI* and lower values for *API* and *EII*, respectively. In particular, BE1 is incomparable with all those regions with a more balanced data profile, i.e., with scores located closely around the center of the distribution as for example ES11 (5.3, 5.3, 4.9).

3.3 Peculiar Regions

To identify some extremely conflicting regions, we apply the peculiarity rule (Sect. 2.2) with $f = 0.05$. Four regions are defined as peculiar (Table 1): the Belgian region hosting the capital Brussels (BE1) (consistent with the finding in the former section), Cyprus (CY0), the western part of Hungary (HU2), and the United Kingdom.

Table 1 The four regions identified as peculiar by the peculiar method with $f = 0.05$

Region/ country	Profile	(0–1) normalized data	Distance from vertex	Interpretation
BE1	[0, 0, 1]	[0.354, 0.011, 0.920]	0.126	This Belgian region, which includes Brussels, features deep pockets of poverty with respect to the rest of the population in the country (very low values of RPI). It is also characterized by very high income levels (EII > 0.9). This is a clear signal of the presence of important levels of inequality
CY0	[0, 1, 1]	[0.322, 0.897, 0.828]	0.144	The region, which coincides with the country Cyprus, is relatively well off in terms of income and relative poverty even if absolute poverty levels are higher that in the other EU regions/countries. This signals an equal society with few very rich but also few very poor people
HU2	[0, 1, 0]	[0.276, 0.851, 0.195]	0.137	The region, which is the western Hungarian macro-region, is relatively worse-off than CY0 because of low income levels. The region's profile indicates an equal society but toward low level of wealth (low *API* and *EII*)
UK	[1, 0, 1]	[0.782, 0.184, 0.770]	0.134	This country[a] stands at the opposite side than HU2, featuring high levels of incomes and low levels of poverty in absolute terms (high *API* and *EII*). The relatively low value of RPI indicates instead the presence of groups of people significantly poorer than the rest of the population (deep pockets of poverty)

[a]Due to data availability issues, the United Kingdom has been described at the national level only

Let us examine two peculiar regions, the Belgian BE1 near to the corner (0,0,1) and the Hungarian HU2, which is found to be in the neighborhood of the corner (0,1,0). Their normalized profiles respectively are [$API = 0.354$, $RPI = 0.011$, $EII = 0.920$] and [$API = 0.276$, $RPI = 0.851$, $EII = 0.195$], so that *RPI* is the most contrasting poverty index. This does not come as a surprise given that relative poverty is by construction a relative concept, which basically captures the level of inequality across the population. The three indicators from which *RPI* is constructed all refer to the poverty line that is the median national income. High values of *RPI* do not always imply that the population is rich; it indicates a low level of heterogeneity of poverty across the population. On the contrary, the earnings and incomes measure is a purely monetary measure of poverty. Relative poverty in BE1 is very high while income levels are within the highest in the European Union. HU2 shows the opposite pattern, quite low values of relative poverty but, at the same time, low wealth. This is a clear indication of the Belgian region being overall rich but highly unequal, while of the Hungarian region being poorer but less unequal (i.e., more homogenous).

The remaining two peculiar regions are briefly discussed in Table 1.

The analysis shows that several regions need attention with respect to compensation effects (Munda 2008). By further aggregating the three poverty indexes, there is the risk to miss relevant information. A population in a region is certainly well-off when low values of relative poverty are associated with low values of monetary and non-monetary absolute poverty. But low values of relative poverty with high values of absolute poverty, either monetary or non-monetary, are a signal of a widespread poverty where, however, there are low levels on inequality. Instead high values of relative poverty and low values of absolute poverty indicate the presence of deep and severe, but not widespread, pockets of poverty.

4 Conclusions

The chapter aim is to show the potential of partial order tools to analyze multi-indicator systems, MIS. When the indicators included in the MIS are carrying different messages, aggregation should be avoided. Partial order tools can be employed to assess how much information is lost due to compensation effects, always hidden in any aggregation process even if to different degrees according to the actually used aggregation technique.

On the flip side, by definition the partial order approach does not solve the incomparabilities between objects due to contrasting indicators. However, the analysis of these incomparabilities can be highly informative. To illustrate the value that a posetic analysis can add, we take as case study the MIS on the multidimensional measure of poverty in the EU regions. The MIS consists of three poverty indexes, Absolute Poverty, Relative Poverty, and Earnings and Incomes, estimated for 88 regions/countries across the European Union.

The analysis of incomparabilities in the poverty case sets a flag on particular regions where the aggregation can hide important contrasting patterns across the poverty measures. In our case, the relative poverty index is mostly contrasting with the remaining two. Relative poverty is in fact a "relative" concept that, by definition, captures the level of deprivation of people with respect to those living in the same area. Low values of relative poverty do not necessarily imply that people are well-off; it shows a low level of heterogeneity of poverty across the population. On the contrary, the earnings and incomes measure is a purely monetary one.

Some regions are detected as peculiar if they deviate from the main pattern of poverty profiles, as defined by the peculiarity rule. Four regions belong to this group with interesting profiles featuring contrasting levels of the poverty indexes.

Appendix

Starting multi-indicator system matrix (scores are oriented in order to have: the higher the score, the lower the level of poverty)

Country	Region code	Region name	Absolute poverty index	Relative poverty index	Earnings and income index
AT	AT1	Ostösterreich	5.4	5.7	6.0
AT	AT2	Südösterreich	5.8	6.1	5.6
AT	AT3	Westösterreich	5.8	6.4	5.8
BE	BE1	Région de Bruxelles-Capitale/Brussels Hoofdstedelijk Gewest	4.6	2.1	5.9
BE	BE2	Vlaams Gewest	5.8	6.1	6.0
BE	BE3	Région Wallonne	5.1	4.7	5.4
BG	BG3	Severna i Iztochna Bulgaria	1.6	3.5	2.8
BG	BG4	Yugozapadna i Yuzhna Centralna Bulgaria	2.7	5.0	3.2
CY	CY0	Κύπρος/Kypros	4.5	6.1	5.6
CZ	CZ01	Praha	5.5	6.4	5.7
CZ	CZ02	Střední Čechy	5.5	6.2	4.2
CZ	CZ03	Jihozápad	5.5	6.3	4.1
CZ	CZ04	Severozápad	5.0	5.0	3.8
CZ	CZ05	Severovýchod	5.2	5.9	4.0
CZ	CZ06	Jihovýchod	5.5	5.8	4.1
CZ	CZ08	Moravskoslezsko	4.9	5.0	4.0
DE	DE	Deutschland	5.5	4.4	5.4
DK	DK	Danmark	5.9	5.1	4.6

(continued)

Country	Region code	Region name	Absolute poverty index	Relative poverty index	Earnings and income index
CZ	CZ07	Střední Morava	5.3	5.5	3.9
EE	EE	Eesti	4.9	5.2	3.6
ES	ES11	Galicia	5.3	5.3	4.9
ES	ES12	Principado de Asturias	5.5	5.9	5.4
ES	ES13	Cantabria	5.7	5.9	5.3
ES	ES21	País Vasco	5.7	5.9	6.1
ES	ES22	Comunidad Foral de Navarra	5.8	6.5	6.0
ES	ES23	La Rioja	5.7	5.0	5.3
ES	ES24	Aragón	5.9	5.6	5.5
ES	ES30	Comunidad de Madrid	5.5	5.0	5.8
ES	ES41	Castilla y León	5.5	4.5	5.1
ES	ES42	Castilla-La Mancha	5.4	3.8	4.7
ES	ES43	Extremadura	5.3	3.2	4.5
ES	ES51	Cataluña	5.5	4.9	5.5
ES	ES52	Comunidad Valenciana	5.3	4.8	5.0
ES	ES53	Illes Balears	5.3	4.5	5.3
ES	ES61	Andalucía	5.0	3.3	4.6
ES	ES62	Región de Murcia	5.2	3.3	4.6
ES	ES70	Canarias	4.6	3.6	4.7
FI	FI13	Itä-Suomi	6.0	5.4	4.6
FI	FI18	Etelä-Suomi	5.9	5.9	5.2
FI	FI19	Länsi-Suomi	6.0	5.6	4.8
FI	FI1A	Pohjois-Suomi	6.2	5.7	4.7
FR	FR10	Île de France	5.2	5.9	6.8
FR	FR20	Bassin Parisien	5.5	6.1	5.3
FR	FR30	Nord - Pas-de-Calais	5.3	5.4	5.1
FR	FR40	Est	5.39	6.16	5.36
FR	FR50	Ouest	5.56	6.43	5.25
FR	FR60	Sud-Ouest	5.35	5.59	5.35
FR	FR70	Centre-Est	5.62	6.14	5.54
FR	FR80	Méditerranée	4.99	4.79	5.30
GR	GR1	Voreia Ellada	4.46	2.67	4.64
GR	GR2	Kentriki Ellada	4.42	2.99	4.54
GR	GR3	Attiki	4.51	4.51	5.35
GR	GR4	Nisia Aigaiou, Kriti	4.11	4.38	4.82
HU	HU1	Közép-Magyarország	3.97	6.22	4.90
HU	HU2	Dunántúl	4.34	6.08	3.68
HU	HU3	Alföld És Észak	3.69	5.15	3.40
IE	IE0	Ireland	5.40	5.76	5.32

(continued)

Country	Region code	Region name	Absolute poverty index	Relative poverty index	Earnings and income index
IT	ITC	Nord-Ovest	5.04	5.68	5.72
IT	ITD	Nord-Est	4.95	6.07	5.61
IT	ITE	Centro (I)	4.82	5.52	5.48
IT	ITF	Sud	4.04	3.15	4.49
IT	ITG	Isole	3.73	2.97	4.54
LT	LT0	Lietuva	4.20	4.86	3.76
LU	LU0	Luxembourg (Grand-Duché)	5.93	5.86	8.15
LV	LV0	Latvija	2.99	3.93	3.34
MT	MT0	Malta	5.41	5.94	5.18
NL	NL	Nederland	5.87	5.48	6.02
PL	PL1	Region Centralny	3.90	5.51	4.00
PL	PL2	Region Południowy	3.99	5.34	3.59
PL	PL3	Region Wschodni	3.88	4.55	3.06
PL	PL4	Region Północno-Zachodni	3.91	5.06	3.48
PL	PL5	Region Południowo-Zachodni	3.70	5.23	3.53
PL	PL6	Region Północny	3.68	5.17	3.40
PT	PT	Portugal	4.32	5.19	4.25
RO	RO11	Nord-Vest	2.47	3.03	2.81
RO	RO12	Centru	2.67	3.31	2.79
RO	RO21	Nord-Est	2.18	3.37	2.57
RO	RO22	Sud-Est	2.45	1.84	2.74
RO	RO31	Sud - Muntenia	3.35	5.05	2.78
RO	RO32	Bucureşti - Ilfov	2.94	4.77	4.19
RO	RO41	Sud-Vest Oltenia	3.28	3.23	2.86
RO	RO42	Vest	3.33	4.36	3.01
SE	SE1	Östra Sverige	5.67	5.31	5.62
SE	SE2	Södra Sverige	5.75	5.13	5.18
SE	SE3	Norra Sverige	5.79	4.91	4.99
SI	SI	Slovenija	4.92	6.11	5.23
SK	SK0	Slovenská Republika	5.16	5.70	3.87
UK	UK	United Kingdom	5.56	3.83	5.47

References

Annoni P, Weziak-Bialowolska D (2016) A measure to target anti-poverty policies in the European Union regions. Appl Res Qual Life 11(1):181–207

Annoni P, Bruggemann R, Carlsen L (2015) A multidimensional view on poverty in the European Union by partial order theory. J Appl Stat 42(3):535–554

Birkhoff G (1984) Lattice theory, vol XXV. American mathematical society, Providence

Bruggemann R, Carlsen L (2014) Incomparable—what now? MATCH Commun Math Comput Chem 71:694–716

Bruggemann R, Patil GP (2011) Ranking and prioritization for multi-indicator systems—introduction to partial order applications. Springer, New York

Bruggemann R, Carlsen L, Voigt K, Wieland R (2014) PyHasse software for partial order analysis. In: Bruggemann R, Carlsen L, Wittmann J (eds) Multi-indicator systems and modelling in partial order. Springer, New York, pp 389–423

Carlsen L, Bruggemann R, Kenessova O, Erzhigitov E (2015) Evaluation of analytical performance based on partial order methodology. Talanta 132:285–293

Decancq K, Lugo MA (2013) Weights in multidimensional indices of wellbeing: an overview. Econ Rev 32:7–34

Fattore M (2015) Partially ordered sets and the measurement of multidimensional ordinal deprivation. Soc Indic Res. Published online. doi:10.1007/s11205-015-1059-6

Foster J, Greer J, Thorbecke E (1984) A class of decomposable poverty measures. Econometrica 52:761–765

Munda G (2008) Social multi-criteria evaluation for a sustainable economy. Springer, Heidelberg

Neggers J, Kim H (1998) Basic posets. World Scientific, Singapore

Törmälehto V-M, Sauli H (2010) Distributional impact of imputed rent in EU-SILC. Eurostat Methodol. Work. Pap., pp 1–78

Trotter WT (1992) Combinatorics and partially ordered sets: dimension theory. The Johns Hopkins University Press, Baltimore

Application of Partial Order Theory to Multidimensional Poverty Analysis in Switzerland

Tugce Beycan and Christian Suter

1 Introduction

Poverty is one of the most urgent global issues both for developed and developing countries. There are several debates on the conceptualization and measurement of poverty. In general, poverty is considered as a monetary and material deprivation and is measured by using income or consumption metrics. But over the last few decades, alternative perspectives—including diverse kinds of lack of resources such as 'hunger, unemployment, homelessness, illness and health care, powerlessness and victimization, and social injustice' (Fukuda-Parr 2006, p. 7)—have led to conceptualize poverty as a deprivation in different spheres of life. In the specialized literature, different methodological approaches are proposed to measure multidimensional poverty, notably classical composite index approach, and the Alkire and Foster's counting approach (2007) which has been used for the construction of several multidimensional poverty indices, for example, the Multidimensional Poverty Index (MPI). Classical composite indices provide poverty statistics such as headcount ratio, poverty severity, and intensity of poverty by reducing the conception of poverty to two distinct groups: poor and non-poor. However, in a multidimensional social space, and particularly in the more differentiated, economically advanced society, there are more complex poverty configurations including people living in a 'in-between' situation (neither poor nor prosperous) characterized by ambiguous situations in the different spheres of life (e.g. good monetary situation but witnessing employment insecurity, few social contacts). Ambiguous situations of the people having an 'in-between' status can be easily detected by a statistical method

T. Beycan (✉) • C. Suter
Department of Sociology, University of Neuchâtel, Faubourg de l'Hôpital 27, Neuchâtel 2000, Switzerland
e-mail: tugce.beycan@unine.ch

M. Fattore, R. Bruggemann (eds.), *Partial Order Concepts in Applied Sciences*,
DOI 10.1007/978-3-319-45421-4_9

providing and ranging different profiles. A profile can be seen as an ensemble of the living situation of a statistical unit (e.g. household, individual person).

Traditionally, in poverty studies (e.g. multidimensional poverty indices) an aggregation process leading to a binary classification, i.e. poor or not poor has been adopted by scholars. However, this statistical methodology does not allow to detect 'in-between' profiles. We need a statistical method, respectively, (i) detecting and ranging profiles; (ii) evaluating them as function of the poverty line criterion to assess a deprivation degree for each profile. Then, we will be able to capture people in the most prosperous, the poor, and in-between of prosperous and poor.

Most recently, a new statistical method, the form of partial order theory approach called 'the poset approach' developed by Fattore and his colleagues (Fattore et al. 2011, 2012; Fattore 2014, 2015) allows to detect and analyse these 'in-between' profiles. Hence, one of the main aims of this chapter is to show that this posetic approach derives 'in-between' profiles and contributes to create a synthetic poverty measure by taking into account the multidimensional and complex nature of poverty.

In our study, monetary poverty, material deprivation, and well-being, measured both with objective and subjective indicators, are used to analyse multidimensional poverty in Switzerland. The empirical analysis is based on the Swiss Household Panel data of 2013 and is realized by using R package PARSEC (2015). After a short introduction, this chapter aims at analysing chronologically how poverty was measured and conceptualized in the scientific field. It discusses then methodological questions and data issues. Finally, it provides a detailed analysis of our results and proposes some conclusive remarks.

2 Conceptualizations and Measurements of Multidimensional Poverty

2.1 Evolution of Multidimensional Poverty's Conceptualizations

The conceptualization of multidimensional poverty goes back to the seminar work of Seebohm Rowntree in the early 1900s. Rowntree measured poverty in the city of York by using the indicator of the 'minimum necessaries for the maintenance of merely physical efficiency' (Rowntree 1901, pp. 86–87), an indicator in which the elements of minimum necessities were converted into monetary metrics in order to obtain absolute (monetary) poverty line. In his model, people who had total earnings below the absolute monetary poverty line were considered (primary) poor. His approach was taking into account different items of life's necessities and human needs but operationally it ended up measuring poverty as a monetary deficiency.

Around the mid-1900s various approaches to poverty such as the culture of poverty, relative deprivation, poor workers, and underclass were developed respectively by anthropologists, sociologists, and economists. In the 1980s, in the wake of these contemporary poverty studies, the economist Amartya Sen (1981, 1987) introduced the capability approach to analyse the notions of well-being and poverty. Amartya Sen's capability approach clearly put in evidence the multidimensional aspects of poverty which manifests itself in different spheres of life. Actually, similarly, Peter Townsend's (1979) relative deprivation approach took into consideration the multidimensional aspects of poverty, although in practice, often only relative monetary poverty or material deprivation have been included.

In the 1990s and early 2000s, social exclusion approaches to poverty have emerged (Laderchi et al. 2006). In 1999, Gordon et al. (2000) analysed poverty from a social exclusion perspective by using 'the 1999 Poverty and Social Exclusion Survey of Britain'. However, conceptually, the origins of social exclusion approaches to poverty has to be traced back to the French debate about '*les exclus*' (Lenoir 1974); a marginalized group of people in precarious situation in the labour market, and suffering from social integration problems like social disqualification in their society (Paugam 1991). In parallel to that, in the German-speaking poverty literature we find the concept of '*Lebenslage*' (life situations) (Hauser and Hübinger 1993; Leu et al. 1997), and notably the concept of 'precarious prosperity' suggested by Hübinger to analyse in-between poverty and precarious prosperity situations (Hübinger 1996).

More recently, poverty has been increasingly related to multidimensional well-being aspects, such as well-being indicators on diverse education and employment dimensions. Nowadays, multidimensional well-being approaches are therefore widely used by scholars for several reasons: (i) to reflect subjective perception of poverty (e.g. subjective poverty), (ii) to take into account personal emotions, satisfaction, and personal evaluation of life situations and living conditions, and finally (iii) to make a combined use of objective and subjective indicators to measure multidimensional poverty. In a synthetic manner, it results from our reconstruction of the evolution of poverty's conceptualization that multidimensional perspectives, capability, material deprivation, social exclusion, and well-being approaches are employed to study poverty.

2.2 *Measurements of Poverty*

As Sen already mentioned (1976), a poverty measurement technique typically includes two steps: the first one is to define who is poor (identification process of the poor, i.e. selecting appropriate indicators, setting the poverty threshold line to define each statistical measurement unit as poor or non-poor), and the second one is to aggregate all poverty factors together (i.e. choosing an aggregate function) to obtain an overall poverty statistics such as: poverty headcount, intensity of poverty, severity of poverty—all of which belonging to the Foster-Greer-Thorbecke's family of poverty measures (Foster et al. 1984). According to the different approaches to poverty, unidimensional ones and multidimensional ones, there are different ways to measure this social phenomenon.

On the one hand, from a unidimensional perspective, poverty is reduced to monetary deficiency, and people situated below the poverty line are considered poor.

By definition, the issue with this kind of measurement of poverty is that it only allows to define the monetary dimension of poverty. In these kinds of approaches, the most common way to measure poverty is to use absolute and relative poverty concepts: 1 USD or 2 USD dollars per day usually represent the absolute poverty line for countries of the global south (e.g. World Bank 1990), 60 % of median income usually represent the relative poverty line (e.g. in Europe used by Eurostat 2000).

On the other hand, a multidimensional perspective includes several dimensions of poverty. In 1997, the human poverty index (HPI) was created by (United Nations Development Programme); and in 2010 the HPI was replaced by the Multidimensional Poverty Index (MPI). This composite index approach aggregates different poverty components together to obtain an overall multidimensional poverty headcount ratio and other related derived statistics such as intensity of poverty and severity of poverty. However, the issue with composite indices is that they produce a binary output (e.g. poor or not poor) and that they are more convenient with numerical (quantitative) data[1]. In opposition to this approach, Alkire and Foster (2007) developed another method called 'counting approach' to measure poverty from a multidimensional perspective[2]. Even though, their counting approach allows to treat ordinal data, it can only produce aggregated poverty statistics (with a binary classification) as in composite indices.

Most recently, in opposition to Alkire and Foster's counting method and traditional composite index approach, Fattore developed a new technique based on partial order theory, called 'posetic approach' in order to measure multidimensional deprivation including poverty, well-being, and quality of life (Fattore et al. 2011, 2012, 2015). This approach allows to treat ordinal data, uses the evaluation process instead of the aggregation one, and produces poverty profiles. In addition, Annoni et al. (2011) applied the partial order theory for describing the multidimensional character of poverty focusing particularly on the ranking analysis (in terms of poverty) among the European Union countries.

Moreover, there are also other mathematical and statistical poverty measurement techniques (Kakwani and Silber (eds)., 2008; Alkire et al. 2015) such as the dominance approach (Duclos et al. 2006; Bourguignon and Chakravarty 2009), a fuzzy set approach (Annoni et al. 2014; Lemmi and Betti 2006), axiomatic approach (Sen 1976), and statistical approaches such as cluster analysis, principal component analysis, multiple correspondence analysis, and structural equation models for measuring multidimensional poverty.

---

[1]Ordinal indicators (e.g. 0 = not at all satisfied, 1 = a very little satisfied, ..., 10 = totally satisfied) can be seen as a (quantitative) scale but they cannot be numerically combined (e.g. by mathematical functions).

[2]In 2010, Alkire and Foster's method is used for creating the United Nations Development Programme's MPI (Multidimensional Poverty Index) which replaced the Human Poverty Index (HPI).

3 Data, Method, and Operationalization

This section is composed of two subsections in which we present respectively our data set, the method, and the way we operationalize poverty in our study.

3.1 Data and Method

In our study, we employed the Swiss Household Panel (SHP) data of 2013.[3] The Swiss Household Panel data is a comprehensive multi-domain survey to observe Swiss population living conditions (which includes both objective and subjective questions). Panel data is conducted on an annual basis since 1999 with the same households and is updated twice a year. There are two questionnaires: one for the household and one for each individual. All individuals aged 14 or more (living in the household) are eligible to answer the individual questionnaire. In our data SHP 2013, there are 10,575 total observations with 4,467 households.

In our study, we used the Fattore partial order theory (Fattore 2015; Fattore et al. 2011, 2012) based on multidimensional deprivation measurement technique (called the poset approach) to measure poverty in Switzerland. Fattore and his colleagues developed the poset approach to measure multidimensional socio-economic phenomenon such as deprivation, subjective well-being, or poverty. This approach allows to treat ordinal data, generate profiles, avoids numerical aggregation of the indicators, characterizes poverty profiles in a differentiated manner (see evaluation process in Fattore 2015), and provides a poverty/deprivation degree for each profile (contrary to binary classification). Here, we provide a very brief overview on the procedure consisting of the following steps: firstly, the profiles due to the indicator set selected are obtained; secondly, benchmarks (i.e. poverty thresholds or poverty lines) are defined; thirdly, the set of profiles are ordered leading to a partially ordered set; fourthly, once a partial order is found it can be represented as a set of linear extensions; fifthly, counting process based on linear extensions and the benchmarks provide (i) profiles below benchmarks, (ii) profiles above benchmarks, and profiles neither (i) nor (ii); finally, poverty/deprivation degree is assessed to each profile. All statistical details related to this method can be found in Fattore's methodological and applied papers (Fattore 2014, 2015; Fattore et al. 2011, 2012). In addition, Bruggemann (2011), Patil and Taillie (2004), Bruggemann and Voigt (2008), Bruggemann and Carlsen (2006) provide detailed information on partially ordered sets, the use of Hasse diagram (to illustrate the partially ordered sets), and the application of the partial order theory. Most recently, Arcagni and Fattore (2015) developed a package 'PARSEC' in R environment for the poset approach's implementation (R is a statistical software and a programming language); and they showed in one of their paper (Fattore 2014) how to use this R package in order to evaluate multidimensional poverty.

[3]The SHP is a longitudinal survey with yearly waves conducted since 1999.

3.2 Operationalization

The goal of our research is to investigate multidimensional poverty in Switzerland. We claim that there are different forms of poverty and that poverty is a process which cannot be reduced to a simple static typification and classification of the population into the two categories of 'poor and non-poor'. Poverty conceived as a multidimensional, gradual, and labile condition means that each statistical measurement unit can occupy different positions in our classification according to the different poverty's dimensions considered, such as 'upper' (prosperous), 'middle' (between prosperous and poor), and 'bottom' (poor).

Hence, the first step to start our research was to find an appropriate statistical method to allow us to see clearly the distribution of poverty characteristics' different combinations in our data set and to analyse 'in-between' profiles. In this perspective, to be able to show different forms of poverty, we decided to choose a profile-based approach (with an evaluation process) rather than an aggregated statistics based one. With regard to our theoretical framework and research objectives, the poset approach (Fattore et al. 2011, 2012; Fattore 2015) represented an adequate method to analyse multidimensional poverty.

In our study, poverty is defined by three main dimensions: (i) monetary poverty, (ii) material deprivation, and (iii) (multidimensional) well-being. We included ordinal data and we combined both objective and subjective indicators (for all indicators, see Table A.1). As well-being is a concept including multiple dimensions, we could not take into account all its components so we decided to select only two of them, education and employment, according to our research focus on employment precariousness. We argue firstly that, notably for developed countries (like Switzerland), indicators should be selected from a relativist perspective, i.e. in our study we adopted Townsend's (1979) 'relative deprivation' approach, that is, we do not use a list of basic/absolute needs as they depend on a specific societal situation. A list of relative needs can be various and long, for that reason we don't fix any list. Secondly, following the first argument, poverty takes different forms and is characterized by different degrees of severity. Thirdly, as poverty is a multidimensional phenomenon the list of indicators can be overlong and arbitrary. For these reasons, in our view, researchers should reduce the scope of their analysis to some specific forms of poverty rather than trying to aggregate a high number of indicators (this is why we focus only on employment precariousness). In the line of this argument, we propose a core model which is composed of three dimensions, monetary poverty, material deprivation, education as well as three models including different aspects of employment precariousness. The logic behind this construction is that the core model (model 1) is the standard one and that the other three derived models (model 2a–c) show different employment precariousness profiles including different employment indicators (job stability, unemployment risk, negative impact of work on private life and family obligations) in addition to the core model's three components (Table 1). The main reason why we used relatively few indicators

Table 1 Research models and corresponding research objectives

Models	Objectives
Model 1 (core model): education and monetary poverty and material deprivation	Identification of core model's different poverty degrees, deprivation/poverty severity degrees, and poverty profiles
Model 2a: education and monetary poverty and material deprivation and job stability	Analysis of labour market precariousness by using the indicators' combination of material deprivation, monetary poverty, and educational level
Model 2b: education and monetary poverty and material deprivation and unemployment risk	Compared analysis (with model 2a) of how the results vary when the variable of unemployment risk perception is replaced by the variable of job stability perception
Model 2c: education and monetary poverty and material deprivation and negative impact of work on private life and family obligations	Analysis of labour market precariousness which includes family-work balance indicator

comparing to the index approach is that we wanted to analyse poverty with a profile-based approach—again, using too much indicators would simultaneously hinder concise and clear interpretation of poverty profiles.

In the core models (model 1), the sample is only composed of people aged between 18 and 66 years old, so the total number of observations in our data set amounts to 4,999. Considering the other models (model 2a, model 2b, model 2c), the sample represents the economically active population (16 – 66 years old), so that the total number of observations in our data set amounts to 3,879. The number of observations are lower in models 2a–2c compared to model 1 because these models focus on labour market precariousness and therefore take into account only people who are active in the labour market. In order to apply partial order methodology (Fattore et al. 2011, 2012; Fattore 2015), we reordered variables so that high numbers represent less deprivation, and conversely, low numbers represent more deprivation. In a general manner, the statistical measurement unit is individual, but questions related to relative monetary poverty and material deprivation come from household questionnaire.

4 Partial Order Theory's Application and Results

In this fourth section, we briefly present the statistical software used in our analysis and, then, the results that we obtained for each models.

4.1 Statistical Software

In order to apply the partial order method developed by Fattore and his colleagues (Fattore et al. 2011, 2012, Fattore 2015), we used the R statistical software and notably the software package namely *PARSEC* (PARtial orders in Socio-Economics). This package is available in R environment and is developed by Arcagni and Fattore (2015). PARSEC can be used to measure multidimensional social issues such as poverty, well-being, quality of life, or other deprivations by poset-based evaluation (Fattore and Arcagni 2014).

4.2 Results

In adequation with partial order method, a benchmark is defined based on Alkire and Foster's (2007) dual cutoff system. In this dual cutoff system, there are two thresholds: the first threshold (first cutoff 'k') serves for determining who is poor regarding a particular indicator; the second threshold allows to determine who is poor regarding all indicators. If a unit gets a score inferior to k, then this unit is considered deprived (poor) regarding to this indicator. Then, we count the number of deprivations of all indicators. If the total number of deprivations is equal or superior to the second threshold, then the unit is considered poor. In Table A.1 (see Appendix), the first cutoff thresholds (k) are shown for each indicator. For the variables[4] 'material possession', 'educational level', 'job stability's perception', 'unemployment risk's perception', 'negative impact of work on private life and family obligations', the cutoff k is determined as function of data distribution. This means no *external threshold criterion* has been adopted; rather we looked which modalities are corresponding to the most disadvantageous situation compared to the whole distribution. We set the second cutoff thresholds from model 1 to model 2c as follows: 2, 3, 3, 3. For instance, in model 1, if a person is deprived regarding at least 2 of 3 indicators, then this person is considered poor; similarly, in model 2a if a person is deprived regarding at least 3 indicators of 4, then this person is considered poor. Hence, benchmark profiles are determined as function of the dual cutoff system; profiles positioning in or under the benchmark profiles (threshold profiles) are considered (totally) poor profiles.

[4]The relative monetary poverty variable is a dichotomous (0 = having an income less than 60% of annual median equivalized net household income (OECD criterion); 1 = otherwise), i.e. having an income less than 60 % of the median is considered poor.

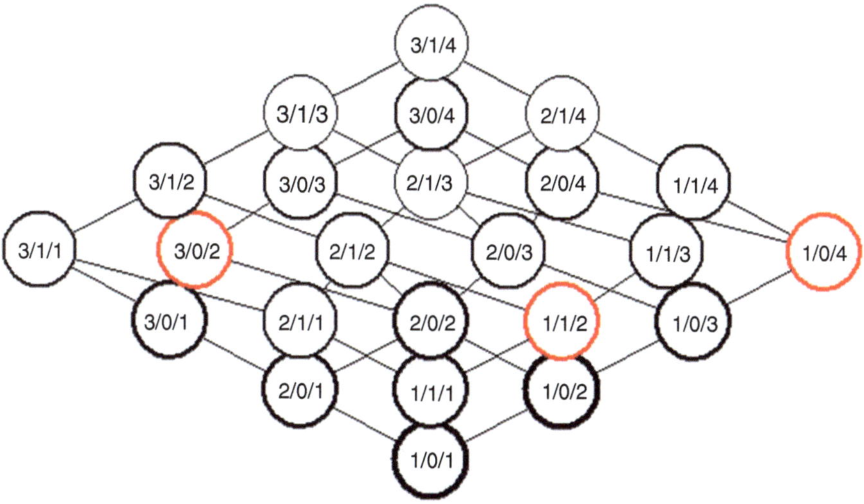

Fig. 1 Profiles (*red*: PARSEC deprived profiles); (*bold*: OPHI deprived profiles) of model 1 (core model) composed respectively by education, monetary poverty, and material deprivation indicators). Source: Authors' calculations in PARSEC

4.2.1 Model 1 (Core Model)

In our data set, we obtained 24 different profiles for model 1. These profiles are illustrated by a Hasse diagram (see Fig. 1; each profile is the ordered set of deprivation scores: for instance, the elements in the profile (3/1/4[5]) designates, respectively, the deprivation scores of educational level, monetary poverty, and material deprivation). As higher numbers signify less deprivation, the profile (3/1/4) at the top of the diagram represents the most prosperous profile and the profile (1/0/1) at the bottom of the diagram corresponds to the poorest profile in our data set. Moreover, the poset approach distinguishes poverty degree through the identification function (see Fig. 2; in the plot of the identification function by profiles, profiles are ordered according to the value of the identification function). This identification function allowed us to interpret poverty through a detailed gradual approach instead of a binary class approach (distinguishing only the categories of poor and non-poor), which is used by Alkire and Foster methodology-OPHI. The poset approach permitted us to define three thresholds (profiles, encircled in red in the Hasse diagram (Fig. 1)). These threshold profiles and other profiles remaining under these thresholds (1/0/4; 1/1/2; 3/0/2) belong to the poor category. Above these thresholds can be defined two categories: the prosperous and the 'in-betweens'.

[5]Generally, profiles can be written in a form such as 3,1,4 instead of 3/1/4. However, as the outputs obtained from the R package PARSEC yield profiles information in a form with '/', so we kept the format with '/' to avoiding confusion.

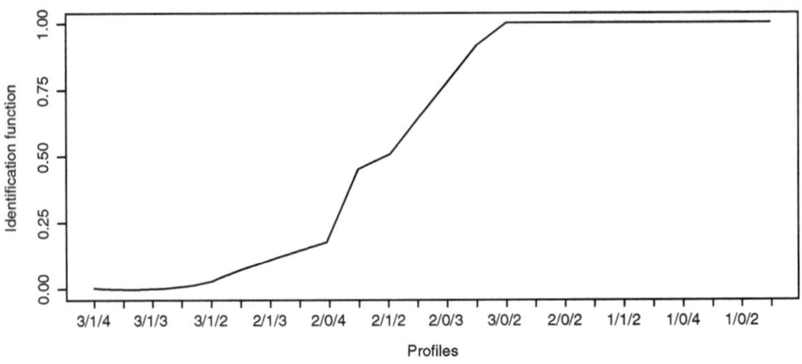

Fig. 2 Identification function (poverty degrees from 0 to 1 for each profile) for model 1 (core model). Source: Authors' calculations in PARSEC

We can define more precisely these two categories as follow: 71 % to the upper prosperous (deprivation degree from 0 to 0.10); 19 % to the bottom prosperous (deprivation degree from 0.11 to 0.49); 4.5 % to the bottom prosperous and the poor ('in-betweens': deprivation degree from 0.50 to 0.99), and 5.5 % to the poor (deprivation degree: 1). The 'in-between' profile deserves a particular attention because people belonging to this category are not situated below the poverty line but still are situated in a risky area; 'in-betweens' suffer from relatively high levels of poverty. In classical composite index approach or in counting approach (Alkire and Foster 2007), 'in-betweens' are placed directly in 'non-poor' category.

In addition to poverty degrees, both deprivation (poverty) severity degrees and relative wealth degrees can also be obtained with a poset approach. In Fig. 3, we can observe that the poorest profile gets a score of 1 for deprivation severity degree and of 0 for wealth degree. Among other examples, it can also be noted that the profile 1/0/4 subjected to deprivation on two items (education and monetary poverty—getting, meanwhile, the highest material possession's score) get a score of 0.3 points for deprivation severity degree and 0 point for wealth degree since it is below the poverty line; or the profile (3/1/1) which suffers from material deprivation get a score of 0.2 point for wealth degree and of 0.10 point for deprivation severity. Moreover, among 'in-between' profiles, the profile (1/1/3) is at a low wealth degree (0.08 point) but get a score of 0.49 point for deprivation severity degree; also the profile (2/1/1) get a score of 0 point for wealth degree but 0.44 point for deprivation severity degree. Hence, we can observe clearly that among the 'in-between' profiles there is a vast heterogeneity of poverty and wealth degrees.

Instead of binary class approach, detailed gradual approach is necessary to understand the complex nature of poverty. The 'in-betweens' particularly encompasses profiles (1/1/3), (2/0/3), and (2/1/1), respectively, (i) low educational level, monetarily non-poor, and not materially deprived; (ii) medium educational level, monetarily poor, not materially deprived; (iii) medium educational level, monetarily non-poor, materially deprived. Among bottom prosperous status profiles, we can identify particularly the following profiles: (i) university degree level, non-monetarily poor

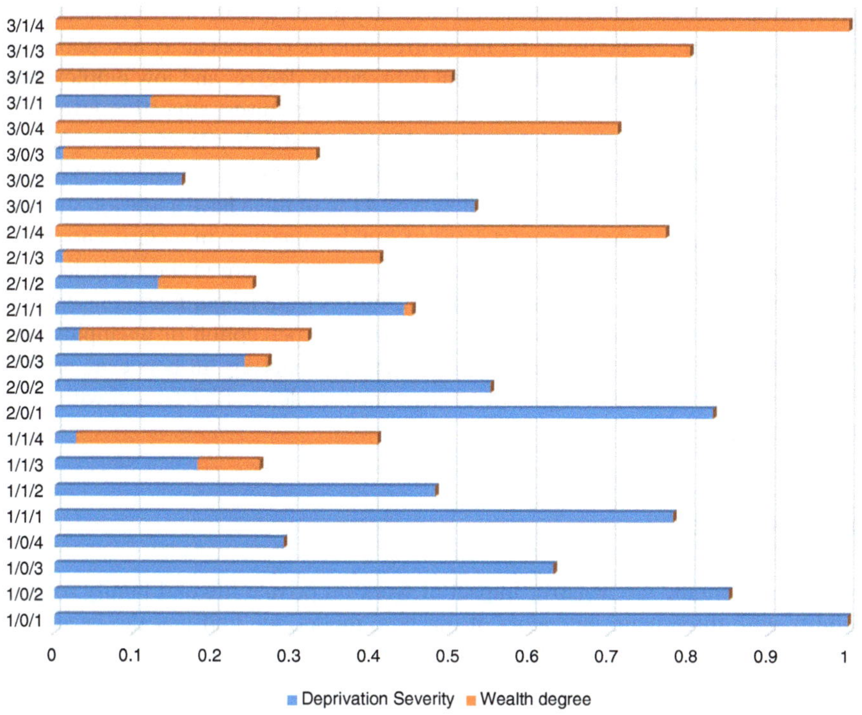

Fig. 3 Deprivation severity and wealth degree for model 1 (core model). Profiles composed respectively by education, monetary poverty, and material deprivation indicators. Source: Authors' calculations in PARSEC

but materially poor; and (ii) medium educational level, non-monetarily poor but materially poor, category which deserves a particular attention since individual belonging to this category gets a score of 0.50 poverty (deprivation) degree point.

4.2.2 Model 2a and Model 2b

We constructed derived models to analyse labour market precariousness by the indicators' combination of material deprivation, monetary poverty, and educational level. Regarding labour market precariousness, the model 2a takes into consideration job stability's subjective perception, while the model 2b includes unemployment risk's subjective perception. In model 2a, 88 profiles are produced from our data set; for instance, the set of 3/1/4/4 designates, respectively, the deprivation scores of educational level, monetary poverty, material deprivation, and job stability. Here, 'in-between' profiles correspond to: (i) monetarily non-poor with the lowest education level, high material deprivation degree, and low employment instability level; and (ii) monetarily non-poor with the lowest education level, low material deprivation degree, and high employment instability level. Among upper prosperous status profiles, we can see notably that 13 % have a low educational level and are not monetarily poor people—however, 41 % of them have a low job stability level. Interestingly enough, around 10 % of highly educated, monetarily non-poor people with a high level of material possession feel insecure in their job. In other words, one out of ten highly educated people suffers from labour market insecurity; and when we replace job stability variable with the perception of unemployment risk variable, we get similar results. However, in model 2b 'in-between' profiles amount to 4 %, which is a higher percentage than model 2a (2.7 %) and with more heterogeneity.

4.2.3 Model 2c

When we combine the negative impact of work on private life and family responsibility with three main indicators of our core model, we find that the headcount poverty rate according to OPHI counting approach is very low, around 1 % (which is equivalent to the poor status' poverty rate in the poset approach). On this point, with a poset approach we saw that despite this very low poverty rate, 6.50 % of the population still belonged to 'in-between' profile (poverty/deprivation degrees from 0.50 to 0.99). Medium educational level graduates form the major part of 'in-between' group, and there is an important heterogeneity among 'in-between' profiles. In particular, it is worth mentioning that 33.3 % of them are suffering from the negative impact of work on their social life even if they are not monetarily poor and are not suffering from material deprivation.

5 Conclusion

Our analysis shows that people belonging to the 'in-between' profiles categories suffer from different deprivation degrees ranged from 0.50 to 0.99. In model 1 'in-betweens' get a score of 4.5 %; in model 2a 2.70 %; in model 2b 4 %; and in model 2c 6.5 %. If we consider the different profiles, we can see that in model 1 (core model) 'in-betweens' encompasses the following ones: (i) low educational level, monetarily non-poor, and not materially deprived; (ii) medium educational level, monetarily poor, not materially deprived; (iii) medium educational level, monetarily non-poor, materially deprived. In labour market precariousness model (model 2a and 2b), we found that one out of ten highly educated people suffers from labour market insecurity; and when we replaced the job stability variable with the perception of unemployment risk variable, similar results were obtained. In model 2c, we found that there is an important heterogeneity among 'in-between' profiles. Notably, in model 2c when we combined the core model, which is only composed of three objective indicators with a subjective employment related variable, we obtained a substantial increase of 'in-between' profiles which are not considered poor but are characterized by a relatively high degree of deprivation. Thus, the poset approach allowed us to see the precarious profiles (active in labour market) when we combined objective variables (education, monetary poverty, material deprivation) with subjective (employment) indicators.

In conclusion, the main contributions of partial order method are (i) to permit the use of ordinal data in order to make multidimensional measurements; (ii) to use evaluation function instead of aggregation function and to measure poverty in terms of profiles; and (iii) to establish what are the poverty degree, wealth degree, and poverty (deprivation) severity for each profile. In our study, we built different models and applied partial order method to show that poverty is a process and a gradual and labile condition which takes different forms, mainly: the prosperous profiles, the 'in-between' profiles and the poor profiles. Developing our analysis further, we could distinguish finally six different poverty profiles according to their poverty degrees: the most prosperous, the upper prosperous, the bottom prosperous, the 'in-betweens', the poor, and the poorest.

Acknowledgment This contribution is based on the project 'Multidimensional well-being: conceptual, methodological, analytical perspectives', financed by the Indo-Swiss Joint Research Programme in the Social Sciences (seed money grants).

This study has been realized using the data collected by the Swiss Household Panel (SHP), which is based at the Swiss Centre of Expertise in the Social sciences FORS in Lausanne. The SHP data are collected within the framework of the research program 'Living in Switzerland', financed by the Swiss National Science Foundation.

Appendix

Table A.1 Poverty indicators by dimension, nature, scale, categories, and first cutoffs

Indicator	Dimension	Nature	Scale	Generated categories; first cutoff (k)
Relative monetary poverty	Monetary poverty	Objective	Cardinal	Two categories: 1 (superior to 60 % of annual median equivalized net household income—(OECD criterion), 0 (inferior to 60 % of it); $k = 1$
Material possession	Material deprivation (10 items[a])	Objective	Ordinal; (1 = deprived at least three or more items; 2 = deprived at two items; 3 = deprived at one item; 4 = no deprivation among 10 items)	Four categories: 1, 2, 3, 4; $k = 3$
Educational level	Education component of well-being	Objective	Ordinal: from 0 to 10 (0 = incomplete compulsory school; 10 = university)	Three categories: 1: ISCED[b] 0–2, 2: ISCED 3–4, 3: ISCED 5–6; $k = 2$
Job stability's perception	Employment component of well-being	Subjective	Ordinal: 4 = very secure; 3 = quite secure; 2 = a bit insecure; 1 = very insecure	Four categories: 1, 2, 3, 4; $k = 3$
Unemployment risk's perception	Employment component of well-being	Subjective	Ordinal: from 0 to 10 (0 = a real risk; 10 = no risk at all)	Six categories: 1: (0), 2: (1–2), 3: (3–4), 4: (5), 5: (6–7), 6: (8–10), $k = 3$
Negative impact of work on private life and family obligations	Employment component of well-being	Subjective	Ordinal: from 0 to 10 (10 = not at all; 0 = very strongly)	Seven categories: 1: (0–1), 2: (2–3), 3: (4), 4: (5), 5: (6–7), 6: (8), 7: (9–10); $k = 3$

[a] We have ten components: savings into third pillar, 1-week holiday away from home per year, meal at a restaurant min. once a month, go to the dentist if needed, having computer at home, having TV at home, having car for private use, having washing machine at home or for exclusive use, invitation of friends min. once a month, having dishwasher at home

[b] *ISCED* International Standard Classification of Education

References

Alkire S, Foster J (2007) Counting and multidimensional poverty measures. OPHI working paper 7

Alkire S, Foster J, Seth S, Santos ME, Roche JM, Ballon P (2015) Multidimensional poverty measurement and analysis. Oxford University Press, Oxford, UK

Annoni P, Fattore M, Bruggemann R (2011) A multi-criteria fuzzy approach for analyzing poverty structure. Stat Appl (special issue): 7–30

Annoni P, Bruggemann R, Carlsen L (2014) A multidimensional view on poverty in the European Union by partial order theory. J Appl Stat 42:535–554

Arcagni A, Fattore M (2015) Parsec: partial orders in socio-economic. CRAN-Package 'parsec'. https://cran.r-project.org/web/packages/parsec/index.html

Bourguignon F, Chakravarty SR (2009) Multidimensional poverty orderings: theory and applications. In: Basu K, Kanbur R (eds) Arguments for a better world: essays in honor of Amartya Sen, I: ethics, welfare, and measurement. Oxford University Press, Oxford, pp 337–361

Bruggemann R (ed) (2011) Special issue: partially ordered sets. Stat Appl Spl Iss: 3–6

Bruggemann R, Carlsen L (2006) Partial order in environmental sciences and chemistry. Springer, Berlin

Bruggemann R, Voigt K (2008) Basic principles of Hasse diagram technique in chemistry. Comb Chem High Throughput Screen 11:756–769

Duclos J-Y, Sahn DE, Younger SD (2006) Robust multidimensional poverty comparisons. Econ J 116(514):943–968

Eurostat (2000) European social statistics: income, poverty and social exclusion. Office for Official Publications of the European Communities, Luxembourg

Fattore M (2014) Partially ordered set, entry title. In: Michalos AC (ed) Encyclopedia of quality of life and well-being research. Springer, Dordrecht, pp 4627–4631

Fattore M (2015) Partially ordered sets and the measurement of multidimensional ordinal deprivation. Soc Indic Res. doi:10.1007/s11205-015-1059-6 (first online: 05 August 2015):http://link.springer.com/article/10.1007%2Fs11205-015-1059-6#/page-1

Fattore M, Arcagni A (2014) PARSEC: an R package for poset-based evaluation of multidimensional poverty. In: Bruggemann R, Carlsen L, Wittmann J (eds) Multi-indicator systems and modelling in partial order. Springer, Berlin, pp 317–330

Fattore M, Maggino F, Greselin F (2011) Socio-economic evaluation with ordinal variables: integrating counting and poset approaches. In: Statistica & Applicazioni, partial orders in applied sciences (special issue). Vita e Pensiero, Milano, pp 31–42

Fattore M, Maggino F, Colombo E (2012) From composite indicators to partial orders: evaluating socio-economic phenomena through ordinal data. In: Maggino F, Nuvolati G (eds) Quality of life in Italy: researches and reflections. Social indicators research series. Springer, Amsterdam, pp 41–68

Foster JE, Greer J, Thorbecke E (1984) A class of decomposable poverty measures. Econometrica 52(3):761–766

Fukuda-Parr S (2006) The human poverty index: a multidimensional measure. In: United Nations Development Program (UNDP), International Poverty Centre (eds) What is poverty? Concepts and measures. UNDP International Poverty Centre, Brazil, pp 7–9

Gordon D, Pantazis C, Levitas R (eds) (2000) Poverty and social exclusion in Britain: the millennium survey. Joseph Rowntree Foundation, New York

Hauser R, Hübinger W (1993) Arme unter uns. Teil 1: Ergebnisse und Konsequenzen der Caritas-Armutsuntersuchung. Lambertus, Freiburg i.B

Hübinger W (1996) Prekärer wohlstand. Lambertus, Freiburg i.B

Kakwani N, Silber J (eds) (2008) Quantitative approaches to multidimensional poverty measurement. Palgrave Macmillan, Basingstoke

Laderchi CR, Saith R, Stewart F (2006) Does the definition of poverty matter? Comparing four approaches. In: United Nations Development Program (UNDP), International Poverty Centre (eds) What is poverty? Concepts and measures. UNDP International Poverty Centre, Brazil, pp 10–11

Lemmi A, Betti G (eds) (2006) Fuzzy set approach to multidimensional poverty measurement. Springer, New York

Lenoir R (1974) Les exclus: Un Français sur dix. Seuil, Paris

Leu RE, Burri S, Priester T (1997) Lebensqualität und armut in der Schweiz. Verlag Paul Haupt, Bern

Patil GP, Taillie C (2004) Multiple indicators, partial order sets, and linear extensions: multi-criterion ranking and prioritization. Environ Ecol Stat 11:199–228

Paugam S (1991) La disqualification sociale. Puf, Paris

Rowntree BS (1901) Poverty, a study of town life. Macmillan, New York

Sen A (1976) Poverty: an ordinal approach to measurement. Econometrica 44(2):219–231

Sen A (1981) Poverty and famines: an essay on entitlement and deprivation. Harvard University Press, Cambridge

Sen A (1987) Commodities and capabilities. Oxford University Press, Oxford, UK

Townsend P (1979) Poverty in the United Kingdom: a survey of household resources and standards of living. Penguin, Harmondsworth

World Bank (1990) World development report: poverty. Oxford University Press, New York

Analysis of Social Participation: A Multidimensional Approach Based on the Theory of Partial Ordering

Stefania Della Queva

1 Social Participation: Sense of Community and Well-Being

Historically, the surge of non-profit organizations is tied to the industrial revolution during the XIXth century and the social issues that it brought. A second wave of growth followed during the 1970s with the first reforms of modern welfare. Since then, the phenomenon of associationism covered an important role both in aggregating an atomized society and in integrating the deprived groups. From this point of view, participation itself is a process that leads to overcoming the isolation of individuals and groups (through aggregation processes) and the reduction of subordination through the distribution of power (through processes of equalization). Associations, made up of economic, cultural and social bonds not directly under the control of traditional organizations, also became a fundamental intermediary between citizens and public institutions, thus giving shape to what will be called civil society.

Today, non-profit institutions form a fabric of relationships and networks that underpins the political system and the state.

In the literature, it is widely recognized that such social fabric, a generalized climate of interpersonal trust, high involvement in associative networks and widespread civic culture, increases individual well-being and social cohesion, allowing a better performance, greater efficiency of public policies and a lower cost of economic transactions. In Italy, the formal networks, which include all the relationships that gravitate around individuals, are of particular interest: environmental association, cultural, recreational associations, volunteer groups or associations and political parties. These networks put in place human and material resources to

S. Della Queva (✉)
ISTAT—Italian National Institute of Statistics, Viale dell'Oceano Pacifico 171, Rome 00144, Italy
e-mail: dellaqueva@istat.it

M. Fattore, R. Bruggemann (eds.), *Partial Order Concepts in Applied Sciences*,
DOI 10.1007/978-3-319-45421-4_10

provide support and protection to individuals both in everyday life and in critical moments and periods of discomfort, representing an essential element of social cohesion (Frisanco 2013).

According to the 2011 Census results, at the 31st of December 2011 Italian non-profit institutions were composed of a heterogeneous set of 301,191 private organizations that represented 6.4 % of the Italian labour market and employed 4.6 % of the paid workers (ISTAT 2014).

A modern definition of social participation should express a very broad concept. One accepted definition is the process of taking decisions concerning the life of an individual and that of the community in which he lives (Hart 1992). The participation is characterized by:

- An ongoing process of self-responsibility and choice;
- Being an interactive part of something, somewhere, some group.

Participating is a place of non-neutrality; it is the awareness that our being in the world is called to take an ethical line according to principles of justice and equity.

Studying this concept is really important because social participation, involving vulnerable and excluded groups, should seek the empowerment of those groups, increasing their effective control over decisions that influence their quality of life.

Because of the complexity of this concept, the study of social aspects increasingly requires the use of new tools that better render the reality and model the multidimensionality of the phenomenon. In this chapter, we will explore dynamic and multidimensional measures of social participation in Italy, leveraging properties of poset theory (Fattore and Arcagni 2013; Fattore 2015).

2 Why Partial Order Theory?

Previous analyses have focused on a particular aspect of social participation (e.g. political); instead, in this chapter we will explore dynamic and faceted aspects of social participation in Italy.

To obtain a synthetic representation of the forms of social participation, we adopt a fuzzy approach. Our purpose is not to give a measure of the phenomenon but to bring out its structure in terms of profiles. In this framework, a profile represents an aggregate of variables that identifies a 'stereotypical individual': we want to identify not only the socially active profiles, but also the borderline forms of participation highlighting the relationships and ranking of these stereotypes.

Because of the multidimensionality of the concept under examination, we cannot assert that there is any single expression of social participation, so we need to use a different language that better exposes the nuances and complexity of the phenomenon.

Furthermore, we need to consider that our data are categorical (nominal variables) with no single ordering between the categories. With this approach, the nature of the data is fully respected, avoiding any kind of scaling and aggregation procedure (Fattore et al. 2011). The adopted methodology will comprise the following steps:

- Selection of variables of interest;
- Casting data in the right mathematical form;
- Building the natural profiles poset;
- Adding relationships between profiles;
- Deciding which profiles have the minimum sets of characteristics to be considered socially active (threshold);
- Computing a ranking of profiles by 'linearizing' the poset;

While preserving the multidimensionality, we can infer a complete ordering of the profiles in the final step. This process is known as linear extension, it is a classical counting approach. We select all paths in the poset from top to bottom and count how many times a profile is above at least one of the threshold profiles. This number will generate an identification function that represents a ranking of all profiles.

The analysis was conducted using R and more specifically the package Parsec (Fattore and Arcagni 2014).

3 Official Statistics

The analysis uses the Multipurpose Survey on households: aspects of daily life, conducted by ISTAT. The population under survey comprises 59 million individuals, but after removing records with missing values it reduces to 51.716 million people aged 14 years and over (this is the effectively considered population).

Official statistics view the participation as taking part in collective actions and classify these actions in different categories. We have considered a subset of the survey, focusing on the four dimensions of social participation listed in Table 1.

Table 1 The dimensions of social participation

Selected variables	Dimensions
Meeting of cultural, recreational or other type of associations	Meeting
Meeting of political parties or trade unions	Meeting
Participation in political rallies or demonstrations/parades	Demonstration
Money given to associations or political parties/trade unions	Donation
Free-of-charge activity for volunteering groups or associations	Activity
Free-of-charge activity for political parties or trade unions	Activity

Selected variables are all binary and have all been modified to show the same polarity towards the participation level, recoded as 0 = not active, 1 = active

4 The Multidimensionality of Social Participation

Having 6 binary variables, our first step produces $2^6 = 64$ different possible profiles, all of which have been observed in our population. The best profile, P64, the most socially active, is a sequence of six 1 digits, while the worst profile, P01, the one that is not active in any of the selected categories, is a sequence of six 0s, as shown in the first Hasse diagram reported in Fig. 1. The first profile poset is composed by partially ordered profiles in a natural way and it represents the logical ordering (Fattore 2008; Fattore and Arcagni 2014).

Table 2 reports the frequency distribution of the generated profiles given by the statistical units belonging to it.

The basic structure is modified by adding relationships between profiles that we consider to be comparable in the specific research and by removing relationships that, while being present in the logical structure, do not have any meaning in the studied subject. Providing this information consists of adding new edges when the profiles are comparable and removing edges in the other case. Five order relationships have been added to our poset, while none have been removed:

- P05 < P17
- P05 < P33
- P13 < P17
- P13 < P33
- P09 < P17
- P09 < P33

Profiles P33 and P17 represent people who are volunteering in non-profit organizations or who are active in political parties or trade unions. We assume that these profiles are more engaged in social participation than profiles P05, P09 and P13. P05 represents people who participate in mass meetings and demonstrations, P09 represents people who donate money to associations or parties and P13 is

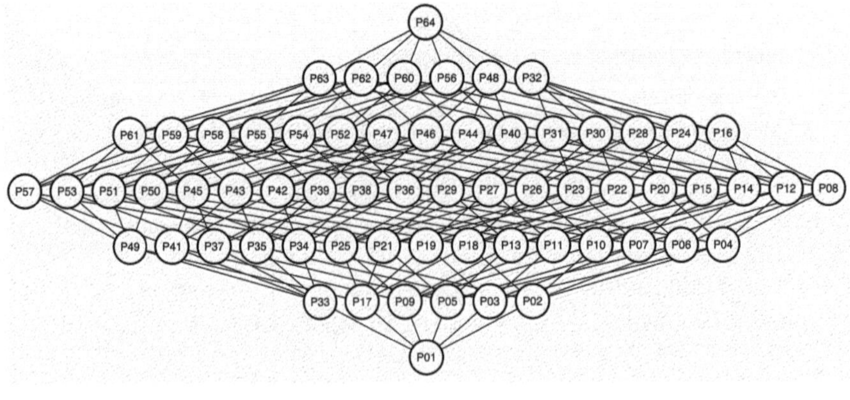

Fig. 1 Hasse diagram of the profiles poset

Table 2 Frequency distribution on elements of the profiles poset

Profiles	Weights[a]	Profiles	Weights[a]	Profiles	Weights[a]
P01	35.446	P23	4	P45	45
P02	1.029	P24	49	P46	19
P03	1.709	P25	8	P47	279
P04	234	P26	31	P48	154
P05	1.767	P27	1	P49	10
P06	506	P28	37	P50	7
P07	281	P29	8	P51	6
P08	140	P30	86	P52	17
P09	2.354	P31	2	P53	4
P10	320	P32	82	P54	7
P11	822	P33	905	P55	8
P12	184	P34	61	P56	25
P13	259	P35	1.339	P57	9
P14	158	P36	136	P58	5
P15	133	P37	97	P59	9
P16	150	P38	24	P60	69
P17	40	P39	174	P61	7
P18	87	P40	76	P62	28
P19	5	P41	306	P63	7
P20	20	P42	35	P64	221
P21	11	P43	1.315		
P22	88	P44	259		

[a]The frequencies of each profile are reported in million

the union of the preceding behaviours. Our assumptions are based on the effort and continuity of the commitment and the time spent on these actions. The new relationships give relevance to some profiles over others in the natural poset, relaxing its shape. The 'relaxed' Hasse diagram is shown in Fig. 2.

Once the structure has been defined, we need to identify a set of profiles that can be considered socially active. The minimal set of such profiles defines a threshold, such that any profiles below one of its elements is classified as not socially active. Imposing this threshold, we give 'social meaning' to the profile poset identifying two subsets, profiles above all the elements of the threshold, which are certainly active, and profiles below all the elements of the threshold, the non-active profiles (Fattore and Maggino 2014). An important feature of this kind of tool is that, besides obviously active profiles and non-active profiles, we also find the 'partially (or ambiguously) active participants' and we will be able to weight their 'activeness' once the poset is linearized.

The selected threshold is composed of the following profiles, highlighted in Fig. 3:

- P33—people belonging to this profile exclusively perform free-of-charge activity for volunteering groups or associations.

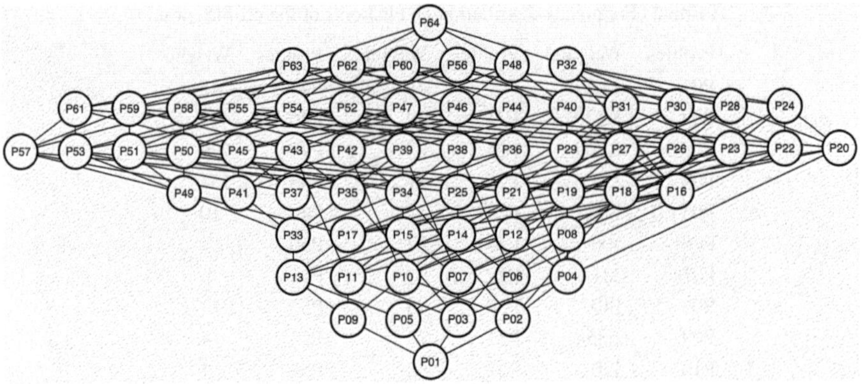

Fig. 2 Hasse diagram of the cover relation of poset

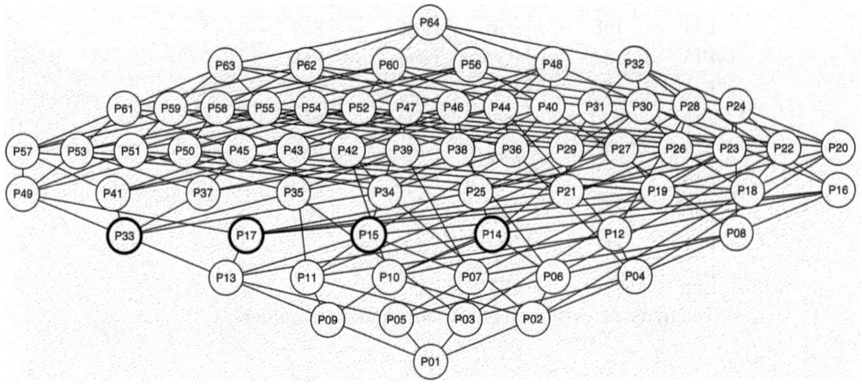

Fig. 3 The profiles poset highlighting the threshold: P33, P17, P15, P14

- P17—people belonging to this profile exclusively perform free-of-charge activity for political parties or trade unions.
- P15—people belonging to this profile participate in the meetings of associations, parades or political rallies and give donations to associations, parties or trade unions.
- P14—people belonging to this profile participate in the meetings of political parties or trade unions, parades or political rallies and give donations to associations, parties or trade unions.

Imposing new relationships and providing a threshold is a value judgement of the researcher. In the many-faceted participation concept, we have included in the threshold profiles that cut the poset across all dimensions. We have chosen instead to emphasize activism and direct participation to non-profit institutions over other forms of activities when we added the relationships. The choice of the threshold backs up the relevance of donations when paired with meetings, as forms of future planning, and parades/rallies.

Fig. 4 Identification function

At the end of our analysis, we can calculate an indicator of the importance of each profile in terms of degree of participation. This indicator is the identification function (Fattore and Arcagni 2014). We can imagine identifying the n possible paths in the poset from profile P64 to profile P01 and unrolling the poset; this process is called linear extension. In each of these paths, a profile can be either above or below the threshold. Counting the number of times that a profile is above the threshold returns an indicator of how much a profile is socially active[1] (Fig. 4).

Profiles are sorted by their level of participation; on the Y-axis we can see their exact ranking, as calculated by the identification function. Notice that the ranking is not a binary condition (0 or 1), but a continuous value determined by the counting approach.

In Fig. 5, we combine the ranking with the weight of each profile.

We can identify three groups of profiles:

- Socially active profiles identified by an identification function value greater than 0.9;
- Partially active profiles with an identification function value between threshold profiles (in red in Fig. 5) and 0.8;
- Non-active profiles for which the identification function evaluates to zero and are not in the threshold set;

The analysis shows that only 14 % of Italians are active in at least some relevant way.

[1]The computations are based on a sample of linear extensions.

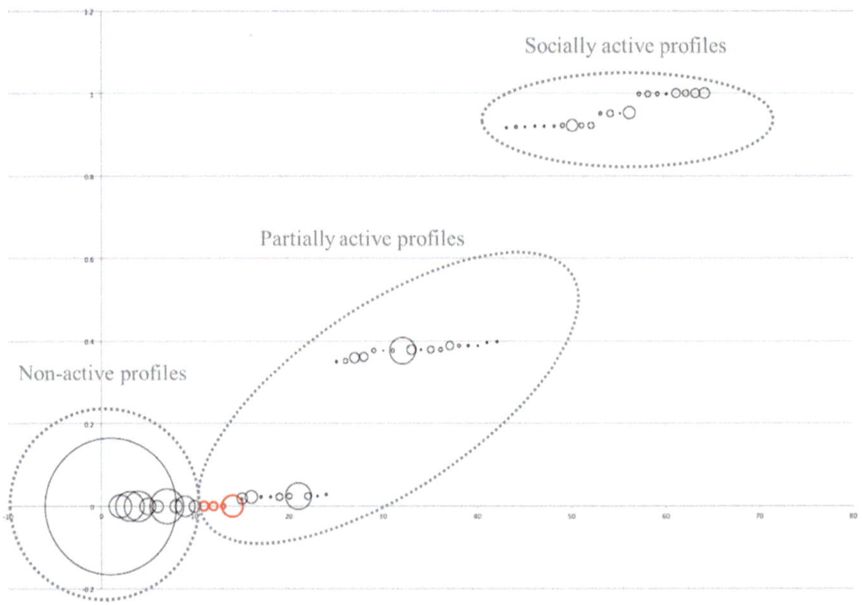

Fig. 5 Weighted profiles

The socially active profiles cover 3.1 % of Italians (nearly 1.6 million people); they are active in many dimensions. In this group, 3 profiles are of particular interest for their weight and manifold activity, all of them contain more than 200,000 units. Profile P44, while not taking part in parades/political rallies and activities for parties/trade unions, participates in the meetings of associations, political parties or trade unions and performs free-of-charge activity for volunteering groups or associations. Profile P47 represents the quintessential volunteer, unbound from the political dimension, in fact participates in the meetings of associations and demonstrations, gives money to non-profit institutions and performs free-of-charge activity for volunteering groups or associations. Profile P64 is the most active profile; in fact it represents the union of all possible activities being socially active in any dimension.

The partially active profiles are very differentiated and represent 10.9 % of Italians. The most significant profiles in terms of units are P12 and P43. The first one represents 184 thousand people who participate in meetings in associations or parties/trade unions and donate, while P43 is composed by 1.3 million people who attend meetings (associations, political parties/trade unions), donate and perform free-of-charge activities for associations.

The remaining 86 % of Italians are not socially active (more than 44 million people). The first profile, P01, includes people who responded 'no' to all questions and covers 35 million Italians, being the largest profile of the whole poset. The other profiles in this group describe people that mainly perform only one action that is by definition occasional compared with other kinds of activities (participating in political rallies or mass meetings and donations).

Analysing some characterizations of profiles, in particular gender, educational level, occupational status and territorial localization, we notice that, in general, men are slightly more participative than women: 15.9 % of men are above the threshold in contrast to 12.2 % of women. Furthermore, 72.5 % of women belong to profile P01, the totally inactive one, while only 64 % of men do not act in any way. Some different behaviours between genders arise: considering the most active profiles, men are more involved in activities tied to the political dimension (meetings, free-of-charge activities for political parties) coupled with participation in mass meetings and political rallies, while women are more interested in volunteering exclusively in non-profit institutions. Educational level has a big impact on participation: 27 % of people with a Ph.D. and 26 % of people with a bachelor's degree are very active in comparison with 5.5 % of people with only a basic education and 2.6 % without any educational level. Employed people belong to profiles that are more active (18 % belong to the high participative group) followed by students (17.9 %) and retired (12 %), mostly present in the partially active group. Housewives, people looking for their first job and people that are in other conditions come last, mostly belonging to the non-participative group. Housewives, in particular, are relevant in the profile of people who only give donations.

Finally, with regard to local distribution, the presence of active profiles is particularly prevalent in Northern Italy: North West contains 15.3 % of active profiles, North East 19.1 % while Islands and South Italy together contain only 20.1 %. The best performing territorial units are Bolzano 30.7 %, Trento 28.3 % and Veneto 20 %. The long tail of participation is closed by Campania, Calabria and Sicily.

5 Conclusions

Multidimensional social participation is a complex phenomenon. In this chapter, we have shown how poset theory provides an effective setting for fuzzy modelling of multidimensional categorical data as the subject under study. Interesting results come out from the analysis of the different profiles active in social participation. The poset methodology could represent the starting point to analyse in a different way the data about population. The poset could serve as a tool to guide the research in addition to obtaining a final output, for example research for further analysis about specific population subsets.

Subsequent experimentation and feedback from usage on more datasets, referring to different times and pertaining to different countries, could lead to improving the methodology, better refining the threshold and ordering relationships to obtain a more conformant model.

References

Fattore M (2008) Hasse diagrams, poset theory and fuzzy poverty measures. Riv Int Sci Soc 1:63–75

Fattore M (2015) Partially ordered sets and the measurement of multidimensional ordinal deprivation. Soc Indic Res: 1–24

Fattore M, Arcagni A (2013) Measuring multidimensional polarization with ordinal data. In: SIS 2013 statistical conference, BES-M3.1—the BES and the challenges of constructing composite indicators dealing with equity and sustainability

Fattore M, Arcagni A (2014) PARSEC: an R package for poset-based evaluation of multidimensional poverty. In: Bruggemann R, Carlsen L, Wittmann J (eds) Multi-indicator systems and modelling in partial order. Springer, Berlin

Fattore M, Maggino F (2014) Partial orders in socio-economics: a practical challenge for poset theorists or a cultural challenge for social scientists? In: Bruggemann R, Carlsen L, Wittmann J (eds) Multi-indicator systems and modelling in partial order. Springer, Berlin

Fattore M, Bruggemann R, Owsiński J (2011) Using poset theory to compare fuzzy multidimensional material deprivation across regions. In: Ingrassia S, Rocci R, Vichi M (eds) New perspectives in statistical modeling and data analysis. Springer, Berlin

Frisanco R (2013) Volontariato e nuovo welfare. La cittadinanza attiva e le amministrazioni pubbliche. Carocci editore, Rome

Hart R (1992) Children's participation. From tokenism to citizenship. UNICEF, Florence

ISTAT (2014) Non-profit institutions profile based on 2011 census results

POSET Analysis of Panel Data with POSAC

Enrico di Bella, Matteo Corsi, and Lucia Leporatti

1 Introduction

In the last two decades, data-driven policymaking has gained more and more importance due to the larger availability of data (and, more recently, Big Data) for designing proper and timely economic and social policies. This larger availability of data has let decision makers have a deeper insight of complex socio-economic phenomena (e.g. unemployment, deprivation, crime, social care, healthcare) but, at the same time, it has drastically increased the number of indicators that can be used to monitor these phenomena. Decision makers are now often in the condition of taking decisions with large batteries of indicators whose interpretation is not always easy or concordant. In order to simplify the decisional process, a large body of literature suggests to use synthetic indicators to produce single measures of vast, latent phenomena underlying groups of indicators. Unfortunately, although simple, this solution has a number of drawbacks (e.g. compensation between components of synthetic indicators could be undesirable; subjective weighting of the components could lead to arbitrary results; mixing information about different phenomena could make interpretation harder and decision-making opaque). Moreover, with operational decisions, it is necessary to distinguish between those situations when decisions can be embedded in automated processes, and those that require human intervention. Under certain conditions, the use of synthetic indicators may bring to a misleading interpretation of the real world and to wrong policy decisions. In order to overcome all these limitations and drawbacks of synthetic indicators, the use of multi-indicator systems is becoming more and more important todescribe and

E. di Bella (✉) • M. Corsi • L. Leporatti
Department of Economics and Business Studies, University of Genoa, Via Vivaldi 5, Genoa 16123, Italy
e-mail: edibella@economia.unige.it; matteo.corsi@unige.it; lucia.leporatti@unige.it

M. Fattore, R. Bruggemann (eds.), *Partial Order Concepts in Applied Sciences*,
DOI 10.1007/978-3-319-45421-4_11

characterize many phenomena in every field of science, as they keep the valuable information, inherent to each indicator, distinct (see, for a review: Bruggemann and Patil 2011).

An effective approach to multi-indicator systems can be found in the theory of Partially Ordered Sets and its applications like Hasse diagrams. A Partially Ordered Set (POSET) is a generally adequate representation of our units of observation if we assume that they are affected by a latent phenomenon which is described by some, not entirely concordant, indicators. In that case, which is the common setting of multi-indicator systems, it is very often impossible to rank all observations according to the intensity of the latent phenomenon, as each indicator is likely to lead to a different ranking of the observed unit. Typically, that is the case in which some total order is ultimately imposed to the observations through a synthetic indicator or through any common data-driven dimensional reduction based on a generalized notion of distance between observations. However, if preserving whatever partial order can be found in the original data is more important than preserving relative distances between observations, partial order methods have considerable advantages over the typical approach.

In our opinion, actual POSET literature has not yet sufficiently analysed two topics that are of absolute relevance for policymakers. First of all, POSET theory and its outputs are still too difficult to be understood by most policymakers, and graphical outputs easier to understand than Hasse diagrams are desirable. Secondly, decision makers are interested not only in comparing statistical units (e.g. neighbourhoods, cities, hospitals, regions, and natural reserves) at one specific time to address resources and different policy interventions, but also in evaluating the effects of policy actions over time. In this chapter, we discuss how Partial Order Scalogram Analysis with base Coordinates (POSAC) can be used to give a bi-dimensional representation of multi-indicator systems while preserving their partial order, consistently with the POSET principles. Afterwards, we apply this technique to two sets of panel data in order to show how the time dimension can be included in POSAC analyses.

The rest of this chapter is organized as follows. First we review the basics of the POSAC technique and we discuss its output in some toy examples in order to evaluate its real accuracy as a graphical tool for decision-making support. Then, we propose a way to analyse panel data in multi-indicator systems via POSAC and we provide two case studies in environmental and economics frameworks. Finally we discuss pros and cons of this technique and we suggest some further developments in this direction.

2 The POSAC Procedure

The principal output of the POSET analysis is the Hasse Diagram, a visual representation of comparability and incomparability across the elements of the set (see Bruggemann and Patil 2011 for a full description of the method). Despite its relevance, the Hasse diagram is not of immediate interpretation, especially for what concerns the level of incomparability. In addition, if the aim is to provide an instrument from which policymakers can directly benefit, it is important to keep the results' interpretation as simple as possible and this is not always the case with the Hasse diagram, which is not a standard and widely used instrument. As a consequence, other partial order methods can be used to simplify the interpretation of results. In this context, Partial Order Scalogram Analysis with base Coordinates (POSAC) provides a two-dimensional representation of a partially ordered set $\underset{(n \times m)}{P}$ composed by n elements and their m attributes $(a_1, a_2, \ldots a_m)$ with $m > 2$ (Shye and Amar 1985; Shye 1985). In practice, POSAC is an iterative algorithm that builds a two-dimensional graph from a multidimensional dataset preserving order relations and incomparabilities as much as possible (Bruggemann and Patil 2011). The POSAC methodology works through different steps:

1. First the m attributes of the n elements are converted into ranks; this reduces random incomparabilities and makes the method more robust to outliers and indifferent to the scales of the original variables (Raveh and Landau 1993; Bruggemann and Patil 2011).
2. The algorithm computes the weak monotonicity coefficients of all attributes and identifies the two attributes that have the least positive correlation.
3. A profile made of the ranking in each attribute is given to each element so that the profile for the k element is given by: $a^{(k)} = a_1^{(k)} \ldots a_i^{(k)} \ldots a_j^{(k)} \ldots, a_m^{(k)}$. The maximum and the minimum profiles are defined as those profiles having the maximum and minimum observed scores in all the m attributes and they are placed, respectively, in the top right corner (1,1) and in the bottom left corner (0,0) of the two-dimensional Cartesian space.
4. The other profiles are then positioned in the two-dimensional graph starting from the initial placement of the $a^{(k)}$ profile (x_k, y_k) that is obtained by solving the simultaneous linear equations:

$$\begin{cases} x_k + y_k = \sum_{t=1}^{m} a_t^{(k)} \\ x_k - y_k = a_i^{(k)} - a_j^{(k)} \end{cases} \tag{1}$$

   and translating the coordinates to fit [0; 1].
5. The final base coordinates of the profiles are afterwards obtained through an iterative procedure that maximizes the correct representation of profile pairs in their order or incomparability such that:

$$a^{(k)} > a^{(h)} \leftrightarrow x_k > x_h \wedge y_k > y_h \tag{2}$$

$$a^{(k)} \,||\, a^{(h)} \leftrightarrow \begin{cases} x_k \geq x_h \wedge y_k \leq y_h \\ x_k \leq x_h \wedge y_k \geq y_h \end{cases} \tag{3}$$

As a consequence, those profiles characterized by high scores in all the attributes will be placed close to the top right corner of the (x, y) plane while those profiles recording lower scores will be located close to bottom left corner of the (x, y) plane. Profiles with large scores for some attributes and small scores for others will instead be placed in the incomparability areas which are located in the bottom right and top left area of the graph.

The POSAC graph identifies two diagonals of particular interest, the main diagonal (called Joint axis: $J = X + Y$) and the off-diagonal (called Lateral axis: $L = X - Y$). By construction, if one profile has higher score in the J coordinate than another, this means that the first profile is better (or worse, accordingly with the direction of the indicators) in all the original attributes than the second profile. Therefore, comparable profiles will be plotted along the J axis in a consistent ranking. On the contrary, those profiles plotted along the L axis cannot be ordered and they represent incomparable situations.

2.1 A Few Toy Examples

We believe that the use of a two-dimensional graph to plot a partial ordered set has undeniable benefits over the use of a Hasse diagram in terms of accessibility of results even for policymakers without a strong quantitative knowledge. Due to these properties, the POSAC procedure has been applied in several fields of research (Pagani et al. 2015; Tzfati et al. 2011; Sabbagh et al. 2003) including studies concerning crime (di Bella et al. 2015; Canter 2004; Raveh and Landau 1993), European countries comparison (Annoni and Bruggemann 2009; di Bella et al. 2016) or education (Eisenberg and Dowsett 1990).

The POSAC procedure is relatively old but no efforts have been made to update and improve the current procedure and overcome some of the shortcomings of the method. Indeed, despite of its benefits in terms of simplicity and interpretability, the POSAC procedure has also a series of limitations which have not been properly addressed in the past but which can, if overcome, make the POSAC procedure a suitable instrument for policymaking decisions. First off, POSAC can be currently run only using one software for statistical analysis (i.e. SYSTAT[1]) or using the original FORTRAN code, but both the two options are not as flexible or widespread as other software solutions (e.g. R, STATA, MATLAB). Moreover, POSAC analysis can't be easily adapted to the researcher's needs; for instance, it is not possible to define a weighting of profiles according to their frequencies: if two statistical units have the same profile, the POSAC procedure simply drops all the duplicates. More generally, the POSAC procedure itself can't be customized and the actual reliability of POSAC analysis is largely unexplored. For this reasons we decided to analyse

[1]All the POSAC plots given in this chapter were done in *R* environment using squared POSAC (x, y) coordinates from SYSTAT 11.0 output.

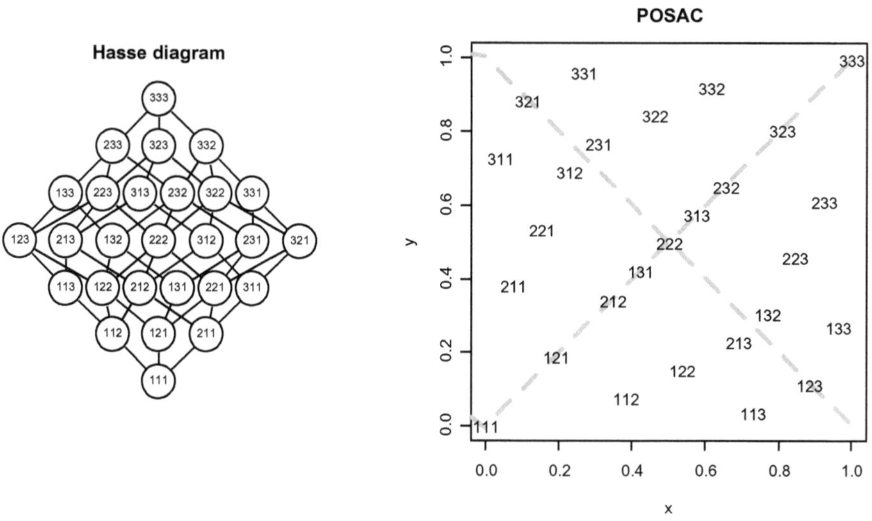

Fig. 1 Hasse diagram and POSAC representation of a complete set of possible profiles obtained from three indicators defined on three levels. *Dashed lines* in the POSAC plot are the J-axis (from *bottom left* to *top right*) and L-axis (from *top left* to *bottom right*)

the POSAC output in a series of toy examples with different degrees of correlation among sets of indicators to detect how the positioning of profiles in the POSAC (x, y) plane, the reliability of the model and the interpretability of the results change under different correlation structures.

In the first toy example, we consider the complete set of profiles that can be obtained from three indicators defined on three levels. The Hasse diagram and the POSAC representation[2] of this set are given in Fig. 1: whereas the Hasse diagram points out the complex incomparability structure of the poset, the POSAC representation shows a large number of violations of conditions (2) and (3) and the proportion of profile pairs correctly represented is 69.2 %.[3] This low performance of POSAC is due to the regular tri-dimensional structure of the poset that POSAC can't obviously reproduce in a bi-dimensional plane and the result given in Fig. 1 is only one of the possible equivalent POSAC projections. So, POSAC does not seem to be a good tool to plot symmetric posets but these cases are quite rare and often some profiles, albeit theoretically possible, are not found in real datasets.

To address this problem and to explore the POSAC capability of mapping profiles we considered that profiles occurrences in real data are a result of the indicators

[2]Each point is plotted accordingly to the (x, y) POSAC coordinates and the labels represent the indicators ranking from 1 (the lowest) to 4 (the highest). So point 111 represents the profile associated to the lowest scores in the three indicators and 444 to the highest.

[3]This measure, equivalent to the STRESS measure used in Multidimensional Scaling, evaluates the correct representations of conditions (2) and (3) for all the profiles pairwise comparisons.

distributional form and correlation. For example, if we have indicators that are highly correlated with positively skewed distributions we should expect to have more occurrences for profiles that are plotted in the top-right part of the POSAC plane; if we have symmetrically distributed positively correlated indicators, the most common profiles will lay along the J-axis, mainly in the centre of the POSAC plane; as long as the correlation among the indicators decreases, the more the most common profiles will be found along the L-axis. Therefore, being not possible to weight the profiles for the POSAC representations, we implemented a procedure to represent various sets of indicators with different correlation structures among the indicators. The procedure we followed is described by the following steps:

1. We generated 100,000 random values from a three-variate standardized normal distribution having correlation structure R.
2. We split the three vectors of randomly generated indicators values into four buckets (`ntile` function in R environment) assigning the appropriate bucket number to each value obtaining 100,000 profiles from 111 to 444.
3. We computed the profiles frequencies distribution
4. We selected the profiles that cover the 75 % of the cases (i.e. occurrences)

In other words, we simulated the distribution of the profiles associated to three correlated indicators (I_1, I_2, I_3) recoded on four ranks and we selected only the most common profiles. The different correlation matrices we used are the following:

$$R_1\,(I_1, I_2,\ I_3) = \begin{pmatrix} 1 & 0.8 & 0.8 \\ 0.8 & 1 & 0.8 \\ 0.8 & 0.8 & 1 \end{pmatrix} \tag{4}$$

$$R_2\,(I_1, I_2,\ I_3) = \begin{pmatrix} 1 & 0.2 & 0.9 \\ 0.2 & 1 & 0.2 \\ 0.9 & 0.2 & 1 \end{pmatrix} \tag{5}$$

$$R_3\,(I_1, I_2,\ I_3) = \begin{pmatrix} 1 & 0.4 & 0.9 \\ 0.4 & 1 & 0.6 \\ 0.9 & 0.6 & 1 \end{pmatrix} \tag{6}$$

R_1 identifies a situation with three highly and equally correlated indicators. The simulation procedure selected 17 profiles (out of 64) that account for at least the 75 % of the total occurrences. The resulting POSAC output is given in Fig. 2a: most of the profiles are plotted along the J-axis because, being the indicators strongly correlated, the profiles with evident incomparabilities have lower frequencies and therefore they are not included in the 75 % selection. But, although the proportion of profile pairs correctly represented is 0.941, even in this case, there are some projection issues due to fact that being the indicators equally correlated the POSAC procedure can't find a predominant indicator according to which the projection should be done and therefore there are various equivalent projections just like the case of Fig. 1.

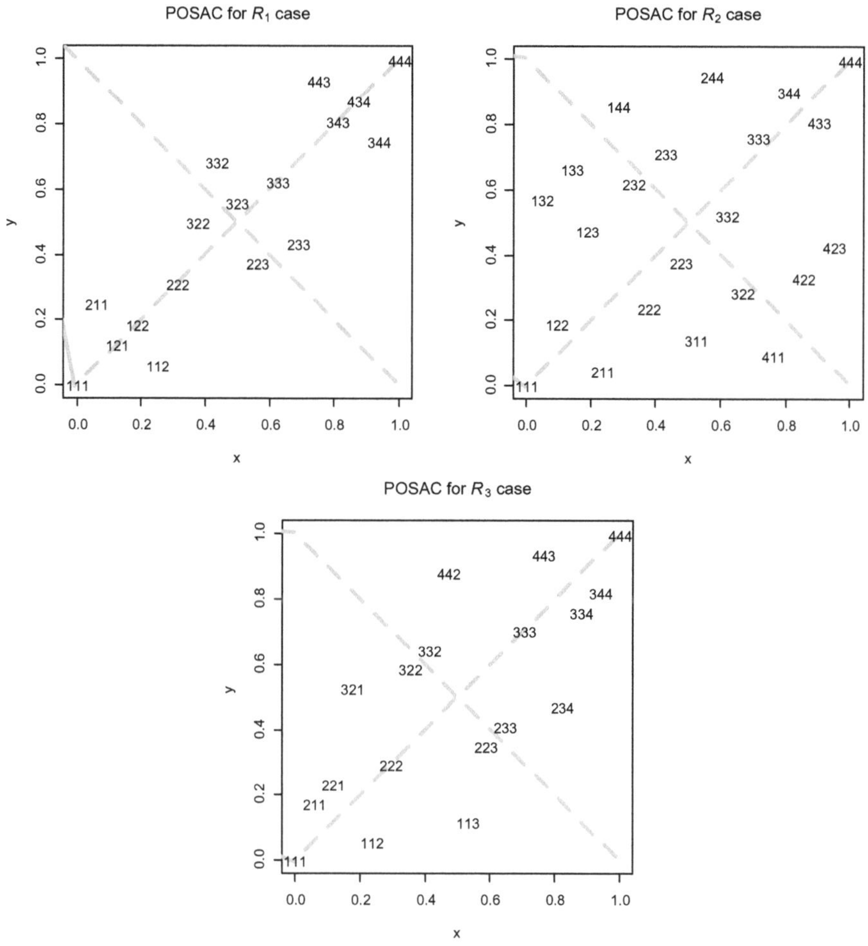

Fig. 2 POSAC representation of toy examples for the correlation structures among indicators specified by (4), (5) and (6)

R_2 identifies a situation with two highly correlated indicators (I_1 and I_3) and one (I_2) weakly correlated with other two. The POSAC output for the 22 selected profiles is given in Fig. 2b: the profiles still basically follow the J-axis but the low correlation that one indicator has with the other two strongly correlated indicators introduces an incomparability element that causes a number of profiles to be plotted far from the J-axis. The POSAC plot puts in evidence the incomparability role of indicator I_2 whose values increase along the L-axis from the right to the left. There are still a number of incorrect representations (the proportion of profile pairs correctly represented is 0.896) such as profiles 123 and 232 or 311 and 411, but generally it seems possible to see how the comparability structure due to indicators I_1 and I_3 is generally respected.

R_3 identifies a last toy example in which the correlation levels vary from medium-weak to strong. This may be quite realistic in real data when consistent indicators are selected. The POSAC output for the 18 selected profiles is given in Fig. 2c. This is best result among the battery of toy examples: all the pairs are correctly represented (100 %); the L-axis is mainly driven by I_1 whereas the other two indicators drive the J-axis component.

In light of these last three toy examples, it seems quite evident that the first one shown in Fig. 1 may closely correspond to the representation of three uncorrelated and uniformly distributed indicators, situation that is a nonsense for policy evaluation.

As a concluding remark for this section, POSAC seems to be a very useful tool to support decision-making but the lack of any form of weighting of profiles or indicators is a real issue that should be addressed in the future. Mediating among various incomparabilities without any weighting of profiles (for instance considering their occurrences) but considering all of them with the same importance, POSAC output may be ambiguous or difficult to interpret for policymakers if not unreliable. This issue seems to be mitigated when a limited number of indicators with a not perfectly balanced correlation structure (e.g. R_3) are given.

3 POSAC in the Context of Panel Data

Partial order methods and POSAC have an important field of application in multi-indicator systems that support policymaking. Once effectively represented, a partial order of government agencies based on last year's performance scores, or one of public universities based on success indicators of a certain enrollment class, may help taking strategic decisions. In many cases, though, a partial order based on a single time period is not sufficient to support informed decision-making: annual scores may be volatile and trends may be as important as outcomes. For example, a partial order of the neighbourhoods of a city based on well-being indicators, if built on a single time period, could not reveal if an area with good scores was the only one to have regressed from last year, compared to significant improvements in all other areas. In order to overcome this limitation, POSAC can be adapted to represent at least certain sets of panel data and the evolution of a certain partial order over some periods of time (di Bella et al. 2016; Schneider et al. 2005).

In a POSET of panel data, our set is described by a matrix $P_{nt \times m}$, where n elements are defined by m attributes $a_1 \ldots a_m$ measured at t time periods. Consequently, our POSAC will be a representation of $n \times t$ profiles and each element k will correspond to t profiles $a^{(kt_1)}, a^{(kt_2)} \ldots a^{(kt)}$. The maximal and minimal profiles a^M and a^m will also be determined according to this expanded population.

A POSAC of panel data has a few distinguishing features. The maximal profile is built with the highest observed scores of each attribute as usual, but in this case these are looked for across several time periods. As a consequence, this maximal may combine observed scores from different time periods. In spite of this difference, the interpretation is about the same: the maximal profile is a benchmark of the best possible scores that we know can be achieved (based on the fact that someone actually achieved them, at some point in time). Another distinguishing feature has greater analytical implications. All the profiles between the maximal and the minimal can now be compared according to three criteria (instead of just one): position relative to all other profiles, position relative to profiles at the same time period and position of profiles relative to the same element in different time periods. The second case can obviously be treated as an ordinary POSAC. The first case provides a rather immediate visualization of time lag in the development of certain performance levels when profiles belonging to different elements and different time periods are comparable and close to each other. For example, using the appropriate indicators, a POSAC of panel data could represent the lag of some countries in developing their industrial sector. Finally, the third case is an opportunity to use POSAC to track change in a partial order context and then to compare change trajectories. The trajectories can be visualized in POSAC by connecting a succession of profiles of the same element in consecutive time periods with directed segments. Segments that are broadly parallel to the J-axis show that the profiles of a certain element are getting closer to the maximal or to the minimal profile; segments parallel to the L-axis indicate changes between incomparable situations.

In order to illustrate the potential use of POSAC of panel data, we present two cases. The first is focused on environmental data concerning six different greenhouse and pollutant gas emissions measured across nine European countries every 10 years starting in 1990. The second deals with five well-known parameters (Maastricht parameters) that regulated the convergence of European countries in the transition from national currencies to the Euro. This POSAC represents 15 European countries, some transitioning to Euro with the first group, some transitioning later, and some not transitioning at all. The parameters are measured in 1997, 1999 (at the introduction of the Euro as a virtual currency) and in 2001.

3.1 *Case Study 1: Greenhouse and Pollutant Gas Emissions*

The international community has produced, since 1992 and the Rio conference, a number of documents, treaties and agreements on climate change and greenhouse gas reduction, with the goal of stabilizing the interference of human activity with the planetary climate system. Several other pollutant gas emissions related to anthropic activities and constituting a danger for human health and the ecosystem are measured and regulated as well. Relatively long-time series measuring these kinds of emissions are available on the data portal of the United Nations Environment Programme.[4]

[4]http://geodata.grid.unep.ch/results.php

There are several factors that make this dataset interesting for POSAC, all ultimately implying that measuring the performance of countries based on their levels of emissions makes more sense in a partial order context. Consistently with the definition of orderability in a multidimensional system, all the emission indicators refer to a clearly identifiable latent variable (air pollution and greenhouse effect). This makes a ranking of countries meaningful even in the presence of multiple indicators, particularly if they are measured *per capita*. Furthermore, such ranking makes sense because these indicators are correlated with each other, but correlation is definitely less than perfect, meaning that we expect several cases of non-comparability. Finally, we have theoretical bases (e.g. Fulton et al. 2011) for the presence of non-comparability and good reason not to collapse this ranking into a total order. The first reason is that we expect some trade-offs between different emissions and we want to identify them: different technological and process arrangements of the same activity may produce more emissions of one kind and less of another. The second reason is that some characteristics and conditions of a country that cannot be modified in the near future may imply high emissions of some kind and low emissions of some other. This is the case with different economic structures (like prevalence of industry or agriculture), different levels of urbanization and mobility systems, different climates and so on. In this case as well, the ranking is more informative if provided in terms of partial order and with POSAC, thus allowing the identification of groups of countries with similar issues. Given the focus of international institutions on reducing emissions and on identifying quantitative targets for countries subscribing climate change agreements, this dataset is also particularly relevant in a panel data context. The ranking at a certain time is not as important as improvement in consecutive time periods and relative to other countries, as these parameters are going to provide the framework for policy decisions, evaluations of compliance to commitments and, possibly, for sanctions against non-compliance.

Our dataset includes nine European countries for which data is complete: differences in size are neutralized by using emissions *per capita* and, in spite of other differences (industrial structure, population density, market economies or transition economies in 1990, etc.), their recent history makes them broadly comparable. In fact, based on many sets of common rules shared by the nine countries, we expect some degree of convergence, and based on changes in technology and structure of the manufacturing sector in Europe and increased awareness about emissions, we expect visible improvements. Each country is represented by three profiles, one for year 1990, one for year 2000 and one for year 2010.

Each profile consists of observed emissions *per capita* of six greenhouse and pollutant gas emissions (Table 1): methane (CH_4), carbon monoxide (CO), nitrous oxide (N_2O), different non-methane volatile organic compounds (NMVOC), different nitrogen oxides (NOX), and sulphur dioxide (SO_2).

Figure 3 presents the POSAC output. Most countries in our dataset are now more virtuous in terms of emissions than they were in 1990 and improvements are more unequivocal from year 2000 onwards. The only country in our dataset that is not comparable (and better) than in 1990 is Portugal, which has an increase of 8.5 %

Table 1 Greenhouse and pollutant gas emissions *per capita* (tonnes)—years 1990, 2000 and 2010

	CH_4			CO			N_2O			NMVOC			NOX			SO_2		
	1990	2000	2010	1990	2000	2010	1990	2000	2010	1990	2000	2010	1990	2000	2010	1990	2000	2010
Belgium	0.97	0.80	0.61	0.13	0.09	0.05	1.09	1.08	0.77	0.03	0.02	0.02	0.04	0.03	0.02	0.04	0.02	0.01
France	1.07	1.04	0.86	0.20	0.12	0.07	1.64	1.36	0.99	0.06	0.04	0.02	0.03	0.03	0.02	0.02	0.01	0.01
Germany	1.35	0.90	0.60	0.15	0.06	0.04	1.07	0.74	0.66	0.04	0.02	0.01	0.04	0.02	0.02	0.07	0.01	0.01
Hungary	1.15	0.91	0.82	0.12	0.07	0.04	1.24	0.83	0.66	0.02	0.02	0.01	0.02	0.02	0.01	0.08	0.04	0.00
Italy	0.80	0.82	0.62	0.16	0.10	0.05	0.67	0.70	0.45	0.04	0.03	0.02	0.04	0.03	0.02	0.03	0.01	0.00
Netherlands	1.73	1.26	0.96	0.08	0.05	0.04	1.34	1.10	0.57	0.03	0.01	0.01	0.04	0.02	0.02	0.01	0.00	0.00
Portugal	1.05	1.19	1.14	0.09	0.07	0.04	0.60	0.60	0.46	0.03	0.03	0.02	0.02	0.03	0.02	0.03	0.03	0.01
Spain	0.68	0.79	0.70	0.10	0.07	0.04	0.69	0.78	0.57	0.03	0.02	0.01	0.03	0.04	0.02	0.06	0.04	0.01
UK	1.91	1.51	0.91	0.16	0.10	0.04	1.22	0.82	0.61	0.05	0.03	0.01	0.05	0.03	0.02	0.07	0.02	0.01

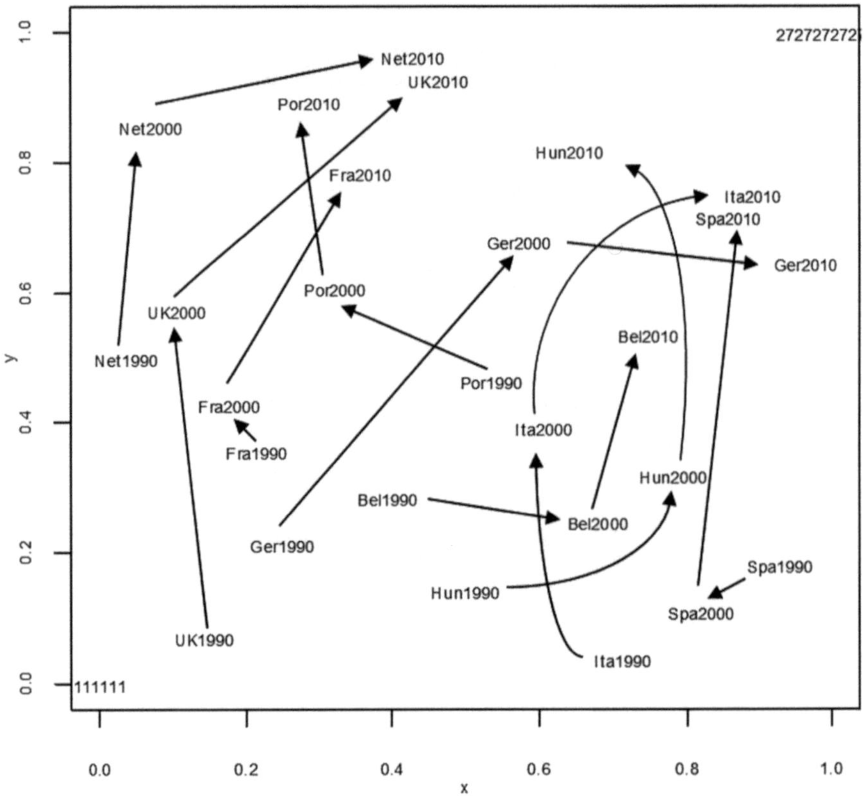

Fig. 3 POSAC representation for greenhouse gas emissions. *Arrows* represent the countries' relative position over the years

of its *per capita* emissions of methane in the last quarter century. Interestingly, our countries are progressively splitting into two distinct groups: in the top left corner are countries that have high emissions of methane and low emissions of sulphur dioxide (Netherlands, the UK, France and, recently, Portugal). This is a circumstance that has become relatively common in recent years. On the bottom right corner are countries which, at the opposite, have low methane emissions but high emissions of sulphur dioxide (Spain, Germany, Italy and Hungary). This kind of unbalance was more frequent in 1990 and in 2000, before European countries had to comply with EU norms concerning the desulphurization of fuels and the abatement technologies of these kinds of emissions.[5] The divide between countries with high methane emissions and countries with high sulphur dioxide emission has nonetheless become more evident.

[5]http://www.eea.europa.eu/data-and-maps/indicators/eea-32-sulphur-dioxide-so2-emissions-1/assessment-3

3.2 Case Study 2: Euro Convergence Criteria

In 1992, the member countries of the European Community signed the Maastricht treaty, which would lead to the creation of the European Union and the common currency, the Euro. At article 121, countries were required to comply with a set of "convergence criteria" before entering the final stage of the Economic and Monetary Union and adopting the common currency. The criteria were all meant to ensure price stability after the adoption of the Euro and revolved around four indicators and one rule: the harmonized index of consumer prices (HICP), the government budget deficit to GDP ratio (BDFtGDP), the government debt to GDP ratio (DBtGDP) and the average yield of 10-year government bonds (BOYE), plus the requirement of not devaluing their national currency in the 2 years prior to the adoption of the common currency. Once again the dataset is an ideal testing ground for POSAC with panel data. The latent variable behind all criteria is explicit: favourable environment for price stability.

These indicators are given in Table 2 and are correlated only to a degree and the actual evolution of the monetary union around year 2000 offers plenty of opportunities to look at different situations, with countries struggling to comply with the criteria and finally adopting the common currency, others still distant from compliance and some more that qualify but decide to keep their national currency anyway. In order to magnify these differences, we chose years 1997, 1999 and 2001 as time periods and 15 EU countries. The complete time series of the indicators for all EU current member states are published online and free of charge on the website of Eurostat.[6]

We straightforwardly converted the rule into a fifth quantitative indicator by counting the number of months in the 2 years prior to each time period during which the national currency remained pegged to the common currency under the European Monetary System (ECU/ERM/EURO).

In Fig. 4, the POSAC output shows rather high levels of incomparability and a tendency of 1999 profiles to be as good as those measured in 2001 in terms of preconditions for price stability. A group of profiles in the bottom right corner represents countries that, in the reference time period, had high yields on long-term government bonds but very low inflation. That happened in the UK, in Belgium, in Sweden and a few more countries including Italy in 1997. At the opposite, countries in the top left corner had relatively high inflation and yet low long-term bond yields. Remarkably, the first group consists mostly of 1997 and some 1999 profiles, whereas the second group includes almost exclusively 2001 profiles. In general, a good number of countries improved their compliance before 1999 and the adoption of the common currency but, a couple of years later, we find that all countries in our dataset, including some of those that kept their national currency, have higher HICP inflation and lower government bond yields. Based on these results, one could

[6]http://ec.europa.eu/eurostat/data/database

Table 2 Euro convergence criteria—years 1997, 1999 and 2001

	BOYE			DBtGDP			BDFtGDP			HICP			ECU/ERM/EURO		
	1997	1999	2001	1997	1999	2001	1997	1999	2001	1997	1999	2001	1997	1999	2001
Austria	5.68	4.68	5.08	64.1	66.8	66.8	−2.4	−2.6	−0.6	1.2	0.5	2.3	12	24	24
Belgium	5.75	4.75	5.13	122.5	113.6	106.5	−2.1	−0.6	0.2	1.5	1.1	2.4	24	24	24
Denmark	6.25	4.91	5.08	65.4	58.1	49.6	−1.2	0.9	1.1	2	2.1	2.3	24	24	24
Finland	5.96	4.72	5.04	53.9	45.7	42.5	−1.2	1.7	5	1.2	1.3	2.7	2	24	24
France	5.58	4.61	4.94	59.4	58.9	56.9	−3.6	−1.6	−1.4	1.3	0.6	1.8	24	24	24
Germany	5.64	4.49	4.8	59.8	61.3	59.1	−2.9	−1.7	−3.1	1.5	0.6	1.9	24	24	24
Greece	9.92	6.3	5.3	97.5	94.9	104.7	−6.1	−5.8	−5.5	5.4	2.1	3.7	6	24	24
Ireland	6.29	4.71	5.01	63.5	47	35.2	1.3	2.4	1	1.3	2.5	4	24	24	24
Italy	6.86	4.73	5.19	117.4	113	108.2	−3	−1.8	−3.4	1.9	1.7	2.3	1	24	24
Luxembourg	5.6	4.66	4.86	7.4	6.4	6.3	3.7	3.6	6	1.4	1	2.4	24	24	24
Netherlands	5.58	4.63	4.96	68.2	61.1	50.7	−1.3	0.3	−0.3	1.9	2	5.1	24	24	24
Portugal	6.36	4.78	5.16	55.5	51.4	53.8	−3.7	−3	−4.8	1.9	2.2	4.4	24	24	24
Spain	6.4	4.73	5.12	66.2	62.4	55.6	−3.9	−1.3	−0.5	1.9	2.2	2.8	24	24	24
Sweden	6.62	4.98	5.11	71.2	64.3	54.7	−1.6	0.8	1.4	1.8	0.5	2.7	0	0	0
United Kingdom	7.13	5.01	5.01	49.4	43.6	37.8	−2.2	0.8	0.4	1.8	1.3	1.2	0	0	0

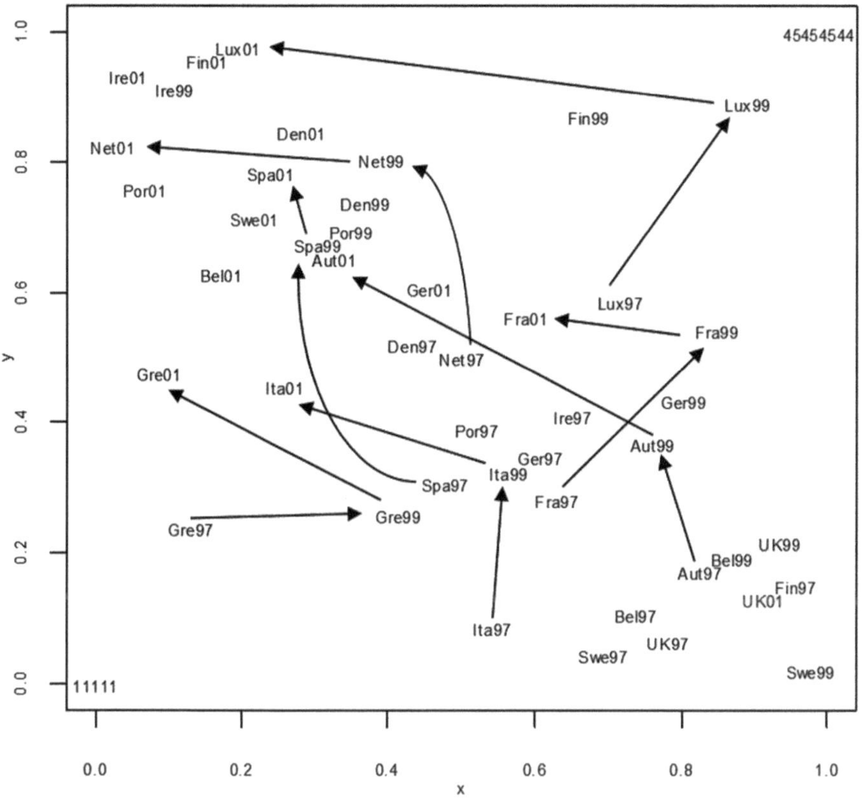

Fig. 4 POSAC representation for Euro convergence criteria. *Arrows* represent the countries' relative position over the years

argue that, between year 1999 and year 2001, the adoption of the Euro was followed by an improvement of sovereign credit risk, which was certainly a goal of the common currency, but also by an unexpected increase in consumer prices, which the convergence criteria were meant to avoid.

4 Conclusions

Although POSAC shows some evident limitations, it is a tool of absolute interest to support decision-making for a number of reasons: the POSAC output is much simpler to understand than Hasse diagrams because the bi-dimensional Cartesian space is much more familiar for decision makers; moreover, in this Cartesian space profiles are clearly positioned from "good" to "bad" cases along the J-axis and the reading of the POSAC is straightforward; it allows a meaningful representation of panel data. Nevertheless, it has some limitations and drawbacks: the percentage of correct representation of the partial orderings drops quite rapidly even with a low number of indicators and their levels. Secondly, the routine for POSAC computation is actually available only in the SYSTAT environment or through the original FORTRAN routines published in Shye (1985). Consequently, the code is not easily editable and, actually, it is not easily possible to improve this technique, as, for instance, giving a different priority to the correct representation of certain profiles because they are observed multiple times.

Nevertheless, it is actually the sole alternative to Hasse diagrams for a visual representation of set of partially ordered indicators and it is therefore of extreme interest for the applicatory contexts herein discussed although its limitations. All these considerations (based on purpose on extreme examples) make us believe that, despite its undeniable potentiality, POSAC procedure may benefit from a revision of the original algorithm.

References

Annoni P, Bruggemann R (2009) Exploring partial order of European countries. Social Indic Res 92(3):471–487

Bruggemann R, Patil G (2011) Ranking and prioritization for multi-indicator systems. Springer, New York

Canter D (2004) A partial order scalogram analysis of criminal network structures. Behaviormetrika 31(2):131–152

di Bella E, Corsi M, Leporatti L (2015) A multi-indicator approach for smart security policy making. Soc Indic Res 122(3):653–675

di Bella E, Corsi M, Leporatti L, Cavalletti B (2016) Wellbeing and sustainable development: a multi-indicator approach. Agric Agric Sci Procedia 8:784–791

Eisenberg E, Dowsett T (1990) Student drop-out from a distance education project course: a new method of analysis. Dist Educ 11(2):231–253

Fulton M, Mellquist N, Bluestein J (2011) Comparing life-cycle greenhouse gas emissions from natural gas and coal. Environ Prot 6:1–29

Pagani RN, Kovaleski JL, Resende LM (2015) Methodi Ordinatio: a proposed methodology to select and rank relevant scientific papers encompassing the impact factor, number of citation, and year of publication. Scientometrics 105:2109–2135

Raveh A, Landau S (1993) Partial order scalogram analysis with base coordinates (POSAC): its application to crime patterns in all the states in the United States. J Quant Criminol 9(1):83–99

Sabbagh C, Cohen EH, Levy S (2003) Styles of social justice judgments as portrayed by partial-order scalogram analysis: a cross-cultural example. Acta Sociol 46(4):323–338

Schneider V, Fink S, Tenbücken M (2005) Buying out the state: a comparative perspective on the privatization of infrastructures. Comp Pol Stud 38(6):704–727

Shye S (1985) Multiple scaling. The theory and application of partial order scalogram analysis. Elsevier, New York

Shye S, Amar R (1985) Partial order scalogram analysis by base coordinates and lattice mapping of items by their scalogram roles. In: Canter D (ed) Facet theory: approaches to social research. Springer, New York, pp 277–298

Tzfati E, Sein M, Rubinov A, Raveh A, Bick A (2011) Pretreatment of wastewater: optimal coagulant selection using partial order scaling analysis (POSA). J Hazard Mater 190(1–3):51–59

Partially Ordered Set Theory and Sen's Capability Approach: A Fruitful Relationship

Giulio Guarini

1 Introduction

The aim of this work is to analyse the epistemological and methodological aspects of the links between the Partial Order Set (POSET) theory and Sen's Capability Approach (CA). CA is one of the best-known approaches to well-being and development analysis, founded by the Nobel Laureate Amartya Sen. If the theoretical bases of CA are sound, the empirical aspects have yet to be fully explored. The complexity of CA empirical verifications involves the requirement of statistical and econometric instruments to tackle: "a plurality of evaluative spaces; a plurality of dimensions and a multiplicity of indicators and scales of a quantitative or qualitative nature, and objectively or subjectively measured; a plurality of units of analysis (individuals, households, subgroups of population) and personal heterogeneities and a plurality of environmental contexts, including socio-economic, geographical, cultural and institutional variables" (Chiappero-Martinetti and Roche 2009, p. 5).

Operationalisation of CA is a major challenge, and Sen stresses that the empirical evidence as well as the theoretical aspects of CA is crucial for the purpose of bringing about improvements in the realities studied, through the supply of results and instruments serving for effective policies. The CA should have practical impact. One of the difficult aspects lies in the subjective and qualitative variables. For this reason, it may be useful to introduce the POSET method in the CA framework. POSET theory is a field of mathematics, recently applied to empirical socio-economic analyses, involving multidimensional ordinal data. Makingreference to

G. Guarini (✉)
Dipartimento di Economia e impresa, Università degli studi della Tuscia, Via del Paradiso n.47, Viterbo 01100, Italy
e-mail: giulioguarini@unitus.it

M. Fattore, R. Bruggemann (eds.), *Partial Order Concepts in Applied Sciences*,
DOI 10.1007/978-3-319-45421-4_12

the main studies on the potential applications of POSET theory to socio-economics (Fattore 2015; Fattore et al. 2011, 2012, 2015), the work will illustrate how its features fit very well with CA.

Although Amartya Sen has never referred directly to the POSET theory, the contribution will quote him extensively to show that his thought is actually rich in interesting indications to develop and enhance the role of POSET theory in the socio-economic sciences.

The contribution is structured as follows: Sect. 2 outlines the basic conceptual framework of the CA according to Amartya Sen; Sect. 3 explains that the two main aspects of POSET theory, namely the multidimensionality and ordinality of the database, are central to CA; Sect. 4 argues the relevance of POSET theory as a method for evaluation in ordinal multidimensional databases, following a perspective à la Sen; finally, Sect. 5 draws the conclusions.

2 The Basic Elements of Sen's Capability Approach

In the CA framework, human life is a set of beings and doings. Well-being is conceived as the being and doing that individuals "value and have reason to value" (Sen 1999 p. 18), e.g. being educated, being healthy, resting, and working. And well-being is the final output of a process starting from resources. Individuals can convert resources (goods and/or services, public and/or private) into capabilities that are what resources enable agents to do and to be. Moreover, people can choose to realise these real opportunities by transforming them into achieved functionings. Thus, well-being is a set of capabilities and in this perspective "development is freedom" (Sen 1999). The conversion factors are what positively influence the conversion of resources into capabilities: they may be personal (e.g. metabolism, physical condition, sex, reading skills, intelligence), social (e.g. public policies, social norms, discriminating practices, gender roles, societal hierarchies, power relations), and environmental (e.g. climate, geographical location) (Robeyns 2005, p. 99). The choice factors may depend upon personal characteristics, personal history, or psychology, which are all directly connected to the social context.

According to the CA definitions, the *standard of living* is the individual well-being related to oneself, while *well-being* integrates standard of living with the well-being deriving from what a person does to increase the standard of living of the others. Finally, *agency* is the ability to enlarge one's own capability set and the capability sets of others, while *empowerment* corresponds to extension of the agency. All kinds of concepts may be evaluated at the individual as well as the collective level (Stewart 2005) (Fig. 1).

Inequality is a relevant issue in capability approach because in the well-being process individuals and groups diverge in their abilities to convert resources into capabilities. One of the types of inequality most analysed lies in education because

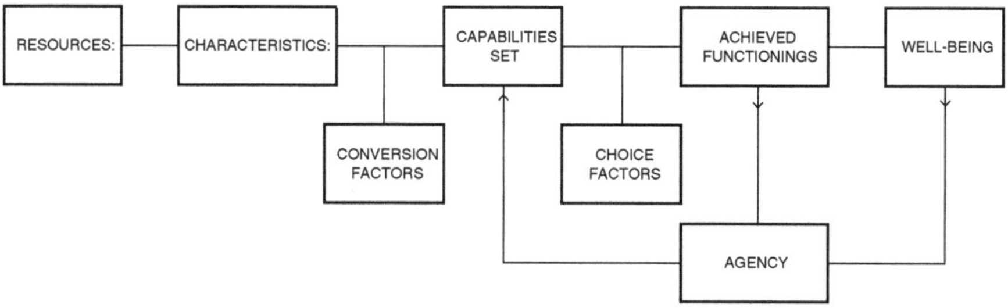

Fig. 1 A well-being process according to the Sen's Capability approach

it is an important element for inclusion in a social and economic system (Robeyns 2006). Formalisation of the above-mentioned capability processes is carried out by Sen (1985 pp. 7–14, 26–27), in these terms

$$\{b_i \,|b_i = f_i(\,c\,(x_i)\,, z_i\} \qquad \forall f_i \in F_i, \quad \text{and} \quad \forall z_i \in Z_i. \tag{1}$$

For each i person (or group), there exists an individual conversion function f_i that converts the resources x_i, with specific characteristics c, into a set of capabilities b_i, some of which can become acquired functionings (Agee and Crocker 2013) according to the individual (or collective) choice. Variable z_i represents the i conversion factors.

3 Multidimensional Ordinal Data and Sen's Capability Approach

3.1 Multidimensionality

In the CA, multidimensionality has an intrinsic and instrumental value. The CA is a people-centred approach, in the sense that people are both the beneficiaries of the process of determination of well-being, and the agents of this process by choosing which capabilities achieve as functionings. In this perspective, the individual is not *homo oeconomicus*, as in the neoclassical view, according to which he is influenced only by economic and material aspects, but is of a multiform nature, according to the Classical Political Economy, starting with Adam Smith (Martins 2011, 2012):

> "The capability perspective involves, to some extent, a return to an integrated approach to economic and social development championed particularly by Adam Smith (both in the Wealth of Nations and in The Theory of Moral Sentiments). In analyzing the determination of production possibilities, Smith emphasized the role of education as well as division of labor, learning by doing and skill formation. But the development of human capability in leading a worthwhile life (as well as being more productive) is quite central to Smith's analysis of the 'wealth of nations'." (Sen 1999, 295f)

In this sense, multidimensionality has an *intrinsic value*, since this is the main aspect of the object of the analysis. Moreover, multidimensionality has an *instrumental value*, since only the multivariate statistical and econometric analyses have the heuristic capacity to represent and evaluate the whole phenomenon studied. At the macro level, Sen explicated this perspective by replacing the Gross Domestic Product, used as the only index of development, with the Human Development Index measured as from 1990 and presented in the UNDP Reports, with ul Haq. The index is composed of Income index, Health index, and Education index. Thus, economics represents only one of the dimensions of development.

At the micro level, another significant example of the bearing of multidimensionality is the concept of poverty and deprivation. Poverty is intrinsically multidimensional. In fact, the capability to live in minimal acceptable conditions may depend upon many factors that influence the conversion of income into basic capabilities. A person with an income above the poverty-line may in practice be poor, for example, on account of an illness entailing high costs, personal characteristics (personal heterogeneity), residence in a problematic area (environmental diversities), variations in the social and environmental context, and differences in the consumption behaviour (Sen 1999). From this viewpoint, relative income poverty determines an absolute deprivation in basic capabilities. Therefore, poverty is deprivation of capabilities. This observation is crucial since it points out that the choice of space of evaluation is fundamental. Indeed, "more income may be needed to buy enough commodities to achieve the same social functioning" (Sen 1992, p. 115). The same social functionings "impose commodity requirements that vary with what others in the community standardly have" (Sen 1992, pp. 115–116). Consequently, inequality analysis should also be multidimensional. At the macro level, since 2010 the UNDP Reports have calculated the Inequality-Adjusted Human Development Index with a method that takes into account the rate of inequality within and across dimensions. However, the multidimensionality of inequality analysis continues to be somewhat undervalued at both the academic level and policy level:

> "First, the conflict between inequality in different "spaces" has often been neglected in the academic as well as policy literature. Indeed, if you announce that you are working on economic inequality, it is quite commonly assumed that you are studying income distribution. The fact that economics has much to say about factors other than income that influence people's wellbeing, or freedom or quality of life, is largely ignored in this narrowing of the understanding of economic inequality. (Sen 1997, p. 156).

3.2 *Ordinality*

As a people-centred approach, in the CA persons are the agents of the well-being process, both by achieving the functionings and by enlarging the capabilities set. In this perspective, the concepts of capability and functioning are a combination of subjectivism and objectivism; in fact, they represent "what people value" (subjective element) and "have reason to value" (objective element). In this way, *subjectivism*, vagueness and nuance are part of the CA, which implies contemplating variables of a strictly ordinal nature. Thus, personal satisfaction is a concept consistent with the capability approach. Amartya Sen criticises the utility principle of neoclassical economics, where satisfaction is the only measure of evaluation of well-being.

> "[...] Consider a very deprived person who is poor, exploited, overworked and ill, but who has been made satisfied with his lot by social conditioning (through, say, religion, political propaganda, or cultural pressure). Can we possibly believe that he is doing well just because he is happy and satisfied? Can the living standard of a person be high if the life that he or she leads is full of deprivation? The standard of life cannot be so detached from the nature of the life the person leads." (Sen 1991, pp. 7–8).

While he inserts satisfaction/happiness in the well-being process in two main ways (Sen 1985), firstly, by considering happiness function h that measures satisfaction deriving from the capabilities b_i, namely $u_i = h_i(b_i)$ (Anand and van Hees 2006), and secondly, "being happy" may be seen as a capability and a functioning that contributes to subjective well-being (Binder 2013; Anand et al. 2011).

Moreover, the indefiniteness and nuance of the measurement may derive from the fact that capabilities are latent variables, making every measurement an indirect and approximate calculation. To overcome this setback, some empirical analyses estimate the achieved functionings directly. However, there is another source of ambiguity and nuance: the meaning of the some relevant capabilities. In the case of poverty analysis, the deprivation of major capabilities concerns social functionings or capabilities like "appearing in public without shame" (Sen 1992, p. 115), which is one of the Smithian necessary goods; "taking part in the life of the community" according to Townsed, (Sen 1992, p. 115), and the concept of self-respect, which Sen recalls from Rawls (Sen 1983). All these elements are evidently not cardinalisable. Within the CA, Nussbaum (2003) draws up a list of capabilities that are not all measurable with a cardinal scale: life bodily health; bodily integrity; the senses, imagination, and thought; emotions; practical reason; affiliation; other species[1]; play; control over one's environment.

Cardinal variables may be intensively correlated with ordinal variables. Indeed in the CA, ordinality not only concerns variables directly linked with the phenomenon analysed, but can indirectly be connected to a type of analysis traditionally described by cardinal measurements. The case of the labour market is emblematic. According to Sen, unemployment influences many subjective meta-economic factors:

> "loss of freedom and social exclusion; psychological harm ("unemployment can play havoc with the lives of the jobless, and cause intense suffering and mental agony); ill health and mortality ("through dejection and a lack of self-respect and a collapse of motivation generated by persistent unemployment); motivational loss and future work ("motivational decline and resignation"); loss of human relations and family life ("unemployment can be very disruptive of social relations, decline of self-confidence"); loss of social values

[1]"Being able to live with concern for and in relation to animals, plants, and the world of nature" (Nussbaum 2003, p. 42).

and responsibility ("there is also evidence that largescale unemployment has a tendency to weaken some social values. People in continued unemployment can develop cynicism about the fairness of social arrangements, and also a perception of dependence on others. These effects are not conducive to responsibility and self-reliance." (Sen 1997, p. 160–161).

4 Evaluation in Ordinal Multidimensional Setting by POSET Theory According to a Perspective à la Sen

According to Fattore (2015) and Fattore et al. (2015), evaluation of well-being using multidimensional ordinal data may be achieved with three different approaches: the Composite Indicator (CI) approach; the Alkire-Foster Counting (AFC) approach (Alkire and Foster 2011a, b) and the Partial Ordered Set (POSET) approach. Let us see why use of the POSET theory is more in line with Sen's viewpoint, integrating the critiques of Fattore (2015) and Fattore et al. (2015) of the first two approaches.

4.1 Critical Analysis of Mainstream Evaluation Approaches

The CI approach is the most common method applied in socio-economic literature. With this approach, after the choice of dimensions, the variables are normalised and aggregated to build a synthetic index through different techniques. With reference to the cardinal variables, there is some debate in the literature as to which mean can better represent the relationship among dimensions. A number of studies, following the CA, criticise the use of the arithmetic mean, which implies a perfect substitutability across them, favouring other kinds of functions with a sort of the offsetting of unbalances (Casadio Tarabusi and Guarini 2013, 2016). In the micro context with ordinal data, within the CA the main methods assume linear correlations among latent variables and between these and explanatory variables (Krishnakumar 2007). Reference here is to factor analysis, principal components, the Multiple Indicators Multiple Causes model (MIMIC) and the Structural Equation Model (SEM). In factor analysis, it is assumed that some observed variables depend linearly upon latent variables representing capabilities. In principal components analysis, observed variables are aggregated in a linear way taking into account the variance of each variable, to obtain a synthetic index that should indirectly represent capabilities. In the MIMIC model, the factor analysis is integrated by introducing the exogenous variables that explain the latent variables. Finally, in the SEM, the MIMIC approach is integrated by inserting equations that consider the interdependence across latent variables. These operationalisations of CA have three major limitations when applied to the multidimensional ordinal data. Firstly, they transform ordinal data into cardinal data, without respecting the ordinal nature and complexity of data. Secondly, "a complex concept comprises many different dimensions, logically related, but not necessarily statistically correlated"

(Fattore et al. 2015 p. 3). This argument is relevant to the CA because statistical instruments substitute the open scrutiny of capabilities, which remains one of the pillars of this approach given its ethical, theoretical and political value. Thirdly, in this context, too, there is a matter of trade-off across components, because every assumption of linear correlation implies perfect substitutability, which is a strong assumption hardly consistent with the CA. Indeed, the relationships across capabilities/functionings are complex and the CA mainly implies discussion and justification of the choice of method, without any implicit a priori hypothesis.

The AFC approach has been used mainly within the CA for deprivation indexing. It avoids aggregating by using a "two-step" procedure. In the identification step, it identifies the deprived persons by dual cut-off procedure: it establishes cut-off both for each attribute and for the number of attributes. Secondly, it counts deprived persons thanks to the dichotomisation and it builds two indexes of deprivation: the Head Count Ratio i.e. the fraction of dissatisfied individuals within the population; the Average Deprivation Share, namely the average fraction of "dissatisfactions" suffered by dissatisfied individuals. This method has two shortfalls: dichotomisation and the absence of attribute relevance. The former aspect does not allow for calculation of the distribution across the poor, while it is relevant according to Sen. Sen's poverty index takes into account the two above-mentioned aspects: the intensity of poverty and the distance from the poverty-line (Sen 1992).[2] This is not only an absence of information but it can also "deflect in anti-poverty policy by ignoring the greater misery of the poorer among the poor" (Sen 1992 p. 105). With reference to this problem, what Sen argues for the headcount index of income poverty is valid *a fortiori* for multidimensional poverty:

> "The aggregation exercise done through simple head counting pays no attention to the fact that people could be a little below the line, or a lot, and also the distribution of income among the poor may or may not be itself very unequal. It is this lacuna in using *H* [headcount ratio] as a measure of poverty [...]" (Sen 1992, p. 102).

By contrast, the POSET theory overcomes the shortfalls of Alkire-Foster and, according to the analyses of Fattore (2015) and Fattore et al. (2015), it can, with opportune techniques, take into account both aspects: intensity of poverty and distribution across the poor. Another limitation of AFC approach is the absence of attribute relevance, which is analogous to weighting in the cardinal context. According to Sen, weighting is central to the study of subjective standard of living. With linear extensions, POSET theory applications can take into consideration the different values of attributes in order to achieve the same functionings; the CA "goes to acknowledge the enormous variability that exists in the commodity requirements of capability fulfilment. [...]" (Sen 1983, p. 164). This variability is not only among countries or among communities, but also within a country or a community concerning, e.g. ethnicity, gender.

---

[2]Sen's poverty index is equal to $P_S = P_0\Big(1 - (1 - G)\,(\mu/z)\Big)$ where P_0, G, μ, z represent the headcount index, the mean income of the poor, the Gini coefficient of inequality across the poor and the poverty line, respectively (Sen 1976).

Weighting is relevant in CA because it represents a transparent procedure to open up evaluation of well-being to public debate: "Any choice of weights should be open to questioning and debating in public discussions (Anand and Sen 1997)". In every socio-economic analysis, the results should be the consequence of specific theoretical and political assumptions, expressed with particular weights. In the POSET theory, attribute relevance by linear extensions settles some incomparabilities, as well as the weighting in the cardinal variables according to Sen. This do not mean eliminating all the incomparabilities, but being able to manage them, with a degree of flexibility.

4.2 Partial Ordering and POSET Theory

Amartya Sen (1992) states that there are two main reasons for choosing partial ordering. The *fundamental reason* for incompleteness concerns the ambiguity and nuances of the phenomena studied. The ambiguities regard the nature of human life and the content of human freedom, and also the objects of value and consequently the characteristics of capabilities. The *pragmatic reason* is inherent to the difficulty for researchers to avail themselves of instruments that can exactly measure all this complexity. Sen observes that it is better to say something than say nothing. In other words, CA is not an "all or nothing exercise".

POSET theory, as a fuzzy analysis, is consistent with an opportune use of mathematics in economics. Sen (2004, p. 597) writes:

> "Certainly, the belief that mathematical expressions must be more precise in capturing what we wish to say can be badly mistaken, since subtleties may be eschewed in choosing particular formal statements. Indeed, the precise expression of a complicated truth cannot possibly take a simple form, however precise-looking the simple formula might be. A formal expression can be extremely precise without being at all a precise representation of the underlying concept to be captured".
>
> Thus, in a context where the objects of value are of an ordinal nature, "in social investigation and measurement, it is undoubtedly more important to be vaguely right than to be precisely wrong" (Sen 1989, p. 45).

Since method should be consistent with theory, if the theoretical elements have an unclear nature, measurement should incorporate this fact, which is a relevant aspect of the object of analysis. In this respect, the transformation of an ordinal variable into a cardinal variable eliminates an important aspect, denaturalising the underlying concept. In the POSET theory, the main basic starting rule is that: given two profiles p and q with two attributes, respectively, p is better than q if at least one attribute is at a higher level. Sen defines this rule as *dominance partial ordering*. It corresponds to the Pareto principle used in the welfare economics. From Sen's point of view, this principle is inappropriate to evaluate policy effects, but it constitutes a first step to analysing the standard of living in a multidimensional way, with integration of weights and threshold (Sen 1987). Partial ordering can offer judgements in some

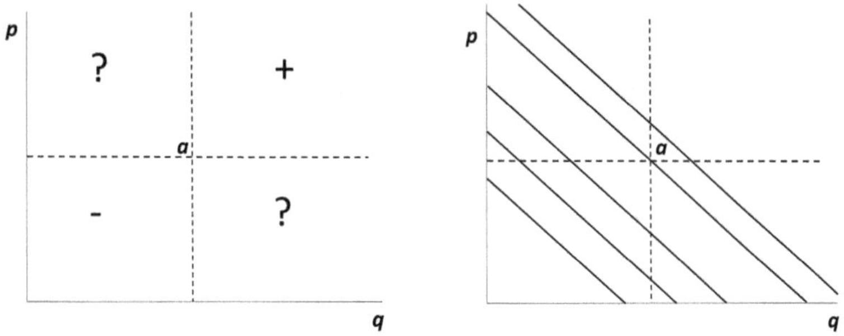

Fig. 2 A comparison between dominance partial ordering (**a**) and complete ordering (**b**) with continuous variables

general questions, like the main contrast between very rich and very poor people (Sen 1987).[3] In this regard, Sen argues:

> "[...] in the evaluation of living standard, there are many intermediate positions between a *complete* ordering of all alternatives and the dominance partial ordering, which may be very incomplete, of the valued functionings and capabilities. As was mentioned earlier, the relative weights may not be precisely determined but fixed over wide ranges, yielding partial orderings more extensive than the dominance partial order, but short of a complete ordering. There is nothing particularly embarrassing in not being able to compare any two life styles in terms of living standard." (Sen 1987 p. 44)

Figure 2 shows a very simple example of dominance partial ordering (a) and complete ordering (b) with continuous variables, given two profiles p and q of well-being, with two attributes for each. With respect to point a, in panel (a) there are no functions of aggregation; every point in areas tagged with "+" ("−") represents an improvement (worsening), while in areas tagged with "?" no evaluation is possible. Instead, in panel (b) the lines represent linear combinations generated by aggregations with arithmetic mean; with reference to point a, the improvements (the worsenings) are represented by shifts of lines to the top (to the bottom); finally, the points in the same line of a have the same level of well-being. Consistently with Sen, in POSET analysis with appropriate attribute relevance and thresholds it is possible to calibrate the ordering and obtain intermediate orderings, which can be more appropriate to the issues studied, and to the theoretical basis of analysis.

In Senian terms, POSET theory is a good method that takes into account the *relevance* of the standards of living in the sense that it respects their ambiguities, as well as considering the *usability* of welfare analysis. In other words, the study should be able to lead to real improvement in the quality of life as a useful instrument

[3]"In comparing across class barriers, or in contrasting the living conditions of the very rich with those of the very poor, or in assessing social change accompanied by progress (or regress) in all fronts, the dominance partial order may indeed give many unequivocal judgments of the ranking of overall living standard" (Sen 1987 p. 40).

for policymakers.[4] With regard to profiles like (0, 0), it can use the *intersection approach* which is shared by all possible alternatives (Sen 1992). With POSET theory, starting from multidimensional ordinal data indexing with weighting is possible, without aggregating and dichotomising. On the analysis of standards of living, Sen observes:

> "The ambiguities in evaluation (even in the identification of "contemporary standards") may require us to be silent on some comparisons while being articulate on others. There is no great failure in the inability to babble" (Sen 1987, p. 44).

The important thing is to have a flexible method that allows for ranking in a flexible way consistently with the theoretical assumptions and the nature of the data. The indexing helps the *usability* of well-being analysis, because it helps policymakers to understand and use the results of analysis to fix targets and monitor specific initiatives. In this respect, Sen's contribution in the human development report 1999 is enlightening. The following passage makes it clear why Sen decided to build the Human Development Index. Initially, he was very critical of the indexing for its crudeness and loss of information, and because it cannot contain all the complexity of human development. But ul Haq was able to persuade him. Sen asserts:

> "We need a measure", Mahbub demanded, "of the same level of vulgarity as GNP—just one number—but a measure that is not as blind to social aspects of human lives as GNP is." Mahbub hoped that not only would the HDI be something of an improvement on—or at least a helpful supplement to—GNP, but also that it would serve to broaden public interest in the other variables that are plentifully analysed in the Human Development Report. Mahbub got this exactly right, I have to admit, and I am very glad that we did not manage to deflect him from seeking a crude measure. By skilful use of the attracting power of the HDI, Mahbub got readers to take an involved interest in the large class of systematic tables and detailed critical analyses presented in the Human Development Report. The crude index spoke loud and clear and received intelligent attention and through that vehicle the complex reality contained in the rest of the Report also found an interested audience (UNDP 1999, p. 23).

In Senian terms, POSET theory may be evaluated with reference to the capability process. Given a database (resources) with multidimensional ordinal data (characteristics), POSET theory can represent both a conversion factor and a choice factor. Firstly, it can transform a database into capability of, e.g. "better understanding poverty" because it takes into account the relevance of "poverty" by avoiding both the aggregation (in order to respect the nature of data) and the dichomtomisation (in order to describe the whole phenomenon studied).

Secondly, thanks to its usability in terms of attribute relevance and indexing, it can promote the use of empirical results on the part of policymakers to implement policy initiatives to reduce poverty. That is, the analysis may actually be an

[4]"These two considerations—relevance and usability—pull us, to some extent, in different directions. Relevance may demand that we take on board the inherent complexities of the idea of the living standard as fully as possible, whereas usability may suggest that we try to shun complexities if we reasonably can. Relevance wants us to be ambitious; usability urges restraint." (Sen 1987 p. 27).

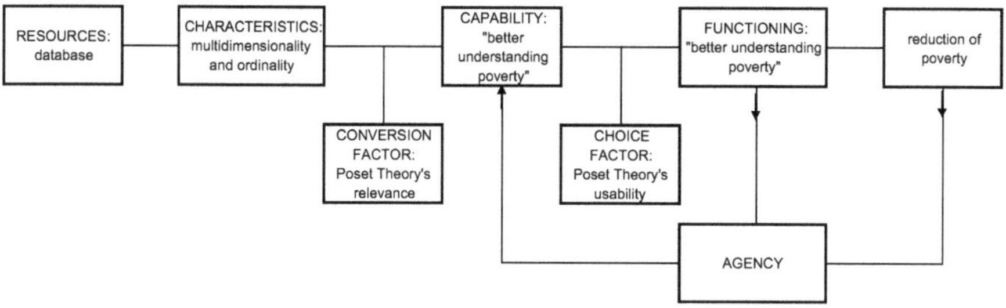

Fig. 3 Evaluation of POSET theory with reference to the Sen's well-being process

instrument to improve the socio-economic conditions that researchers intend to measure, in line with Sen's argument that analysis should have not only theoretical interest, but also practical implications. The results of analysis may stimulate new, more thorough research projects enlarging the capabilities set thanks to the openness and flexibility of POSET theory. The collective agency can be sustained by POSET theory implementations: the inherent POSET theory discretionality may be motivated and made transparent within the CA framework and may help in understanding the phenomenon studied and build consensus on social and political initiatives (Fig. 3).

CA is a partial theory of justice, well-being and development. Sen defines his theory "partial" because it is a framework with minimal elements that fix the main aspects of analysis, but it needs to be adapted according to the specificities concerning: the particular topic, the beneficiaries, the policy, the territory of application, and the time. Thus, it is inherently flexible. Also in this respect, POSET theory is a method consistent with the CA perspective. The following passage by Sen on the list of capabilities makes his concept of partial theory clear:

> "pure theory cannot 'freeze' a list of capabilities for all societies for all time to come, irrespective of what the citizens come to understand and value. That would be not only a denial of the reach of democracy, but also a misunderstanding of what pure theory can do, completely divorced from the particular social reality that any particular society faces." (Sen 2005, p. 160)

Finally, the fact that it is "partial" and "flexible" does not make the CA and POSET theory less "rigorous" than other "complete" theories, since "rationality is an exercise of reasoning, valuation, and choice, not a fixed formula with a pre-specified maximand" (Rothschild and Sen 2006, p. 358).

5 Concluding Remarks

The chapter illustrates the fruitful relationship between the POSET theory and Sen's CA. Epistemological and methodological issues are discussed to bring out the potential benefits from the integration of the two fields. According to the present work, the POSET theory is a mathematical instrument that can be well integrated in the CA framework from various points of view. Firstly, POSET theory concerns multidimensional and ordinal data, and these two properties are significantly appropriate to the theoretical framework of CA, characterising the database of many CA studies regarding (subjective) well-being, quality of life, and inequality. Secondly, POSET theory allows for attribute relevance and indexing without running into aggregation (the limitation of the composite index approach) or dichotomisation (the limitation of the Alkire-Foster Counting approach).

From the point of view of CA researchers, implementation of POSET theory overcomes the shortfalls of CA empirical analyses and makes their studies more consistent with the theoretical assumptions. While for POSET theory users, the CA framework can reinforce the important role of this method for the socio-economic sciences, and it can have many fields of application. This contribution could stimulate the multidisciplinary perspective and incentivise new paths for applied research in socio-economic topics, such as building a Sen Poverty Index for ordinal variables with the POSET method.

Acknowledgements I wish to thank Enrico Casadio Tarabusi and Marco Marini for their precious suggestions and the referee for his important comments. Any errors or inaccuracies are my responsibility alone.

References

Agee TD, Crocker MD (2013) Operationalizing the capability approach to assessing well-being. J Socio-Econ 46:80–86

Alkire S, Foster JE (2011a) Counting and multidimensional poverty measurement. J Public Econ 95(7–8):476–487

Alkire S, Foster JE (2011b) Understandings and misunderstandings of multidimensional poverty measurement. J Econ Inequal 9:289–314

Anand S, Sen A (1997) Concepts of human development and poverty: a multidimensional perspective. Human development report background paper

Anand P, van Hees M (2006) Capabilities and achievements: an empirical study. J Socio-Econ 35:268–284

Anand P, Krishnakumar J, Tran NB (2011) Measuring welfare: latent variable models for happiness and capabilities in the presence of unobservable heterogeneity. J Public Econ 95:205–215

Binder M (2013) Subjective wellbeing capabilities: bridging the gap between the capability approach and subjective wellbeing research. Papers on economics and evolution 2013-02

Casadio Tarabusi E, Guarini G (2013) An unbalance adjustment method for development indicators. Soc Indic Res 112(1):19–45

Casadio Tarabusi E, Guarini G (2016) Level dependence of the adjustment for unbalance and inequality for the human development index. Soc Indic Res 126:527–553

Chiappero-Martinetti E, Roche JM (2009) Operationalization of the capability approach, from theory to practice: a review of techniques and empirical applications. In: Chiappero-Martinetti E (ed) Debating global society: reach and limits of the capability approach. Fondazione Giangiacomo Feltrinelli, Milan

Fattore M (2015) Partially ordered sets and the measurement of multidimensional ordinal deprivation. Social Indic Res. doi:10.1007/s11205-015-1059-6

Fattore M, Maggino F, Greselin F (2011) Socio-economic evaluation with ordinal variables: integrating counting and POSET approaches. Statistica & Applicazioni, special issue, pp 31–42
Fattore M, Maggino F, Colombo E (2012) From composite indicators to partial orders: evaluating socio-economic phenomena through ordinal data. In: Maggino F, Nuvolati G (eds) Quality of life in Italy: researches and reflections. Springer, Amsterdam, pp 41–67
Fattore M, Maggino F, Arcagni A (2015) Exploiting ordinal data for subjective wellbeing evaluation. Statistics in transition new series—special issue, pp 1–20
Krishnakumar J (2007) Going beyond functionings to capabilities: an econometric model to explain and estimate capabilities. J Hum Dev 8(1):39–63
Martins NO (2011) The revival of classical political economy and the Cambridge tradition: from scarcity theory to surplus theory. Rev Pol Econ 23(1):111–131
Martins NO (2012) Sen, Srafaa and the revival of classical political economy. J Econ Methodol 19(2):143–157
Nussbaum MC (2003) Capabilities as fundamental entitlements: Sen and social justice. Fem Econ 9(2–3):33–59
Robeyns I (2005) The capability approach: a theoretical survey. J Hum Dev 6(1):93–117
Robeyns I (2006) The capability approach in practice. J Polit Philos 14(3):351–376
Rothschild E, Sen A (2006) Adam Smith's economics. In: Haakonssen K (ed) The Cambridge companion to Adam Smith. Cambridge University Press, Cambridge
Sen A (1976) Poverty: an ordinal approach to measurement. Econometrica 44(2):219–231
Sen A (1983) Poor relatively speaking. Oxford economic papers, new series 35(2):153–169
Sen A (1985) Commodities and capabilities. North-Holland, Amsterdam
Sen A (1987) The standard of living. Tanner lectures. Oxford University Press, New Delhi, pp 1–51
Sen A (1989) Development as capability expansion. J Dev Plan 19:41–58
Sen A (1991) The standard of living. Cambridge University Press, Cambridge
Sen A (1992) Inequality reexamined. Clarendon, Oxford
Sen A (1997) Inequality, unemployment and contemporary Europe. Int Labour Rev 136(2):155–171
Sen A (1999) The possibility of social choice. Am Econ Rev 89:349–378
Sen A (2004) Economic methodology: heterogeneity and relevance. Soc Res 71(3):583–614, originally published in Social Research 56(2), 1989
Sen A (2005) Human rights and capabilities. J Hum Dev 6(2):151–166
Stewart F (2005) Groups and capabilities. J Hum Dev 6(2):185–204
UNDP (1999) Globalisation with human face. Human Development Report, New York

Part III
Partial Order Theory in Environmental Sciences

Ranking Chemicals with Respect to Accidents Frequency

Ghanima Al-Sharrah

1 Introduction

The classical science-based approach to risk management focuses on the integrated numerical assessment of the two basic risk criteria, frequency of occurrence and extent of damage or consequences. People seriously misjudge accident risk because they routinely neglect relevant information regarding these two criteria. Such risk judgment affects both personal and public policy decision such as assessment of recklessness. This study investigates the frequency part of risk with its different forms. Frequency is commonly defined as the number of times an accident occurs over a period of time. In many cases, the frequency is divided by the size of a facility in order to normalize it. Accident frequency depends on the type of work an individual or company does and the dangers inherent to certain actions. Note that this research does not suggest using accident frequencies as a replacement for risk assessment, nor does it suggest ignoring accident consequences in analysis. This research aims to show some misleading representations of frequency in health, safety, and environmental literature due to size negligence and then suggests guidance criteria to correct the errors. Many published accident analyses give information on the number of accidents; however, they do not reveal information on other important variables, like size or scale of a firm's operation. In other words, they focus on the numerator of the frequency calculation much more than the denominator. This gives a great advantage to small-scale industries over large ones. Examples from the literature are presented below with the relevant comment from the author:

G. Al-Sharrah (✉)
Department of Chemical Engineering, Kuwait University, P.O. Box 5969, 13060 Safat, Kuwait
e-mail: g.sharrah@ku.edu.kw

M. Fattore, R. Bruggemann (eds.), *Partial Order Concepts in Applied Sciences*,
DOI 10.1007/978-3-319-45421-4_13

Examples

1. "*Hydrocarbons with 3 or 4 carbons atoms account for 40 % of known incidents*" (Khan and Abbasi 1999). Comment: Hydrocarbons with 3 or 4 carbon atoms are a range of compounds such as propane, propene, and butane and are widely used in industry. The most common is LPG (liquefied petroleum gas). Because of its high usage, it has the highest percentage of incidents.
2. "*In China, the accident rate in developed southeast coastal areas, e.g., Guangdong, Zhejiang and Jiangsu, was far higher than that in the northwest regions, e.g., Xizang, Xinjiang, and Qinghai*" (Duan et al. 2011). Comment: The southeast costal area of China is more industrialized than the northwest area. Robb et al. (2012) reported the number of manufacturing firms in these regions to be 38,995 in Guangdong, 49,450 in Zhejiang, and 39,997 in Jiangsu. On the other hand, there were only 1157 factories in Xinjiang and 303 in Qinghai. These data show that a clear difference in industry size can justify a high accident rate.
3. "*In the post-World War II period, the incidence of major industrial accidents was only one every 5 years or so, until 1980. Since 1980, the incidence has risen to two major accidents per year. Thus, according to the OECD, the frequency of major chemical accidents has increased 10-fold in recent years*" (Rachel's 1994). Comment: After World War II, industry expanded greatly. For Europe alone, economic development outstripped growth of any period of comparable length before or since (Eichengreen 1994). This can explain the high rate of accidents without any blame of negligence on industry people.
4. "*A list of the top 15 states shows California as number one (with 100,000 incidents during the period 1987–1996), nearly double the number of incidents recorded for Texas (55,209), the second leading state. Incidents were most frequently reported for chemical manufacturing and fuel companies*" (The 600 Report 1999). Comment: The rate of incidents is highly consistent with the size of industry; at this time, California had 48,478 manufacturing plants while Texas had 21,450 (U.S. Census Bureau 2005).

Clearly, statistical data on accident frequency often leaves out important variables. Often times, this misrepresentation of the data skews results and leads to conclusions that do not show actual accident frequency based on all appropriate factors.

2 Materials and Method

2.1 Denominator Selection

When evaluating the actual risks of different hazardous chemicals, it is necessary to know which chemicals have a high frequency of accidental releases. In this case, incidents are commonly normalized by dividing the number of incidents by some measure of the number of opportunities for an accident to occur. However, it is scientifically acceptable to report accidents frequency in number of incidents per year without referring to size or number of opportunities if no comparison or ranking is desired. For example, when considering a specific plant (see for example Khan and Abbasi (2001) and Meel et al. (2007)) or for socio-demographic (re)development to know how accidents are distributed across metropolitan regions (see Schweitzer 2008). Other than these cases, plant size is essential to correctly present accidents frequency.

The U.S. Department of Labor calculated occupational injury and illness rates (frequency) by dividing the number of occupational injuries at a facility by the total number of person-hour worked at the facility over a given period. This allows large and small facilities to be fairly compared, assuming that all else is equal. The overall number of occupational injuries at a workplace over a given time period will generally be directly proportional to the number of employees working there. Similarly, Squire (2001) selected an exposure hour for studying process safety. In this definition, exposure hour is defined as the number of people multiplied by the number of hours they work. Smith et al. (2006) used three denominators for evaluating work injury frequency: (1) per US$1 million payroll, (2) per 100 workers, and (3) per 100,000 h worked. He determined that the US$1 million payroll was a poor indicator of risk and that the rate per hours worked is a better denominator. Burgherr and Hirschberg (2008) selected the unit of electricity production for comparing frequencies in fossil energy chains.

To focus attention on chemicals, the size of the facility should be related to the existence of hazardous chemicals. Hazardous chemical facilities vary greatly in size, number of processes, chemical quantities stored and produced, operating schedules, and other characteristics. This study focuses on the number of processes and aggregate chemical quantity as a denominator (normalization factor) for frequency calculation. In choosing these factors, the assumption implied is that a chemical contained in a large number of processes or in large quantities has more opportunities to be accidently released than does a chemical contained in fewer processes or smaller quantities. While these divisors are certainly not perfect, they appear to be reasonable (Belke 2000). Clearly, the numerical value of an accident's frequency is different depending on the denominator. When the frequency of chemical accidents is presented as rate of accidents per the number of processes, this means that the basic bulk properties of chemicals (density, viscosity, molecular weight, and compatibility with process materials) affect these accidents. The principle is that the chemical environment in the pipe or plant equipment controls the severity of deterioration by corrosion on metal surfaces, and hence increases the failure rates of a process. On the other hand, when the frequency is presented by rate of accidents per quantity, other reactive properties affect this rate. Properties, such as maximum heat of decomposition and overall energy release potential, are related to the reactivity of the chemical, and consequently, the existence of high quantities in a chemical plant can result in a higher energy release and/or then violent self-reaction followed by an explosion.

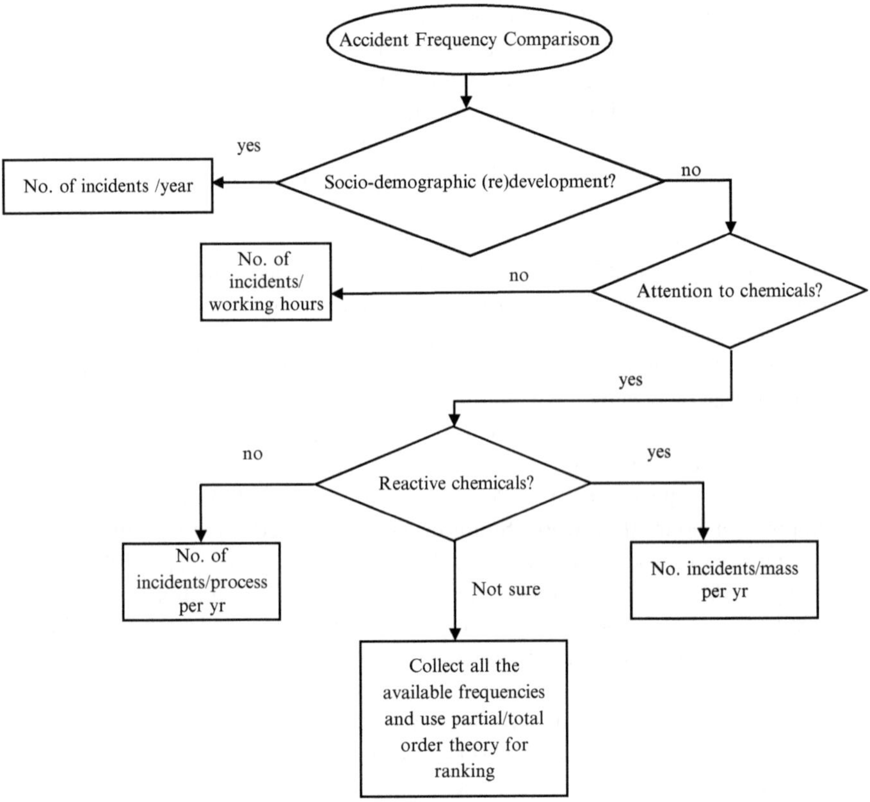

Fig. 1 Flowchart for frequency selection for ranking

Situations happen when the decision on a specific frequency is difficult and/or data available facilitates the calculation of different types of frequencies. In this case, it is advantageous to collect as many data as possible on the available frequencies and calculate judgments based on a suitable combination of them. Judgment usually comes in the form of ranking facilities or chemicals from the most hazardous to the least hazardous. This procedure can be done using partial order theory with the objective to rank chemicals according to their multiple accidents frequencies. To sum up, a simplified chart can be drawn summarizing denominator selection from the above discussion (Fig. 1).

2.2 *Ranking*

Ranking methods can be classified as relative ranking or categorical methods. Relative ranking means that an overall rank or score is derived for the objects relative to one another. A categorical ranking means that groups of objects are assigned a high, medium, or low rank, or they are determined to be either selected or nonselected objects using different comparisons among their indicators. An example of categorizing methods are the Hasse diagram and the method described by Ashby et al.(1990). The Hasse diagram is a visual representation of partially ordered sets, and details of this method are presented in Halfon and Reggiani (1986).

For relative ranking, the available methods are numerous and have different levels of complexity. Some examples are the Copeland method, the Hasse average rank and linear extensions. Using the concept of partially ordered sets and social choice theory, the Copeland score ranking methodology was modified and applied outside of its usual political environment (voting) by Al-Sharrah (2010) to rank objects in the science field. This method assumes neither linearity nor any numerical relationship among indicators and is therefore defined as a nonparametric method. The Hasse average ranking is calculated as an approximation based on a local partial order model (Bruggemann et al. 2004) using the total number of objects, the number of objects ranked below the i-th object in the Hasse diagram, and the number of objects incomparable with the i-th object. The linear extension of a partial order set is a permutation of objects that does not contradict the order of objects in the original data. Each linear extension assigns a rank to the object, and then all of these ranks are used to obtain a total rank of the object (see Patil and Taillie 2004).

2.3 Confidence Limits on Frequency Estimates

Statistical confidence limits may be attached to frequency estimation which reflects the uncertainty in estimating the underlying frequency from a small sample of events. This is usually needed since some accident databases provide limited case studies for frequency analysis. For quantitative risk assessment, a 90 % confidence range is usually adequate (OGP 2010). This confidence value extends between a lower (5 %) and an upper (95 %) confidence limit. In addition, it implies a 90 % chance that the true frequency lies within the stated range, a 5 % chance of it being lower than the lower limit, and a 5 % chance of it being above the upper limit.

These confidence ranges only take into account uncertainty due to estimating the frequency from a small number of random events, assuming the underlying frequency is constant. They do not take into account the numerous other sources of uncertainty, such as incomplete event data or inappropriate measures of exposure. Therefore, the total uncertainty in the frequency may be much higher than indicated, and the confidence limits estimated may be misleading. If the confidence is from normally distributed data, then the confidence is proportional to the square root of the data size. Therefore, it is clear that increasing data size will increase the confidence in frequency estimation. In other words, higher data size is a good indicator for frequency estimation.

3 Data and Results

To illustrate the denominator selection for accident frequency, a case study of chemical accidents is presented. Table 1 shows some data for accidents taken from Belke (2000) from 1994 to 1999. The table lists:

(a) Number of recorded accidents
(b) Accident frequency presented as the number of accidents per process per year
(c) Accident frequency presented as the number of accidents per Mlbs per year

Belke's (2000) research is a preliminary characterization of the RMP*Info database, and is therefore only a first step toward investigating the data for clues to the nature and causes of catastrophic chemical accidents. Although the study presented an analysis, much work remains to be done to analyze the data. Therefore, Belke concluded that there was a need for further research. He posed the following:

Table 1 Data for frequency analysis (Belke 2000)

No.	Chemical name	Number of accidents	Number of accidents per process per year	Number of accidents per Mlbs per year
1.	Ammonia	656	0.016	0.014
2.	Chlorine	518	0.022	0.16
3.	Hydrogen fluoride	101	0.064	0.27
4.	Flammable mixture	99	0.007	0.00075
5.	Chlorine dioxide	55	0.155	1.97
6.	Propane	54	0.006	0.0012
7.	Sulfur dioxide	48	0.013	0.011
8.	Ammonia (aqueous)	43	0.017	0.018
9.	Hydrogen chloride	32	0.060	0.25
10.	Hydrogen	32	0.031	0.24
11.	Methane	30	0.027	0.0064
12.	Butane	26	0.011	0.00089
13.	Ethylene oxide	19	0.027	0.045
14.	Hydrogen sulfide	19	0.067	0.50
15.	Formaldehyde	17	0.009	0.024
16.	Isobutane	17	0.010	0.011
17.	Pentane	17	0.013	0.0052
18.	Titanium tetrachloride	15	0.056	0.090
19.	Phosgene	12	0.044	2.49
20.	Nitric acid	12	0.038	0.047
21.	Ethane	12	0.014	0.00071
22.	Oleum	11	0.022	0.011
23.	Ethylene	11	0.014	0.00089
24.	Vinyl chloride	11	0.042	0.0051
25.	Trichlorosilane	11	0.034	0.10

- Do the data reveal the need for any policy, practice, or regulatory changes with regard to particular chemicals, industrial sectors, processes, or equipment?
- Does the database constitute a large enough sample of chemical facilities to determine risk distributions with significant confidence to make decisions about low-frequency or/and high-consequence events?
- Do the accident history data contain enough information to identify any trends or patterns in accidents, or is more data needed?

Some of these questions can be answers and the chemicals can be ranked according to their accident frequencies without any knowledge of their properties (reactive or corrosive) as explained in the following paragraphs.

The sets of frequency data listed in Table 1 together with the number of accidents used to evaluate these frequencies are used as the data set. The two frequencies are the main indicators that can be used with partial order theory to rank chemicals (objects). In addition, since the number of accidents is also available, these data can strengthen the ranking results, i.e., high frequency resulting from small number of accidence is not that reliable. Note that none of the above indicators should be used alone since they do not guarantee reliable frequency comparison, especially in regard to complex chemicals and facilities. For example, phosgene would be ranked very low if the only variable employed was the total number of accidents. In actuality, phosgene is a classified as a very high-risk chemical (The National Board of Health and Welfare 2000).

3.1 Partial Order

The theory of partial order is suitable for comparing a number of objects using different indicators. The Hasse diagram of the data in Table 1 is shown in Fig. 2a, and was created by using the PyHasse program (Bruggemann et al. 2013). Highly hazardous materials with high accident frequency are at the top level, and the relatively safe chemicals are at the lower level. However, to classify/identify hazardous chemicals it is imperative to classify hazardous chemicals and chemicals that perform worse than the acceptable risk. This requires identification of acceptable chemical risks in terms of accident frequency.

Data are available in literature about normal or acceptable operation of chemical plants:

- Typical chemical industry risk is 8.5×10^{-5} fatality per person per year (Kletz 1980).
- For each fatality, there are 30 lost-time injuries, 300 recordable injuries (Squire 2001).
- From the RMP*Info database analysis an average of 2187 processes will hold 7030 Mlbs of chemicals (Belke 2000).
- An average of 0.425 workers are needed per unit per shift (3 shifts per day) (Peters et al. 2004).

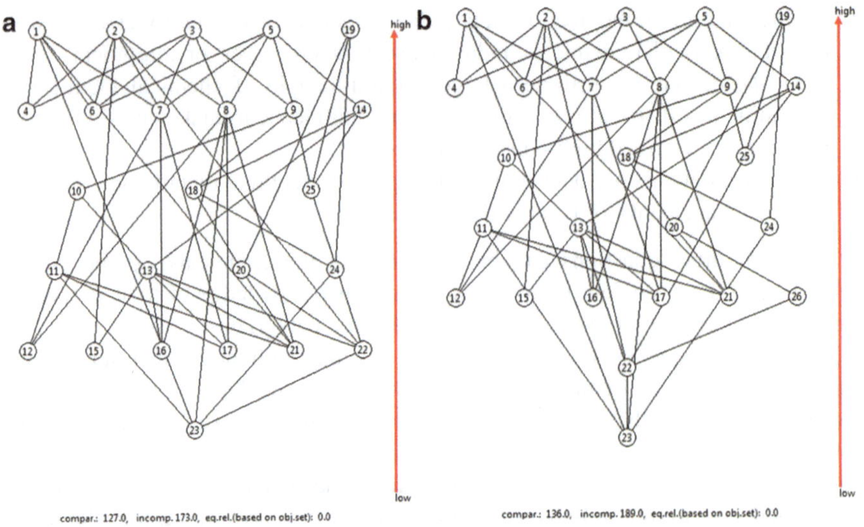

Fig. 2 Hasse diagram for chemicals: (**a**) original data (**b**) with acceptable

Using the above data, the acceptable risk was 0.035 number of accidents per process per year. The number of accidents per Mlbs per year was 0.011. This acceptable risk is attached to Table 1 as an additional object; however, the first indicator of Table 1 is missing for this new object, which is the number of accidents. To be on the conservative side, the lowest number of accidents can be taken, which is 11. Now the acceptable risk is considered as an additional object (chemical 26) that can be included in ranking. The modified Hasse diagram is shown in Fig. 2b. The diagram clearly shows that some chemicals operate higher than the acceptable risk, and they are the chemicals at the top three level of the Hasse diagram. These levels are higher than object 26. A quick and incautious analysis of the Hasse diagram in Fig. 2b indicates that the number of chemicals that needs more attention in accident prevention and which operated higher that an acceptable risk is higher than the number of safe chemicals or with acceptable risk operation.

The percent of risky chemicals, according to Fig. 2a, is high (72 %); however, due to the high incomparability in the diagram, a more careful look is essential. The acceptable risk is object 26. Chemicals that are comparable to this object are obtained by a local Hasse diagram in which only comparable objects to object 26 are plotted. The local Hasse diagram is shown in Fig. 3 and results are more confidently declared, giving only 28 % of the chemicals operating above the acceptable risk. These are chemicals 3, 5, 9, 14, 18, 19, and 20. This percent is acceptable because other studies have reported that 29 % of chemicals do more harm than good (Crowl and Louvar 2011).

Most of the *"above acceptable risk"* chemicals are corrosive chemicals. These chemicals are hydrogen fluoride, chlorine dioxide hydrogen chloride, hydrogen sulfide, titanium tetrachloride, and phosgene. Therefore, more attention should be given to corrosion prevention because corrosion is a serious threat to the industry.

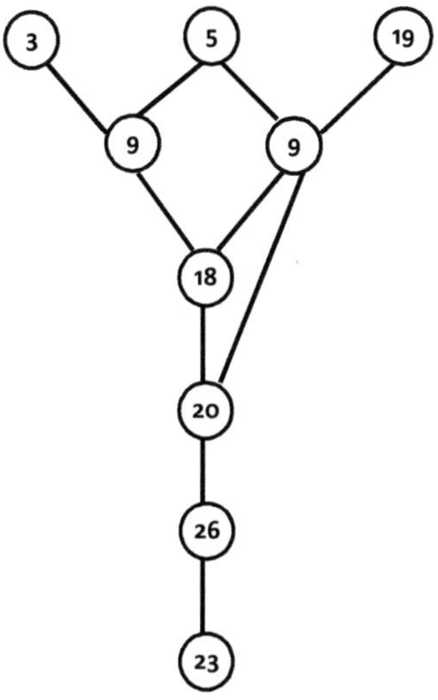

Fig. 3 Local Hasse diagram for object 26

On the other hand, the lowest *"below acceptable risk"* chemical is ethylene, which has no reactivity with common materials and is very stable during transport. Ethylene is mainly flammable but cause fire only after release, particularly when released in large quantities. Above ethylene is oleum (chemical 22). Oleum is a solution of various compositions of sulfur trioxide in sulfuric acid. It is less corrosive than pure sulfuric acid, where low alloy material of construction, such as carbon steel, can be used with oleum (see Table 4.11 Carson and Mumford 2002).

3.2 Sensitivity Analysis

The sensitivity analysis aims to find the most important indicator significantly affecting ranks. It can easily be carried out using the PyHasse program. The most important indicator is based on the distances of the different posets resulting from eliminating one after another column of the data matrix in comparison to the original poset. The program calculates a sensitivity number: the higher the number, the higher the sensitivity. The PyHasse program gave sensitivity numbers of 122, 57, and 25 for the first, second, and the third indicator, respectively, in Table 1. Consequently, ranking is most sensitive for the first indicator, which is the number of accidents. However, it should not be used as a single indicator for simple total ranking since it will misrepresent risk, as stated previously.

From our data, some questions raised by Belke (2000) can be answered:

- 28 % of the chemicals in the industry are operating above acceptable risk.
- Most risky chemicals are corrosive chemicals that need more corrosion control strategies.

- The number of data collected is essential information because it gives more confidence to chemical ranking. Therefore, stricter accident reporting systems should be imposed on all manufacturing firms.

3.3 Total Order

If strict total raking is desired, various total ranking methods can be used. The important question is: Can total ranking tools give contradicting conclusions with the Hasse diagram? Is it possible that some vital information is lost after the aggregation for total order?

Ranking methods applied on the data are: (1) the Hasse average rank from the DART program, (2) the Copeland method (Al-Sharrah 2010) and (3) linear extensions from the PyHasse program; the results are shown in Table 2. Note that in Table 2, the lower the rank, the more hazardous the chemical is. Different tools and different methods are used to get an overview on ranking decisions with the possible available tools.

The Hasse average rank is highly correlated with the Copeland rank (correlated coefficient of 0.97), while the Linear extension rank is relatively low when correlated with the Copeland rank (correlation coefficient of 0.73). The most important concern at this stage is to verify the warning about risky chemicals expressed by the Hasse diagram. Is it still the same, or did any chemical enter/leave the risk alert group? Previously, the Hasse analysis revealed that 28 % of chemicals operate above the acceptable risk. However, now and according to Table 2 this percentage can easily be detected. It has changed to 60, 52, and 92 % with Hasse average rank, Copeland and Linear extensions, respectively. The risky chemicals are now the ones which have a rank lower than the acceptable risk object 26. Evidently, these percentages are very high and unrealistic. For example, some safe chemicals, such are methane and hydrogen, are included in these percentages. Although, linear extension did not violate any of the orders in Hasse diagram, the total order presented by this method was unrealistic.

4 Conclusion

Quantitative risk assessment is widely used for improving safety and land use planning. For this purpose, there is a strong need for accident frequency data that are actual, reliable, and generally acceptable. This chapter has shown the importance of selecting a denominator for chemical accident frequency. A denominator presents the size of a facility and is a type of normalization, which is important to evenly compare large and small industries. Industry officials have to be aware so as not to take risk-related numbers and statistics as they find them; since some of them are flawed, resulting in under- or overestimation of risk. The first step in selecting numbers and statistics wisely is realizing that choosing a normalization

Table 2 Relative total ranking

Chemical No.	Chemical name	Hasse average rank	Copeland rank	Linear extensions rank
1.	Ammonia	4	8	1
2.	Chlorine	3	5	2
3.	Hydrogen fluoride	2	2	3
4.	Flammable mixture	18	19	6
5.	Chlorine dioxide	1	1	4
6.	Propane	19	20	7
7.	Sulfur dioxide	13	15	8
8.	Ammonia (aqueous)	9	10	9
9.	Hydrogen chloride	7	3	10
10.	Hydrogen	8	6	12
11.	Methane	12	14	15
12.	Butane	22	24	19
13.	Ethylene oxide	11	11	16
14.	Hydrogen sulfide	6	4	11
15.	Formaldehyde	21	17	20
16.	Isobutane	23	22	21
17.	Pentane	24	23	22
18.	Titanium tetrachloride	10	9	13
19.	Phosgene	5	7	5
20.	Nitric acid	14	12	17
21.	Ethane	25	25	23
22.	Oleum	20	21	25
23.	Ethylene	26	26	26
24.	Vinyl chloride	17	18	18
25.	Trichlorosilane	15	13	14
26.	*Acceptable risk*	16	16	24

factor is important, and if selection is difficult or nearly impossible, partial order can help in combining all the available data in a reliable manner. Although, normalization is important, the number of accidents can be a useful additional indicator for presenting the quality of the data in the form of confidence. When multiple indicators are used, the Hasse diagram can present a reasonable safety perspective for the chemical industry from a small sample of chemicals; it is better than some total ranking methods, indicating that 28 % of chemicals are operating above acceptable risk.

Results show that corrosive chemicals need more attention. Accident prevention is the most effective strategy for controlling corrosive chemical accidents. This does not, however, imply that response actions are less relevant or valid since chemical accidents, often unreported, take place daily in the world even with advanced protection.

References

Al-Sharrah G (2010) Ranking using the Copeland score: a comparison with the Hasse diagram. J Chem Inf Model 50:785–791

Ashby J, Doerrer N, Flamm F, Harris J, Hughes D, Johannsen F, Lewis S, Krivanek N, McCarthy J, Moolenaar R, Raabe G, Reynolds R, Smith J, Stevens J, Teta M, Wilson J (1990) A scheme for classifying carcinogens. Regul Toxicol Pharmacol 12:270–295

Belke JC (2000) Chemical accident risks in U.S. industry – a preliminary analysis of accident risk data from U.S. hazardous chemical facilities. United States Environmental Protection Agency. Chemical emergency preparedness and prevention Office. http://www.epa.gov/ceppo/pubs/stockholmpaper.pdf

Burgherr P, Hirschberg S (2008) Severe accident risks in fossil energy chains: a comparative analysis. Energy 33:538–553

Bruggemann R, Sørensen PB, Lerche D, Carlsen L (2004) Estimation of average ranks by a local partial order model. J Chem Inf Comput Sci 44:618–625

Bruggemann R, Carlsen L, Voigt K, Wieland R (2013) PyHasse software for partial order analysis: scientific background and description of selected modules. In: Bruggemann R, Carlsen L, Wittmann J (eds) Multi-indicator systems and modelling in partial order. Springer, New York

Carson P, Mumford C (2002) Hazardous chemicals handbook, 2nd edn. Butterworth Heinemann, Great Britain

Crowl D, Louvar J (2011) Chemical process safety: fundamentals with applications, 3rd edn. Prentice Hall, Boston, MA

Duan W, Chen G, Ye Q, Chen Q (2011) The situation of hazardous chemical accidents in China between 2000 and 2006. J Hazard Mater 186:1489–1494

Eichengreen B (1994) Institutional prerequisites for economic growth: Europe after World War II. Eur Econ Rev 38(3–4):883–890

Halfon E, Reggiani M (1986) On ranking chemicals for environmental hazard. Environ Sci Technol 20:1173–1179

Khan FI, Abbasi SA (1999) The world's worst industrial accidents of the 1990s, what happened and what might have been: a quantitative study. Process Saf Prog 18(3):135–145

Khan FI, Abbasi SA (2001) Risk analysis of a typical chemical industry using ORA procedure. J Loss Prev Process Ind 14:43–59

Kletz TA (1980) Benefits and risks: their assessment in relation to human needs. IAEA Bull 22(5/6):2–12

Meel A, O'Neilla LM, Levina JH, Seidera WD, Oktemb U, Kerenc N (2007) Operational risk assessment of chemical industries by exploiting accident databases. J Loss Prev Process Ind 20(2):113–127

OGP, International Association of Oil & Gas Producers, March 2010. Risk Assessment Data Directory: Report No. 434 – 11.1, March 2010, Aviation transport accident statistics. http://www.ogp.org.uk/pubs/434-11.pdf

Patil GP, Taillie C (2004) Multiple indicators, partially ordered sets and linear extensions: multi-criterion ranking and prioritization. Environ Ecol Stat 11:199–228

Peters MS, Timmerhaus KD, West R (2004) Plant design and economics for chemical engineers, 7th edn. McGraw-Hill, New York

Rachel's (1994) Rachel's Environment & Health Weekly # 408.Environmental Research Foundation. www.monitor.net.rachel.r408.html

Robb DJ, Liu F, Richard Lai R, Ren J (2012) Inventory in main land China: historical, industry, and geographic perspectives. Int J Prod Econ 135:440–450

Schweitzer L (2008) Accident frequencies in environmental assessment and land use studies. J Hazard Mater 156:44–50

Smith GS, Huang Y, Ho M, Chen PY (2006) The relationship between safety climate and injury rates across industries. The need to adjust for injury hazards. Accid Anal Prev 38(3):556–562

Squire RH (2001) Zero period process-a description of a process to zero injuries. Process Saf Prog 20(1):17–28

The 600 Report (1999) Commercial chemical incidents in the United States 1987–1996. Chemical Safety and Hazard Investigation Board

The National Board of Health and Welfare (2000) Chemical Accidents and Disasters Medical Care Planning Guidance Modin-Tryck, March 2000

U.S. Census Bureau (2005) 2002 Economic Census. Manufacturing geographic area series. EC02-31A-CA (RV). U.S. Department of Commerce Economics and Statistics Administration. http://www.census.gov//econ/census02/guide/EC02_31.HTM

Formal Concept Analysis Applications in Chemistry: From Radionuclides and Molecular Structure to Toxicity and Diagnosis

Nancy Y. Quintero and Guillermo Restrepo

1 Introduction

One of the advantages of Formal Concept Analysis (FCA) is its clear definition of *context*, entailing *relations* between objects and attributes and the knowledge extracted from those relations. Chemistry is all about objects, their relations and the properties emerging from relations; thus, it is not a surprise that FCA has begun to find chemical applications (Gardiner and Gillet 2015), some of which are here discussed.

Any FCA study entails declaring *objects*, their *attributes* (presence/absence variables) and the relation objects-attributes (*context*). By exploring the context, *concepts* are found and related by subsethood, from which knowledge on the relations among attributes is provided as implication and association rules, i.e. assertions $X \Rightarrow Y$, where X (*antecedent*) and Y (*consequent*) are disjoint sets of attributes (Lenca et al. 2008). An *implication* is an exact rule holding for all objects of the rule, and it is written as $(x)\ X \Rightarrow (x)\ Y$, indicating that for x objects with attribute X, they also have attribute Y. An *association* does not apply to all its involved objects, having the form $(x)X \overset{p}{\Rightarrow} (y)Y$, which reads that there are y objects

N.Y. Quintero (✉)
Laboratorio de Química Teórica, Universidad de Pamplona, Pamplona, Colombia

Universidad de Antioquia, Antioquia, Colombia

Universidad Pontificia Bolivariana, Antioquia, Colombia

Universidad Católica de Oriente, Antioquia, Colombia
e-mail: ytrioradiac@gmail.com; yaneth.quintero@udea.edu.co

G. Restrepo
Laboratorio de Química Teórica, Universidad de Pamplona, Pamplona, Colombia

Bioinformatics Group, Department of Computer Science, Universität Leipzig, Leipzig, Germany
e-mail: guillermo@bioinf.uni-leipzig.de; grestrepo@unipamplona.edu.co

M. Fattore, R. Bruggemann (eds.), *Partial Order Concepts in Applied Sciences*,
DOI 10.1007/978-3-319-45421-4_14

with attribute Y out of x objects having attribute X, or that $p < 100\,\%$ out of x objects having X also have Y. FCA rules may be generated by several algorithms, some of them described in Carpineto and Romano (2004), or by freeware software like Conexp (Yevtushenko 2000) and Lattice Miner (2015), among others (Priss 2008). The results discussed in the following sections use Conexp and Lattice Miner; for a detailed FCA description, the reader is addressed to references (Ganter and Wille 1998; Carpineto and Romano 2004; Gardiner and Gillet 2015).

2 Applications to Structure–Activity Relationships (Mutagenicity)

A chemical question is whether a macroscopic property of substances can be modelled from attributes of molecules, e.g. molecular structure. Attempts to solve this question have led to come up with structure–activity and quantitative structure–activity relationships (QSAR) approaches, which, despite their satisfactory results, have drawbacks in relating properties with structures (Rincón-Villamizar and Restrepo 2014; Greene et al. 2010). It is here that FCA comes into play, allowing finding rules relating structure with activity and vice versa, being the study of mutagenicity of amines a case in point (Franke et al. 2001).

2.1 Substances Attributes

Current chemistry studies substances through molecules (Restrepo and Harré 2015) in a one-to-one relation. Hence, characterising a substance entails characterising a molecule, which is done, e.g. using molecular descriptors, fingerprints or molecular frameworks, to name but a few possibilities (Leach and Guillet 2007; Lounkine et al. 2008; Todeschini and Consonni 2009). Frameworks were used to characterise 95 aromatic amines (Restrepo et al. 2011), whose relationship with mutagenicity was afforded through FCA by setting up mutagenicity as an attribute of substances. To make it suitable for FCA, Restrepo et al. (2011) categorised mutagenicity (Bartel and Bruggemann 1998) by creating six property levels.

2.2 FCA Rules and Interpretation

Early FCA applications to medicinal chemistry (Lounkine et al. 2008) studied formal concepts but did not explore association rules, something performed in Restrepo et al. (2011). As many rules may show up, Restrepo et al. (2011) selected only those where molecular frameworks and mutagenicity were present. These rules, coming out from a context where objects are substances and frameworks and mutagenicity levels attributes, are shown in Restrepo et al. (2011). It was found that all amines with polycyclic frameworks possess high mutagenicity, with ortho- and peri-fused systems having the highest values. Substances with a single six-atom ring system have low mutagenicity and amines having frameworks with two rings and six atoms linked by a molecular bridge have low mutagenicity. Cycles turned out to be important, as while six-membered rings show low mutagenicity, if they are connected (fused) with a five-membered ring, the mutagenicity goes up. This knowledge is useful for chemists, for they can avoid synthesising substances with worrisome frameworks.

3 Applications to Structure–Activity Relationships (Hepatotoxicity)

Besides their benefits, medicines may pose threats, e.g. hepatotoxicity in the form of drug-induced liver injury (Suk and Kim 2012), which is one important reason for drug development failure and for drug removal from the market (Njoku 2014). Thus, knowing whether a drug is hepatotoxic or not is of importance. Traditionally, a drug is hepatotoxic if after its application, concentrations of hepatic enzymes in blood increase (Ozer et al. 2010); but this is an a posteriori determination. It would be better if an a priori method were developed, reducing humane exposure and bringing down production costs of unhealthy medicines (Clark et al. 2004). Several methods have been used to estimate hepatotoxicity (Dambach et al. 2005; Ekins et al. 2010); but just recently FCA has been used to study the relation between structure and hepatotoxicity.

3.1 Drugs Attributes

In Rincón-Villamizar and Restrepo (2014), the hepatotoxicity of 274 drugs (objects) from the liver toxicity knowledge base (Chen et al. 2011) was analysed. This time besides characterising substances by molecular frameworks (Bemis and Murcko, 1996), frameworks were split into terminal ring systems and molecular bridges on the one side, and on rings and structural elements on the other, to attain a finer panorama of the mutual relation between structure and activity. As it has been found that hepatotoxicity depends on atom and bond types as well as electronic factors (Low et al. 2011), molecules were also characterised by 281 molecular descriptors accounting for such factors. By further statistical treatment the aromatic ratio (number of aromatic atoms/number of atoms), number of bonds and of atoms were the selected descriptors for conducting the study. As they are real-valued attributes, they were split into two levels by using tools reported in Quintero et al. (2013b).

3.2 FCA Rules and Interpretation

Many rules were obtained, which required a pruning strategy to look for informative rules relating structure with activity and vice versa. This strategy was based on support and confidence; briefly, the *support* (S) of the rule $X \Rightarrow Y$ is the percentage of objects characterised by attributes X and Y and the *confidence* (C) is the percentage of objects characterised by X that are also characterised by Y. If, for example, the rule $X \Rightarrow Y$ has 20 % of S and 85 % of C, the rule states that when X happens, there is 85 % probability of having Y and that X and Y occur together in 20 % of the cases.

Some rules found were 2TR $\Rightarrow$ hepatotoxicity ($S = 40.1$; $C = 79.7$) and hepatotoxicity $\Rightarrow$ 2TR ($S = 40.1$; $C = 51.2$) (Rincón-Villamizar and Restrepo 2014), indicating that molecules with two terminal rings are hepatotoxic and that the same proportion of hepatotoxic molecules has two terminal rings. Rules 1TR $\Rightarrow$ hepatotoxicity ($S = 30.3$; $C = 83.8$) and hepatotoxicity $\Rightarrow$ 1TR ($S = 30.3$; $C = 38.6$) (Rincón-Villamizar and Restrepo 2014) show relations structure–property and property–structure, illustrating the suitability of FCA as an alternative approach to traditional QSAR methods. Finer contexts showed that substances having either one or two terminal ring systems increase their likelihood of being hepatotoxic if the number of rings in the terminal ring is high (Rincón-Villamizar and Restrepo 2014). Therefore, the presence of at least one terminal ring in molecules having fused rings or spiro connections could increase the chance of hepatotoxicity. It was found that the most worrisome substances are those with a direct link between terminal ring systems, so the inclusion of ring systems in the bridge may reduce the probability of hepatotoxicity, which turns out to be valuable information for drug developers.

4 Applications to Nuclear Chemistry (PET Radionuclides)

Many β^+ radionuclides have been used in Positron Emission Tomography (PET) for diagnosis, including cancer (Pagani et al. 1997; Ziessmann et al. 2007; Raaphorst et al. 2015). In PET a non-invasive tumour visualisation is provided by using positron emitters radionuclides (Carrió et al. 2003), including ^{18}F, ^{68}Ga, ^{89}Zr and ^{86}Y, among others (Pagani et al. 1997). Quintero et al. (2013a, b) analysed 42 β^+ radionuclides currently used in PET (with clinical or preclinical applications) using 46 attributes. They also analysed an expanded set of 54 β^+ radionuclides where 12 radionuclides with potential PET use were added; the analysis was carried out using classification techniques and chemotopology (Restrepo et al. 2004; Restrepo and Mesa 2011). The results show potential applications for radionuclides, e.g. ^{47}V, ^{63}Zn and ^{70}As for studying slow metabolic processes. In the current study, we apply FCA to the extended set of 54 β^+ radionuclides.

4.1 β^+ Radionuclides Attributes

The 54 β^+ radionuclides studied are: ^{11}C, ^{13}N, $^{15}O_a$, $^{15}O_b$, $^{18}F_a$, $^{18}F_b$, $^{38}K_a$, $^{38}K_b$, ^{43}Sc, ^{44}Sc, ^{45}Ti, ^{47}V, ^{48}V, $^{51}Mn_a$, $^{51}Mn_b$, $^{52}Mn_a$, $^{52}Mn_b$, $^{52}Fe_a$, $^{52}Fe_b$, $^{55}Co_a$, $^{55}Co_b$, ^{60}Cu, ^{61}Cu, ^{62}Cu, ^{63}Zn, $^{64}Cu_a$, $^{64}Cu_b$, $^{66}Ga_a$, $^{66}Ga_b$, ^{68}Ga, ^{70}As, ^{72}As, ^{73}Se, $^{75}Br_a$, $^{75}Br_b$, $^{75}Br_c$, $^{76}Br_a$, $^{76}Br_b$, ^{77}Kr, $^{82}Rb_a$, $^{82}Rb_b$, ^{86}Y, ^{89}Zr, ^{90}Nb, ^{106}Ag, ^{115}Sb, ^{116}Sb, ^{118}Sb, ^{120}Sb, $^{124}I_a$, $^{124}I_b$, ^{130}Cs, ^{134}La and ^{140}Pr; where subindices *a*, *b* and *c* indicate differences in the production of the labelled β^+ radionuclide. Nuclear species may be characterised by several attributes, e.g. physico-chemical, physical, nuclear, quantum and dosimetric ones (Quintero et al. 2013a). In the current study, radionuclides are characterised by 36 attributes: mean β^+ energy, spatial resolution loss, percentage of positrons (%β^+), percentage of negatrons (%β^-), percentage of electron capture (EC), 0, +1, −1, +2, −2, +3, −3, +4, +5, +6, +7 oxidation numbers, atomic number (*Z*), mass number (*A*), number of neutrons (*N*), atomic weight, spin, dose, half-life ($t_{1/2}$), mean life ($1/\lambda$), parity, transition metal, p-block metal, noble gas, thermal conductivity, maximum β^+ energy, mass defect, non-metal, alkaline metal, halogen, semi-metal and lanthanide. The procedure for selecting attributes was carried out according to Quintero et al. (2013b), where a cluster analysis was performed; from each cluster a property was selected as representative by calculation of average internal and external Pearson correlations (Sanche and Lonergan 2006; Quintero et al. 2013b). After clustering, 14 β^+ attributes were selected; as most of them were continuous, they were transformed to a presence/absence scale according to Quintero et al. (2013b).

4.2 FCA Rules and Interpretation

High thresholds for *S* and *C* (50–90 %) were set up, but either none or many rules were generated; therefore, thresholds were gradually decreased until generating informative rules. Thus, we analysed different values of minimum support (min*S*) and minimum confidence (min*C*), which led to set up $\min S = 5\,\%$ and $\min C = 80\,\%$. A large number of rules was obtained making necessary another pruning methodology, namely the standardised lift (McNicholas et al. 2008), which ranks association rules (Quintero et al. 2013b). From these rules it arises that the maximum β^+ energy is related to mass defect and odd parity (−1), two useful attributes for production of radionuclides; herein, some of these rules are discussed: (8) Medium maximum β^+ energy, parity +1 $\overset{88}{\Rightarrow}$ (7) high mean life; indicating the relevance of maximum β^+ energy for dosimetric studies, where mean life is important. (18) High maximum β^+ energy, parity +1 $\overset{89}{\Rightarrow}$ (16) low mean life and (11) high maximum β^+ energy, medium mass defect $\overset{82}{\Rightarrow}$ (9) low mean life. In PET, low mean life is a sought attribute to reduce the irradiation dose in patients; on the other hand, a high maximum β^+ energy leads to lose of image spatial resolution, which can be solved by using mathematical methods (Herzog et al. 2008). Hence, given the double relationships between maximum β^+ energy and low mean life, it is necessary to find a compromise between maximum β^+ energy and resolution to select the right β^+ radionuclide.

5 Applications to Biotechnology (Uranium Bioremediation)

Contamination with radionuclides is a global problem (Noyes 1995), for they are deposited on land, air and aquatic ecosystems and are released from different anthropic activities. Depending on the process from which radionuclides are generated, they may occur as oxides, organic and inorganic complexes (van Hullebusch et al. 2005) and in all these forms they continue decaying, posing risks (Smith 2011). Microorganisms may play an important role in removing radionuclides, as they have the ability to accumulate, immobilise and biotransform radionuclides and metal ions by enzymatic or cellular uptake processes (van Hullebusch et al. 2005; Gupta et al. 2002). Uranium is a radionuclide successfully removed using microorganisms, e.g. bacteria (Nakajima and Sakaguchi 1986) and it is the subject of the present study, which intends prioritising suitable bacteria for bioremediation.

5.1 Bioremediation Attributes

We studied 70 bacteria from the following genera: *Arthrobacter*, *Citrobacter*, *Bacillus*, *Micrococcus*, *Escherichia*, *Nocardia*, *Pseudomonas*, *Streptomyces* and *Paenibacillus*, among others (Horikoshi et al. 1981; Nakajima and Sakaguchi 1986; Nakajima and Tsuruta 2002, 2004; Kazy et al. 2009; Reitz et al. 2010, 2014; Tsuruta 2007), which have been studied in uranium bioremediation along with attributes for availability, mobility and binding of metals. From those attributes, we selected:

1. Uptake capacity (Q) (mg U/g dry biomass) that quantifies the ability of microbial biomass to accumulate metals from aqueous solutions (Yi and Lian 2012).
2. Classification of bacteria as Gram-positive or Gram-negative. Both classes are differentiated by the amount of peptidoglycan in their walls, endowing the bacteria with affinity towards metals (Bäuerlein 2003).
3. pH, which influences the oxidation state and may affect the availability of functional groups at bacteria's walls to trap uranium (Gupta et al. 2002; Yi and Yao 2012).
4. Concentration of uranium (mM) that may influence the efficiency of metal removal (Singh and Gadi 2012).
5. Concentration of bacterial biomass (mg biomass/L solution), which quantifies the amount of biomass per unit of volume to recover (Hammaini et al. 2007).

As time of contact between biomass and uranium solutions is important in bioremediation, all information here reported is based on 1-h contact. Gram stain is the only presence/absence attribute, treated as Gram-positive and Gram-negative attributes with 57 and 13 bacteria, respectively; the remaining attributes are continuous, needing to be split into levels (Quintero et al. 2013b). For attributes such as concentration of bacterial biomass and concentration of uranium metal, split into two intervals, the lower interval included values within -3σ and μ and the higher within μ and 3σ. Attributes divided into three intervals were pH, Q and percentage of uranium removal, whose low, medium and high values were in between $[-3\sigma, -1\sigma]$, $(-1\sigma, 1\sigma]$ and $(1\sigma, 3\sigma]$, respectively (Stephens 1998).

5.2 FCA Rules and Interpretation

As most of the rules lacked biotechnological interpretation, different values of min$S$ and min$C$ were explored ranging from 50 to 100 %, which led to few rules. Setting up min$S$ and min$C$ thresholds to 10 % and 90 %, respectively, 38 association rules were obtained (15 implications and 23 associations). Rules highlight Gram-positive bacteria capacity for binding uranium when high biomass concentration (4.5–6.5 mg/L) is in diluted uranium solutions with pH ranging from 5.1 to 6.5, e.g. (38) low U concentration, high biomass, high pH $\overset{97}{\Rightarrow}$ (37) Gram-positive. However, not all Gram-positive bacteria achieve important uranium uptakes: (22) Gram-positive, low U concentration, high biomass concentration, high pH $\Rightarrow$ (22) medium Q, i.e. only 22, out of the 57 Gram-positive bacteria studied, attain medium uptake (21.8–69.8 mg U/g dry weight).

Only few attribute levels resulted with the support and confidences set up, namely low uranium concentration, high and medium pH and medium Q. The absence of rules for high uranium concentrations indicates poor suitability of bacteria for those solutions, which is offset by the absence of rules with low uranium uptake and which indicates the suitability of these microorganisms for bioremediation. There is a group of bacteria worth studying, for their high uranium uptake (69.9–143.5 mgU/g dry weight) with low biomass ((12) high Q $\overset{92}{\Rightarrow}$ (11) low U concentration, low biomass concentration) and even at medium pH (3.8–5.0) ((11) High Q, medium pH $\overset{91}{\Rightarrow}$ (10) low U concentration, low biomass). These bacteria belong to genera: *Actinomyces*, *Bacillus*, *Micrococcus*, *Pseudomonas*, *Streptomyces* and *Zoogloea*.

6 Conclusions and Outlook

When selecting attributes to characterise objects in FCA, there is always a compromise; the finer the attributes, the lower the supports and confidences. Hence, knowledge on the objects and the extent of the desired conclusions have to be considered for attribute selection. A strategy to take the most advantage of FCA is starting with rough attributes, generating rough contexts, which are further explored in subcontexts by using finer attributes. This was the approach, e.g. in the hepatotoxicity case.

FCA leads, normally, to many rules and the researcher needs to explore them to separate trivial from relevant ones; confidence and support are two measurements to assist in this process, but there are others worth exploring (Lenca et al. 2008). It is even possible to devise an ordering of rules based on several of these statistics and semantic and syntactic features (Berardi et al. 2006). There are also graphical methods for exploring association rules (Plasse et al. 2007). Regarding FCA software, one advantage of Lattice Miner (2015) is that each rule has directly associated its support and confidence, something lacking in Conexp. However, for graphical purposes, as depicting and editing the line diagram, Conexp is better than Lattice Miner.

How to categorise continuous attributes is an item worth discussing, which leads to studying the distribution of attribute values. Some approaches include the application of strategies from descriptive statistics (Stephens 1998) and non-parametric tests (Kolmogorov-Smirnov and Shapiro-Wilks) for assessing normality (Quintero et al. 2013b). In some cases, the application of transformation methods of variables like Box and Cox (Osborne 2010) are important. However, further interaction with novel statistical tools for categorisation is needed. As mentioned, chemistry is all about relations and FCA has helped to explore the following relationships of importance to chemists:

- Amines with polycyclic frameworks are more associated to mutagenicity than amines with few cycles. This opens the way to finer studies with finer attributes for the molecular structure. A question that arises is to what extent the amine group is related to mutagenicity or whether mutagenicity is more related to molecular connectivity and molecular shape.
- The hepatotoxicity study allowed finding rules relating activity with molecular features that are of importance for synthetic chemistry, where the researcher has a clear idea, even quantitative, of the sought activity, which thanks to FCA, may lead to particular aspects (frameworks) of the molecular structure associated to substances to synthesise.
- The β^+ radionuclides study shed light on the relationships between physico-chemical attributes and desired properties like low radiation exposure, which led to suggesting potential novel radionuclides.
- For diluted samples of uranium at pH between 5.1 and 6.5, a heavy charge of Gram-positive bacteria leads to uranium uptakes ranging from 21 to 70 g U/g of dry bacteria biomass. These bacteria are: *Actinomyces levoris* HUT 6156, *Arthrobacter* sp. US-10, *Bacillus badius* IAM 11059, *Bacillus cereus* AHU 1030, *Bacillus cereus* AHU 1355, *Bacillus cereus* AHU 1356, *Bacillus stearothermophillus* IAM 11062, *Bacillus subtilis* AHU 1219, *Bacillus subtilis* AHU 1390, *Bacillus subtilis* IAM 11062, *Bacillus thuringiensis* IAM 11064, *Streptomyces albidoflavus* HUT 6129, *Streptomyces albidus* HUT 6129, *Streptomyces albosporeus* HUT 6130, *Streptomyces albus* HUT6132,

Streptomyces antibioticus HUT 6137, *Streptomyces chartreusis* HUT 6140, *Streptomyces cinereoruber* HUT 6142, *Streptomyces novaecaesareae* HUT 6158, *Streptomyces violaceus* HUT 6164, *Streptomyces viridochromogenes* HUT 6166 and *Streptomyces viridochromogenes* HUT 6167.

We hope this chapter, besides bringing knowledge on each particular application, shows the methodological advantages of FCA in chemistry and promote motivates its use and development in this science.

Acknowledgements N. Y. Quintero thanks Colciencias (Colombia) for the Ph.D. grant to conduct the current study. She also thanks Professors A. Nakajima and T. Reitz for their valuable comments regarding U bioremediation. G. Restrepo thanks the Alexander von Humboldt Foundation/Stiftung and the Universidad de Pamplona for the support to conduct this research.

References

Bartel H-G, Bruggemann R (1998) Application of formal concept analysis to structure-activity relationships. Fresenius J Anal Chem 361:23–38

Bäuerlein E (2003) Biomineralization of unicellular organisms: an unusual membrane biochemistry for the production of inorganic nano- and microstructures. Angew Chem Int Ed Engl 42:614–641

Bemis GW, Murcko MA (1996) The properties of known drugs. 1. Molecular Frameworks. J Med Chem 39:2887–2893

Berardi M, Appice A, Loglisci C, Leo P (2006) Supporting visual exploration of discovered association rules through multi-dimensional scaling. In: Esposito F, Râs ZW, Malerba D, Semeraro G (eds) Foundations of intelligent systems, 16th International Symposium, ISMIS 2006, vol 4203, Lecture notes in computer science. Springer, Berlin

Carpineto C, Romano G (2004) Concept data analysis: theory and applications. Wiley, Hoboken, NJ, 07030, USA, pp 1–220

Carrió I, González P, Estorch M, Canessa J, Mitjavila M, Massardo T (2003) Medicina nuclear. Aplicaciones clínicas. Masson. S.A, Barcelona

Chen M, Vijay V, Shi Q, Liu Z, Jang H, Tong W (2011) FDA-approved drug labeling for the study of drug-induced liver injury. Drug Discov Today 16:697–703

Clark RD, Wolohan PR, Hodgkin EE, Kelly JH, Sussman NL (2004) Modelling in vitro hepatotoxicity using molecular interaction fields and SIMCA. J Mol Graph Model 22:487–497

Dambach D, Andrews B, Moulin F (2005) New technologies and screening strategies for hepatotoxicity: use of *in vitro* models. Toxicol Pathol 33:17–26

Ekins S, Williams AJ, Xu JJ (2010) A predictive ligand-based Bayesian model for human drug-induced liver injury. Drug Metab Dispos 38(12):2302–2308

Franke R, Gruska A, Giuliani A, Benigni R (2001) Prediction of rodent carcinogenicity of aromatic amines: a quantitative structure-activity relationships model. Carcinogenesis 22(9):1561–1571

Ganter B, Wille R (1998) Formal concept analysis: mathematical foundations. Springer, Berlin

Gardiner EJ, Gillet VJ (2015) Perspectives on knowledge discovery algorithms recently introduced in chemoinformatics: rough set theory, association rule mining, emerging patterns, and formal concept analysis. J Chem Inf Model 55(9):1781–1803

Greene N, Fisk L, Naven RT, Note RR, Patel ML, Pelletier DJ (2010) Developing structure-activity relationships for the prediction of hepatotoxicity. Chem Res Toxicol 19(23):1215–1222

Gupta R, Saxena RK, Mohapatra H, Ahuja P (2002) Microbial variables for bioremediation of heavy metals from industrial effluents. In: Singh VP, Stapleton JR (eds) Biotransformations: bioremediation and technology for health and environmental protection, vol 36, Progress in industrial microbiology. Elsevier, Amsterdam, pp 189–230

Hammaini A, González F, Ballester A, Blázques ML, Muñoz JA (2007) Biosorption of heavy metals by activated sludge and their desorption characteristics. J Environ Manage 84:419–426

Herzog H, Tellmann L, Scholten B, Cohenen HH, Qaim SM (2008) PET imaging problems with the non-standard positron emitters Yttrium-86 and Iodine-124. Q J Nucl Med Mol Imaging 52(2):159–165

Horikoshi T, Nakajima A, Sakaguchi T (1981) Studies on the accumulation of heavy metal elements in biological systems. Eur J Appl Microbiol 12(2):90–96

Kazy SK, D'Souza SFD, Pinaki S (2009) Uranium and thorium sequestration by a *Pseudomonas* sp.: mechanism and chemical characterization. J Hazard Mater 163:65–72

Lattice Miner (2015) http://sourceforge.net/projects/lattice-miner/. Accessed 25 Apr 2015

Leach AR, Guillet V (2007) An introduction to chemoinformatics. Kluwer, Dordrecht, The Netherlands. doi:10.1007/978-1-4020-6291-9

Lenca P, Meyer P, Vaillant B, Lallich S (2008) On selecting interestingness measures for association rules: user oriented description and multiple criteria decision aid. Eur J Oper Res 184(2):610–628

Lounkine E, Auer J, Bajorath J (2008) Formal concept analysis for the identification of molecular fragment combinations specific for active and highly potent compounds. J Med Chem 51(17):5342–5348

Low Y, Uehara T, Minowa Y, Yamada H, Ohno Y, Urushidani T, Sedykh A, Muratov E, Kuz'min V, Fourches D, Zhu D, Rusyn I, Tropsha A (2011) Predicting drug-induced hepatotoxicity using QSAR and toxicogenomics approaches. Chem Res Toxicol 24(8):1251–1262

McNicholas PD, Murphy TB, O'Regan M (2008) Standardising the lift of an association rule. Comput Stat Data Anal 52:4712–4721

Nakajima A, Sakaguchi T (1986) Selective accumulation of heavy metals by microorganisms. Appl Microbiol Biot 24:59–64

Nakajima A, Tsuruta T (2002) Competitive biosorption of thorium and uranium by actinomycetes. J Nucl Sci Technol 39(Suppl 3):528–531

Nakajima A, Tsuruta T (2004) Competitive biosorption of thorium and uranium by *Micrococcus luteus*. J Radioanal Nucl Chem 260(1):13–18

Njoku DB (2014) Drug-induced hepatotoxicity: metabolic, genetic and immunological basis. Int J Mol Sci 15:6990–7003

Noyes R (1995) Nuclear wastes cleanup technology and opportunities. Noyes, Park Ridge, NJ

Osborne JW (2010) Improving your data transformations: applying the Box-Cox transformation. Practical assessment. Res Eval 15(12):1–9

Ozer JS, Chetty R, Kenna G, Palandra J, Zhang Y, Lanevschi A, Koppiker N, Souberbielle BE, Ramaiah SK (2010) Enhancing the utility of alanine aminotransferase as a reference standard biomarker for drug-induced liver injury. Regul Toxicol Pharmacol 56:237–246

Pagani M, Stone-Elander S, Larsson SA (1997) Alternative positron emission tomography with non-conventional positron emitters: effects of their physical properties on image quality and potential clinical applications. Eur J Nucl Med 24(10):1301–1327

Plasse M, Niang N, Saporta G, Villeminot A, Leblond L (2007) Combined use of association rules mining and clustering methods to find relevant links between binary rare attributes in a large data set. Comput Stat Data Anal 52(1):596–613

Priss U (2008) FcaStone-FCA file format and interoperability software. In: Croitoru M, Jaschkë R, Rudolph S (eds). Conceptual structures and the web. Proceedings of the third conceptual structures and tool interoperability workshop, pp 33–43

Quintero NY, Restrepo G, Cohen IM (2013a) Chemotopological study of positron emitters ra-dionuclides used in PET diagnostic imaging: physical, physico-chemical, dosimetric, quantum and nuclear properties. J Radioanal Nucl Chem 295(2):823–833

Quintero NY, Restrepo G, Cohen IM (2013b) Relating β^+ radionuclides' properties by order theory. J Radioanal Nucl Chem 298(3):1937–1946

Raaphorst RM, Windhorst AD, Elsinga PH, Colabufo NA, Lammertsma AA, Luurtsema G (2015) Radiopharmaceuticals for assessing ABC transporters at the blood-brain barrier. Clin Pharmacol Ther 97(4):362–371

Reitz T, Merroun ML, Rossberg A, Selenska-Pobell S (2010) Interactions of *Sulfolobus acidocaldarius* with uranium. Radiochim Acta 98:249–257

Reitz T, Rossberg A, Barkleit A, Selenska-Pobell S, Merroun ML (2014) Decrease of U (VI) immobilization capability of the facultative anaerobic strain *Paenibacillus* sp. JG-TB8 under anoxic conditions due to strongly reduced phosphatase activity. PLoS One 9(8):1–13

Restrepo G, Mesa H, Llanos EJ, Villaveces JL (2004) Topological study of the periodic system. J Chem Inf Comput Sci 44(1):68–75

Restrepo G, Mesa H (2011) Chemotopology: beyond neighbourhoods. Curr Comput Aided Drug Des 7(2):90–97

Restrepo G, Basak S, Mills D (2011) Comparison of QSARs and characterization of structural basis of bioactivity using partial order theory and formal concept analysis: a case study with mutagenicity. Curr Comput Aided Drug Des 7(2):109–121

Restrepo G, Harré R (2015) Mereology of quantitative structure-activity relationships models. HYLE Int J Phil Chem 21:19–38

Rincón-Villamizar E, Restrepo G (2014) Rules relating hepatotoxicity with structural attributes of drugs. Toxicol Environ Chem 96(4):594–613

Sanche R, Lonergan KF (2006) Variable reduction for predictive modeling with modeling clustering. Casualty Actuarial Society Forum, pp 89–100

Singh N, Gadi R (2012) Bioremediation of Ni (II) and Cu (II) from wastewater by the nonliving biomass of *Brevundimonas vesicularis*. J Environ Chem Ecotoxicol 4(8):137–142

Smith JT (2011) Nuclear accidents. In: Hester RE, Harrison RM (eds) Nuclear power and environment, vol 32, Issues in environmental science and technology. Royal Society of Chemistry, London

Stephens LJ (1998) Theory and problems of beginning statistics, Schaum's outline series. McGraw-Hill, Mexico

Suk KT, Kim DJ (2012) Drug-induced liver injury: present and future. Clin Mol Hepatol 18(3):249–257

Todeschini R, Consonni V (2009) Molecular descriptors for chemoinformatics. Wiley-VCH, Weinheim

Tsuruta T (2007) Removal and recovery of uranium using microorganisms isolated from North American uranium deposits. Am J Environ Sci 3(2):60–66

van Hullebusch E, Lens PNL, Tabak HH (2005) Developments in bioremediation of soils and sediments polluted with metals and radionuclides. 3. Influence of chemical speciation and bioavailability on contaminants immobilization/mobilization bio-processes. Rev Environ Sci Biotechnol 4:185–212

Yevtushenko SA (2000) System of data analysis "Concept Explorer". Proceedings of the 7th national conference on Artificial Intelligence KII, Russia, pp 127–134

Yi Z, Lian B (2012) Adsorption of U (VI) by *Bacillus mucilaginosus*. J Radioanal Nucl Chem 293(1):321–329

Yi Z, Yao J (2012) Kinetic and equilibrium study of uranium (VI) adsorption by *Bacillus licheniformis*. J Radioanal Nucl Chem 293(3):907–914

Ziessmann H, O'Malley J, Thrall JH (2007) Medicina nuclear: Los requisitos en radiología. Elsevier, España

Partial Order Analysis of the Government Dependence of the Sustainable Development Performance in Germany's Federal States

Alexander Hilckmann, Vanessa Bach, Rainer Bruggemann, Robert Ackermann, and Matthias Finkbeiner

1 Introduction

In the year 1992 during the Earth Summit (United Nations Conference on Environment and Development) in Rio de Janeiro the Agenda 21 was established. The Agenda 21 is an action plan with regard to sustainable development that can be executed on local, national, and global level (United Nations 1992). To comply with the requirements of the Agenda 21, the Council of Sustainable Development (Rat für Nachhaltigkeit) was established in 2001 by the German Federal Government followed by the national sustainability strategy in 2002 Federal Ministry for Economic Cooperation and Development (2016). It consists of several goals, quantified by indicators to measure the success of policy implementations and other measures on national and local level (Bundesregierung Deutschland Nationale Nachhaltigkeitsstrategie 2012). To determine meaningful and applicable indicators, the Conference of Environmental Ministers (Umweltministerkonferenz) ordered the Bund/Länder-Arbeitsgemeinschaft Klima, Energie, Mobilität—Nachhaltigkeit (BLAG KliNA—Federal-State working group for climate, energy, mobility, and sustainability) to develop a comprehensive indicator set: the German core sustainability indicators Landesamt für Natur und Verbraucherschutz (2015). This set is divided into four main groups:

1. Climate and energy (A)
2. Nature and countryside (B)

A. Hilckmann • V. Bach • R. Ackermann (✉) • M. Finkbeiner
Technische Universität Berlin, Straße des 17 Juni 135, 10623 Berlin, Germany
e-mail: robert.ackermann@tu-berlin.de

R. Bruggemann
Department of Ecohydrology, Leibniz—Institute of Freshwater Ecology and Inland Fisheries, Müggelseedamm 310, 12587 Friedrichshagen, Berlin, Germany

M. Fattore, R. Bruggemann (eds.), *Partial Order Concepts in Applied Sciences*,
DOI 10.1007/978-3-319-45421-4_15

Table 1 Indicators considered

Indicator number	Description
A2.1	Energy-related carbon dioxide emissions, per inhabitant
A2.2	Fuel-based carbon dioxide emissions generated by traffic, per inhabitant
A3.1	Primary energy consumption, per inhabitant
A4	Share of renewable energy in primary energy consumption
B3	Share of strictly protected nature conservation areas nationwide
B4	Proportion of evidently damaged beech trees (level 2 or higher)
C3.1	Local public passenger transport services, per inhabitant

3. Environment and health (C)
4. Resources and efficiency (D)

Each group consists of 4–9 indicators, which are named after the main group they belong to, for example, "A2 Carbon Dioxide Emissions." Some are divided further into subindicators, for example, "A2.1 Energy related carbon dioxide emissions, per inhabitant" or "A2.2 Fuel-based carbon dioxide emissions generated by traffic, per inhabitant." Overall only seven of the existing 47 indicators were used within the analysis (see Table 1) due to data availability constraints. High indicator values indicate an increasing sustainable development. The selected years for the analysis are 1991 to 2009 as for the considered federal states only this time frame has good data availability.

This chapter aims at answering the following question:

Does a relation between political parties and the sustainable development within Germany's federal states exist?

To answer the question, the introduced seven indicators are analyzed for the federal states Bavaria (BY), Schleswig-Holstein (SH), Baden-Württemberg (BW), Thuringia (TH), Hesse (HN), and Rhineland-Palatinate (RP) for the years 91–09 using Partial Order methodology. The Partial Order methodology can support the analysis of indicator systems as shown in several case studies (Bach et al. 2015; Carlsen and Bruggemann 2013; Bruggemann et al. 2014a).

2 Method

The concept of comparison is the basis of the Partial Order methodology. Given a set of objects, $X = \{a, b, \ldots\}$, the objects a, b, etc., are mutually compared. A Partial Order is present when the three following axioms are valid.

(I) (Reflexivity): $a \leq a$, for all $a \in X$.
(II) (Transitivity): If $a \leq b$ and $b \leq c$, then $a \leq c$, for all $a, b, c \in X$.
(III) (Antisymmetry): If $a \leq b$ and $b \leq a$, then $a = b$, for all $a, b \in X$.

The ≤-relation can be achieved in many different ways. Considering the analyzed objects together with the multiindicator system (MIS) as presented in Table 1, two objects $x, y \in X$ fulfill the ≤-relation, if and only if it is valid: $q_j(x) \leq q_j(y)$ for all $j = 1, \ldots, m$. The natural (weak) order among objects for any indicator $q_j (j = 1, \ldots, m)$ is combined to obtain a Partial Order by considering all indicators at the same time. For Partial Orders a total (or complete or linear) order is often not an option. Partial orders can be visualized by Hasse diagrams, which are often useful representations.

Hasse diagram: The Hasse diagram focuses on individual objects and their relation to each other and visualizes them (Bruggemann et al. 2014a; Bruggemann and Patil 2011).

Cover relation: Given $x < y$. If there is no object z, for which is valid: $x < z < y$ then the relation between x and y is called a cover relation. The object y is covering object x.

Set of linear extensions: Any Partial Order can be represented by a set of linear orders (Bruggemann et al. 2014a), called LE, which are extensions of the partial order. Any object $x \in X$ has a certain position within any linear order of LE. The average of all positions over all linear orders of LE is called the average height, hav.

Level: The drawing of the Hasse diagram is organized in that manner that objects $x \in X$ are as far as possible positioned in the same vertical location. The set of objects in the same vertical position is called a level. The concept of levels allows a valuable weak ordering of the objects (Bruggemann and Patil 2011).

Generalized ranking: In strict terms a ranking is understood as a total ordering of the objects of X. Applying Partial Order often only for subsets of X such a total order can be found. Therefore, the wording 'generalized ranking' is used.

Equivalence: If x and y are different objects and $q_j(x) = q_j(y)$ is valid for all $j = 1, \ldots, m$, then the objects x and y can be classified as equivalent and are declared as $x \cong y$ (Carlsen and Bruggemann 2013; Saxl 1995). An equivalent part can be represented by one element. This element is a representative element.

Incomparability: If $q_j(x) \leq q_j(y)$ cannot be established for all $j = 1 \ldots m$, the two elements x and y are "incomparable," denoted as $x \mid\mid y$. Incomparability indicates a data conflict: With respect to some indicators x is better than y, but with respect to some other indicators y is better than x.

Discretization: Using various attributes can lead to many incomparabilities caused by minute numerical differences among the attribute values of different objects. To minimize these and preserve the meaningfulness of the results, beside other techniques the formation of intervals can be applied. By dividing the range of highest and lowest value for every attribute into intervals (K value), the amount of incomparabilities can be reduced (Carlsen and Bruggemann 2013; Bruggemann et al. 2014b).

Minimal, maximal elements: Elements x for which no relation $y < x$, $y, x \in X$, can be found are called minimal elements. Elements x for which no relation $x < y$, $y, x \in X$ can be found are called maximal elements.

Similarity: When two sets of attributes are available, based on the same set X of objects, two partial orders arise. The similarity between the two partial orders can

Table 2 Description of similarity terms

Aspect	Relations	Remark
Isotone	$x <_1 y, x <_2 y$	
Antitone	$x <_1 y, x >_2 y$	and other combinations
Weak isotone	Combinations with one equality, such as $x <_1 y, x =_2 y$	
Indifference	All combinations with $x \mid\mid y$, *such as* $x \mid\mid 1\ y, x <_2 y$	
Identity	$x =_1 y, x =_2 y$	

Table 3 Applied modules of PyHasse

Name of module	Application
mHDCl	Standard partial order analysis (Hasse diagram)
discretiz	Building intervals within attributes to improve the comparability of objects
Avranks	Calculating the average ranks (heights)
Similarity	Comparison of two partial orders derived from two MIS

be determined in terms of fractions of isotone, antitone, weak isotone, indifferences, and identities (iso, anti, wiso, indiff, ident). Let $x <_i y$ denote an order relation obtained from the ith MIS. Then the five aspects of similarity (i.e., isotone,. . . , ident.) can be explained as presented in Table 2.

Software: The program PyHasse (Bruggemann et al. 2014a) can be used to support the calculation of Partial Order results. Within this chapter, the following modules were applied (Hilckmann 2015) (see Table 3).

3 Results and interpretation

3.1 Comparison of Bavaria and Schleswig Holstein: General remarks

First, the results for the comparison of the federal states Bavaria (BY) and Schleswig-Holstein (SH) are shown. The objects to be compared are years. The indicators characterize each year (of a certain federal state) with respect to the sustainability as far as 7 indicators can be considered as sufficient. Their Hasse diagrams are presented in Fig. 1. BY and SH were chosen as both federal states show different characteristics:

- Bavaria:
  Government: BY was governed by CDU (Christlich Demokratische Union Deutschlands—Christian Democratic Union of Germany) during the period considered 1990. In the year 1990 the FDP (Freie Demokratische Partei—Free Democratic Party) joined the government.

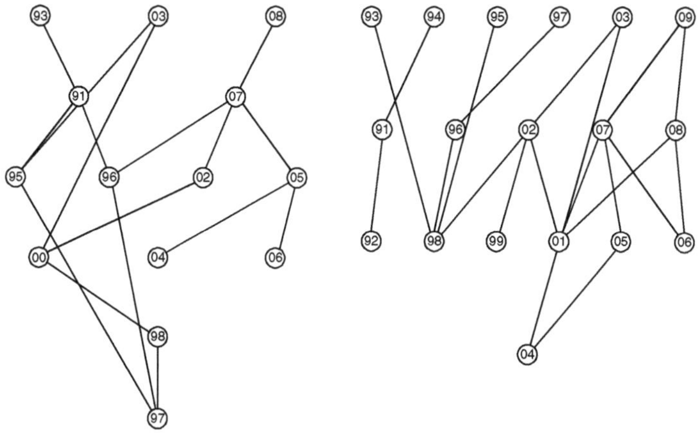

Fig. 1 Hasse diagram of Schleswig-Holstein (*left*) and Bavaria (*right*) for the years 91–09, considering seven indicators. Three equidistant intervals were chosen ($K = 3$). Equivalent classes (SH): {91, 92}, {93, 94} {98, 99}, {00,01}, {08, 09}, Equivalent objects (BY): {99, 00}, PyHasse-Software

Wealth: BY is one of the wealthiest federal state (GDP of 38.429€ per inhabitant and year)

Location: Bavaria is in the south of Germany

- Schleswig-Holstein:
  Government: In SH the SPD (Sozialdemokratische Partei Deutschlands—Social Democratic Party of Germany) governed together with Bündnis 90/Die Grünen (Alliance 90/the Greens) until 2005. After 2005, CDU governed together with SPD (Hilckmann 2015; Landesregierung 1946).
  Location: SH is situated in the north of Germany
  Wealth: SH is one of the poorest (GDP of 27.684€ per inhabitant and year).

The population density is the same in both federal states Statistik Nord (2015); Landesamt für Statistik (2015).

3.2 *The Partial Orders of Bavaria and Schleswig Holstein*

The Hasse diagrams show the generalized ranking of objects (here: years 91–09) based on the seven indicators considered (e.g., "A2.1 Energy related carbon dioxide emissions, per inhabitant"). Both diagrams are based on a discretization by three equidistant intervals, i.e., $K = 3$.

- Schleswig-Holstein (SH):
  Level: The Hasse diagram of SH has six levels from which the following weak order can be deduced: 97 < {98, 99} < {00, 01, 04, 06} < {95, 96, 02, 05} < {91, 92, 07} < {93, 94, 03, 08, 09}

Note, we neglect the set symbols for the singletons.

Cover relation: For example, the years 07 and 08.

Equivalence: Noticeable is the high amount of equivalent objects, which are often consecutive years. Equivalent objects: {91, 92}, {93, 94} {98, 99}, {00, 01}, {08, 09},

Minimal, maximal elements: The year 97 is one of the minimal elements. Thus, it performs worse compared to the other years, e.g., 98 regarding sustainable development. Three maximal representative elements (e.g., 93, 03 and 08) exist; these years perform well regarding their sustainable development. *Incomparability*: Several noncomparable objects occur (e.g., 95 and 96). For them it is hard to say (without checking the data matrix) in which year sustainable development was increasing more, only that they are not comparable with other years and each other, thus indicating a data conflict.

- Bavaria (BY):
  Level: The Hasse diagram of BY has 4 levels. A weak order can be established: 04 < {92, 98, 99, 00, 01, 05, 06} < {91, 96, 02, 07, 08} < {93, 94, 95, 97, 03, 09}

  Cover relation: For instance, the years 99 and 02.

  Equivalent objects: {99, 00}

  Minimal, maximal elements: The year 2004 is one of the minimal elements. Thus, it performs worse regarding sustainable development compared to other years, e.g., 01. Several maximal elements occur (e.g., 95, 97, 03).

  Incomparability: In comparison to SH many more incomparabilities are found. This finding is in coincidence with the lower number of levels of the Hasse diagram of BY. For example, 93 || 94. For these two years the profiles in terms of modes from 1 to 3 exhibit data conflicts. At least one aspect of sustainability for 93 is better than that of 95, whereas there is another indicator, for which 95 has the better value.

3.3 Average Heights

Based on the Hasse diagrams the comparison of the years and their corresponding sustainable development is limited. Hence it is difficult to establish comparison among federal states on the basis of their temporal development. Thus, calculating the average heights allows for a chronological and thus more adequate comparison (see Fig. 2). The horizontal X-axis displays time while the vertical Y-axis shows the average height of the objects. Thus, we introduce the function hav(t) with t taken from {91, 92, … , 09} for different federal states, which is a highly aggregated measure for sustainable development. By the functions hav(t) conclusions regarding the sustainable development of the federal states can be made by analyzing the direction of movement of the graph (Fig. 2).

For the years 92, 94, and 99 the direction of the graphs differs. Whereas for SH the graph heads down, it goes up for BY. For those years the sustainable development differs in the federal states. For the other years both graphs behave

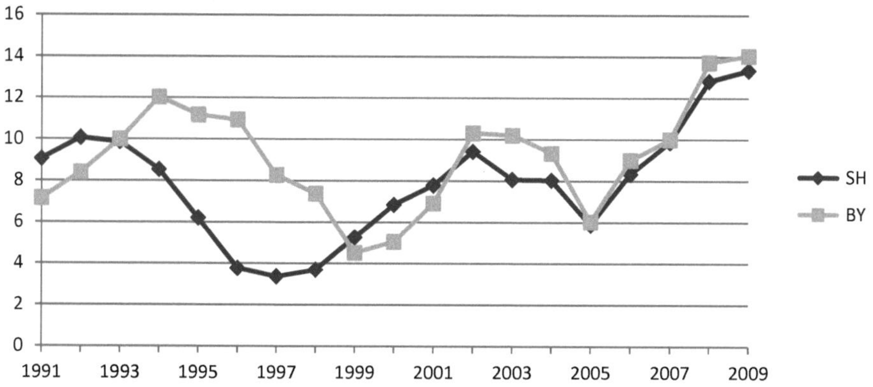

Fig. 2 Average heights of BY and SH for the years 91–09 ($K = 3$)

similar. Starting from the year 04 the development of both federal states is increasing simultaneously. Concluding, the sustainable development of both states is similar considering the chosen time frame and indicators. As the federal states vary in their governance as well as in several other characteristics (e.g., economic background) these results are surprising.

3.4 Similarity Between the Partial Order of SH and BY

When testing the data sets of both federal states regarding their similarity 44 isotone, 4 antitone, 6 wiso, and 288 indifferent relations are identified. As the number of antitone relations only sums up to four a shift in sustainable development hardly takes place, as also indicated in the average heights analysis. The relatively high number of isotone relations points to a similar sustainable development for several indicators. However, as many indifferent objects exist, the sustainable development of both federal states can not be seen as identical.

3.5 Extension to Other Federal States

To make sure that the results of the federal states BY and SH are not artifacts, four additional federal states are analyzed (see Fig. 3). Thereby, federal states are chosen based on a wide variety regarding the considered characteristics: economically strong and weaker regions (e.g., Hesse and Thuringia), high and low population density (e.g., Baden-Württemberg and Thuringia), as well as states of former East Germany (e.g., Thuringia) and former West Germany (e.g., Rhineland-Palatinate) are taken into account.

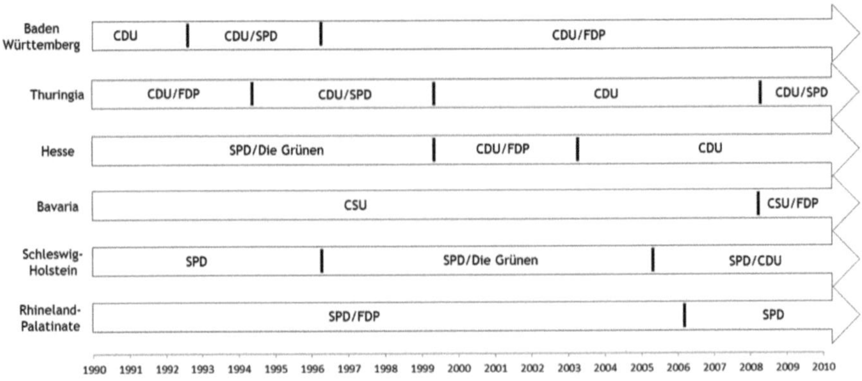

Fig. 3 Overview of governing political parties in the six considered federal states during the years 91–09

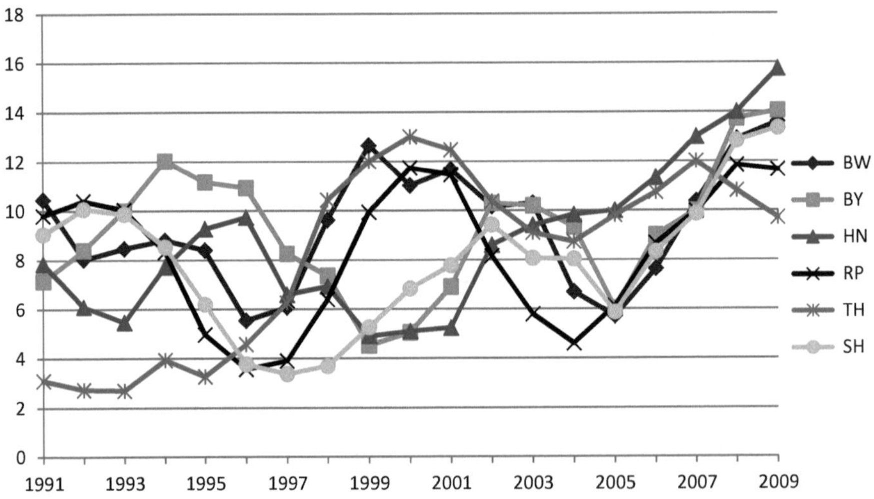

Fig. 4 Average heights of six federal states, $K = 3$

The considered federal states next to Bavaria (BY) and Schleswig-Holstein (SH) are Baden-Württemberg (BW), Thuringia (TH), Hesse (HN), and Rhineland-Palatinate (RP). BW was governed by CDU over the considered time period. For the years 92–96 SPD was supporting the governance. Afterward the FDP joined. In TH the CDU formed the government over the considered time frame. First supported by FDP (90–94) later by SPD (94–99 as well as 09). In HN a more drastic change occurred. First it was governed by SPD and Bündnis 90/Die Grünen (91–99) and then by CDU and FDP until 03 when CDU governed it by themselves. RP was governed by SPD over the considered period of time, joined by FDP from 91 to 06.

The average height analysis of these six states is shown in Fig. 4.

It shows that starting with the year 99, where the change of government (from SPD to CDU) occurs, the sustainable development of HN increases gradually. However, also in the years 93–96 an increase can be observed. For the development of BW (governed by CDU) and RP (governed by SPD) such a trend cannot be detected. The direction of BW's sustainable development changes pretty often. The direction of RP's sustainable development also changes but not as often as BW.

Overall it can be stated, that neither the composition of the government, nor wealth or the geographical location can be identified as a significant paramter for sustainable development. Even though HN and TH have very different gross domestic production, their sustainable development is alike for the most periods of time. Same is true for the federal state BW and TH as well as for TH and RP for the time period 96 to 04. The development of all six federal states is similar but not identical, indicating that some other hitherto not identified characteristics might influence their sustainable development over time.

The main hypothesis of this study, namely, the role of the political parties with regard to sustainable development can be rejected at this point. Reasons for the similar development could be the influence of the federal government, which is not included in the analysis.

4 Conclusion and Outlook

As demonstrated within the chapter, a connection between the sustainable development of federal states and their governing political parties could not be established. This might have methodological reasons like the amount of considered indicators as well as content-related reasons like the influence of Germany's federal government.

Hence, these aspects should be analyzed in more detail. To exclude that the considered indicators and time frames misrepresent the results, a comprehensive assessment of both parameters in several sensitivity analyses should be carried out. If it is confirmed, that neither the amount of indicators nor the considered timeframes lead to a distortion of results, contextual aspects have to be analyzed in more detail. This includes the role of the federal government. As the governing parties and their composition have changed over the last decades, an evaluation similar to the one carried out within this chapter for overall Germany could be established. Furthermore, subsequent changes of sustainable development have to be taken into account as well. As measures to enhance sustainable development will often be visible after considerable time delays, it might happen that successes of one government can only be seen years later when other political parties govern.

References

Bach V, Bruggemann R, Finkbeiner M (2015) Using partial order to analyze characteristics of resource availability indicators. In: Simulation in Umwelt-und Geowissenschaften, pp 11–17

Bruggemann R, Patil GP (2011) Ranking and prioritization for multi-indicator systems. Springer, Berlin

Bruggemann R, Carlsen L, Wittmann J (2014a) Multi-indicator systems and modelling in partial order. Springer, New York

Bruggemann R, Carlsen L, Voigt K, Wieland R (2014b) PyHasse Software for partial order analysis. In: Bruggemann R, Carlsen L, Wittmann J (eds) Multi-indicator systems and modelling in partial order. Springer, New York, pp 389–423

Bundesregierung Deutschland Nationale Nachhaltigkeitsstrategie Fortschrittsbericht (2012) http://www.bundesregierung.de/Webs/Breg/DE/Themen/Nachhaltigkeitsstrategie/1-die-nationale-nachhaltigkeitsstrategie/nachhaltigkeitsstrategie/_node.html;jsessionid=AB7764D74BA79942AF3B8D330034811D.s3t2

Carlsen L, Bruggemann R (2013) An analysis of the "failed states index" by partial order methodology. J Soc Struct 14

Federal Ministry for Economic Cooperation and Development (2016) http://www.bmz.de/en/

Hilckmann A (2015) Partial Order als Tool zur Analyse von Multiindikatorsystemen am Beispiel der Anwendung auf die Kernindikatoren der Nachhaltigkeit, Masterthesis

Landesamt für Natur und Verbraucherschutz (2015) Länderinitiative Kernindikatoren—LIKI, http://www.lanuv.nrw.de/liki/

Landesamt für Statistik (2015) Bevölkerungsstand 2015, https://www.statistik.bayern.de/statistik/bevoelkerungsstand/

Landesregierung S-H (1946) Landesportal Schleswig Holstein—Inhalte—Schleswig-Holstein—Ministerpräsidenten seit

Saxl J (1995) Discrete Mathematics

Statistik Nord (2015) Gemeinsames Datenangebot der Statistischen Ämter des Bundes und der Länder, http://www.statistik-portal.de/Statistik-Portal/

United Nations (1992): Earth summit—Agenda 21, ISBN: 978-92-1-100509-7

Part IV
New Applications of Partial Order Theory

A Matching Problem, Partial Order, and an Analysis Applying the Copeland Index

Rainer Bruggemann, Peter Koppatz, Frauke Fuhrmann, and Margit Scholl

1 Introduction

Often the problem is how needs on one side and offers on the other side can be assigned to each other in order to fulfill some optimization constraints. Sometimes, needs and offers, respectively, can be described by sets of indicators and depending on the values of the indicators the best assignment is of interest. This problem is called a matching problem. Table 1 offers some examples.

Both needs and offers are described by a multi-indicator system (MIS), and it is assumed that there is a finite set of needs and a finite set of offers. Therefore, not only one vector describes the need and one vector describes the offer but there are sets of vectors and the problem is which two vectors fulfill best the needs and offers. As an example, the project iBaMs—Barrier-Reduced Machines in Innovative Interaction—is taken (see also Table 1). It was aimed at promoting social inclusion for people with intellectual disabilities and their integration into labor markets and everyday activities, especially as to how far user interfaces of computer-assisted machines fulfill the needs and abilities of the employees and the supervising people. As details of the iBaMs project are explained in other papers (Wiesner-Steiner et al. 2014; Bruggemann et al. 2015; Fuhrmann et al. 2016), the paper here focuses on some mathematical aspects of the matching problem.

R. Bruggemann (✉)
Department of Ecohydrology, Leibniz—Institute of Freshwater Ecology and Inland Fisheries, Müggelseedamm 310, D-12587 Berlin-Friedrichshagen, Germany
e-mail: brg_home@web.de

P. Koppatz
Sudile Gbr, Jägerstr. 36, D-14467 Potsdam, Germany

F. Fuhrmann • M. Scholl
Technical University of Applied Sciences Wildau, Hochschulring 1, D-15745 Wildau, Germany

M. Fattore, R. Bruggemann (eds.), *Partial Order Concepts in Applied Sciences*,
DOI 10.1007/978-3-319-45421-4_16

Table 1 Examples of matching problems

Problem	Indicator set 1	Indicator set 2	Remarks
Marriage problem	Abilities, preferences of person 1	Abilities, preferences of person 2	See also Hall's marriage theorem (Clark and Holton 1994)
Science enterprises	Project data of scientific institutions	Needs of enterprises	
Logistic	Transport needs	Offers of logistic enterprises	
Human capital	User interfaces of CNC-machines	Skills of employees	See iBaMs-project (Wiesner-Steiner et al. 2014)

2 Materials and Methods

We introduce two sets, A and B, elements of A are denoted as a, elements of B are denoted as b. Both sets are considered as finite; hence, we define $|A| = n_A$ and $|B| = n_B$. As mentioned above, we assume that a and b can be characterized by a multi-indicator system (MIS); hence, any element of A is described by the tuple $q(a)$ and that of B by the tuple $q(b)$:

$a \rightarrow q(a) = (qa_1, qa_2, \ldots, qa_{ma})$ and $b \rightarrow q(b) = (qb_1, qb_2, \ldots, qb_{mb})$. The numbers ma and mb are not necessarily the same. The components of $q(a)$, $q(b)$, etc. are called indicators.

Irrespective of which nature A and B have, there are five steps to find a best matching between an element of A and an element of B:

1. A ranking of n_A tuples is to be performed to find the relative best tuples associated with A. Similarly, a ranking of n_B tuples is convenient to reduce the matching problem as far as possible. As often constraints are to be taken into regard, not necessarily the top ranked elements can be selected. Therefore, the ranking procedure has to be generalized in two aspects:

   (a) A ranking should be performed without additional knowledge, hence many Multi-Criteria-Decision-Aid (MCDA) systems are not suitable because, for example, weights are to be determined.

   (b) The procedure to rank tuples of A and B, respectively, must be flexible enough to regard constraints. Therefore, the concept of heterarchies becomes popular and can successfully be realized by tools offered by the theory of partially ordered sets (Bruggemann et al. 2015; Büschenfeldt and Scholl 2013).

2. A subset of tuples of A and of B (due to step 1) must be combined and optimal pairs of tuples $(q_A(a), q_B(b))$, $a \in A$, $b \in B$ are to be identified. The usual method of constructing a goal function, where the components of q(a) and $q(b)$ are

numerically combined is, in general, not possible because often the scaling level of the indicators is just qualitative and does not allow a combination by field operations as known in conventional arithmetic.

3. A decision for the indicators qa_i and qb_j is needed as to how far they fit. Here, a matrix M is to be defined as follows:

$$M(i,j) = \begin{cases} 1 & qa_i \text{ can contextually be combined with } qb_j \\ 0 & \text{otherwise} \end{cases} \tag{1}$$

   Once again the project iBaMs delivers examples: A user interface of a CNC-machine may beside others be described by an indicator for optical design. The tuple of a worker may consist of an indicator describing the haptical ability. These two indicators cannot contextually be combined as they do not appropriately correspond with each other.

   Clearly, the amount of a good fit could be expressed by fractions taken from the closed interval [0,1]; however, the project iBaMs showed that the crisp decision (yes/no) for any pair of indicator qa_i from A and qb_j from B seems to be sufficient (Wiesner-Steiner et al. 2014; Fuhrmann et al. 2016).

4. With help of M (1), we define a triple and set:

$$\left(qa_i, M(i,j), qb_j\right) = \begin{cases} \left(qa_i, qb_j\right) & \text{if } M(i,j) = 1 \\ (0,0) & \text{otherwise} \end{cases} \tag{2}$$

   When there are ma components of $q(a)$, $a \in A$, and mb components of $q(b)$, $b \in B$, then we arrive at $ma \times mb$ pairs, i.e. the evaluation of (2) leads to a tuple having $2 \times ma \times mb$ elements for each pair (a,b).

5. Find the optimal combination of a tuple $q(a)$ and $q(b)$. A simple way would be just to sum up the components. However, due to the ordinal character of the elements of the two tuples, this kind of operation is not allowed. Instead the algorithm of the Copeland index (Al-Sharrah 2010; Saari and Merlin 1996) is applied. Here, it is only counted how often a certain pair $(q(a),q(b))$ is better than a pair $(q(a'),q(b'))$, where $a \neq a'$ or $b \neq b$'.

3 Example

The example was developed under the premises of the project iBaMs. Here however, beside the configurations selected (see below) we do not refer to that project. See for details (Wiesner-Steiner et al. 2014; Bruggemann et al. 2015; Fuhrmann et al. 2016). The number of tuples of A and B may be an outcome of an a priori ranking as described in step 1.

The configurations are as follows:

$$n_A = 4, \; ma = 4$$
$$n_B = 3, \; mb = 3$$

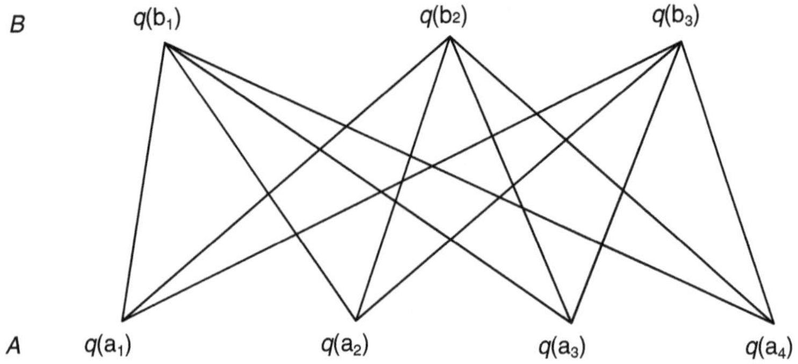

Fig. 1 Without any knowledge any tuple $q(a)$ of A may be combined with any tuple $q(b)$ of B

The coincidences of n_A and ma on the one hand and of n_B and mb on the other hand are artificial. The most general situation for a matching is shown in Fig. 1, where four elements of A, $\{a_1, a_2, a_3, a_4\}$ are combined with three elements of B, $\{b_1, b_2, b_3)$. The matching will be performed on the basis of the two MIS; therefore, in Fig. 1 the tuples $q(a_i)$ and $q(b_j)$ are inserted:

Figure 1 shows that 3 × 4 pairs can be found. To sharpen this result, we perform the remaining four steps, mentioned above. Now, it is needed to specify the seven tuples of A and B, respectively: The numbers are arbitrarily found and could be realistic within the project iBaMs. Note the numbers 0, 1, 2, 3 are thought of as an ordinal representation of qualitative data or even of linguistic expressions.

$$
\begin{aligned}
q(a_1) &= (2, 1, 1, 2) \\
q(a_2) &= (1, 2, 1, 2) \\
q(a_3) &= (1, 0, 3, 1) \\
q(a_4) &= (0, 0, 3, 3) \\
q(b_1) &= (1, 0, 0) \\
q(b_2) &= (0, 1, 0) \\
q(b_3) &= (0, 0, 1)
\end{aligned}
$$

As matrix M the following assignments are done:

$$
M = \begin{pmatrix} 1\,0\,0 \\ 0\,1\,0 \\ 0\,1\,1 \\ 1\,1\,0 \end{pmatrix}
$$

that is $M(1,1)$ the first indicator of $q(a)$ fits well with the first indicator of $q(b)$,

$M(1,2)$ the first indicator of $q(a)$ does not fit well with the second indicator of $q(b)$,

$M(2,1)$ the second indicator of $q(a)$ does not fit well with the first indicator of $q(b)$.

This assignment in a concrete situation is extremely difficult and needs an intensive iterative procedure.

The next step is the application of M as an operator on the two tuples, one arising from A and one arising from B (2). We arrive at $n_A \times n_B$ tuples, having $2 \times ma \times mb$ components. These tuples we denote as $q(ab)$ (or if a specification is needed $q(a_i b_j)$) and call them “pair’s tuple”:

2.0	1.0	0.0	0.0	0.0	0.0	0.0	0.0	1.0	0.0	0.0	0.0	0.0	0.0	1.0	0.0	1.0	0.0	2.0	1.0	2.0	0.0	0.0	0.0
2.0	0.0	0.0	0.0	0.0	0.0	0.0	0.0	1.0	1.0	0.0	0.0	0.0	0.0	1.0	1.0	1.0	0.0	2.0	0.0	2.0	1.0	0.0	0.0
2.0	0.0	0.0	0.0	0.0	0.0	0.0	0.0	1.0	0.0	0.0	0.0	0.0	0.0	1.0	0.0	1.0	1.0	2.0	0.0	2.0	0.0	0.0	0.0
1.0	1.0	0.0	0.0	0.0	0.0	0.0	0.0	2.0	0.0	0.0	0.0	0.0	0.0	1.0	0.0	1.0	0.0	2.0	1.0	2.0	0.0	0.0	0.0
1.0	0.0	0.0	0.0	0.0	0.0	0.0	0.0	2.0	1.0	0.0	0.0	0.0	0.0	1.0	1.0	1.0	0.0	2.0	0.0	2.0	1.0	0.0	0.0
1.0	0.0	0.0	0.0	0.0	0.0	0.0	0.0	2.0	0.0	0.0	0.0	0.0	0.0	1.0	0.0	1.0	1.0	2.0	0.0	2.0	0.0	0.0	0.0
1.0	1.0	0.0	0.0	0.0	0.0	0.0	0.0	0.0	0.0	0.0	0.0	0.0	0.0	3.0	0.0	3.0	0.0	1.0	1.0	1.0	0.0	0.0	0.0
1.0	0.0	0.0	0.0	0.0	0.0	0.0	0.0	0.0	1.0	0.0	0.0	0.0	0.0	3.0	1.0	3.0	0.0	1.0	0.0	1.0	1.0	0.0	0.0
1.0	0.0	0.0	0.0	0.0	0.0	0.0	0.0	0.0	0.0	0.0	0.0	0.0	0.0	3.0	0.0	3.0	1.0	1.0	0.0	1.0	0.0	0.0	0.0
0.0	1.0	0.0	0.0	0.0	0.0	0.0	0.0	0.0	0.0	0.0	0.0	0.0	0.0	3.0	0.0	3.0	0.0	3.0	1.0	3.0	0.0	0.0	0.0
0.0	0.0	0.0	0.0	0.0	0.0	0.0	0.0	0.0	1.0	0.0	0.0	0.0	0.0	3.0	1.0	3.0	0.0	3.0	0.0	3.0	1.0	0.0	0.0
0.0	0.0	0.0	0.0	0.0	0.0	0.0	0.0	0.0	0.0	0.0	0.0	0.0	0.0	3.0	0.0	3.0	1.0	3.0	0.0	3.0	0.0	0.0	0.0

The first row (pair’s tuple $q(a_1 b_1)$ describes the outcome of the operator for $q(a_1)$ with $q(b_1)$, the second row for $q(a_1)$ with $q(b_2)$ until we arrive at $q(a_4)$ combined with $q(b_3)$, according to (2).

The last step includes the formalism of the Copeland index. The results are shown in Table 2.

Table 2 The 12 combinations of A- and B-tuples are shown evaluated due to the Copeland procedure. Note: we are using a simplified notation $a_1 b_2$ for expressing that we consider the pair’s tuple $q(a_1,b_2)$

Combination	Copeland
$a_1 b_1$	0
$a_1 b_2$	12
$a_1 b_3$	−12
$a_2 b_1$	−3
$a_2 b_2$	9
$a_2 b_3$	−15
$a_3 b_1$	−12
$a_3 b_2$	0
$a_3 b_3$	−24
$a_4 b_1$	15
$a_4 b_2$	27
$a_4 b_3$	3

Due to the numbers in Table 2, we arrive at a certain sequence (weak order):

$$\begin{aligned}&(a_4b_2) >> (a_4b_1) > (a_1b_2) > (a_2b_2) > (a_4b_3) >> (a_1b_1)\\ &\cong (a_3b_2) > (a_2b_1) > (a_1b_3)\\ &\cong (a_3b_1) > (a_2b_3) > (a_3b_3).\end{aligned} \tag{3}$$

Several principles can be helpful to select a suitable subset from the ranking sequence (3):

1. All tuples of A must appear at least once in the subset of pairs.
2. All tuples of B must appear at least once in the subset of pairs.
3. The values of the Copeland index may be considered as random numbers. Therefore, the set of values constitutes a distribution. Then, the percentiles of the distribution may be a guiding information.

Here we follow advice 2, because in iBaMs it is necessary to find for all tuples of B (describing the skill profiles of the people with intellectual disabilities) a good counterpart (a good user interface for CNC-machines). Therefore, we selected the first three pair's tuples from the top.

$$b1 - a4,\ b2 - a1,\ b3 - a4.$$

This constellation can be graphically summarized by a matching diagram (Fig. 2).

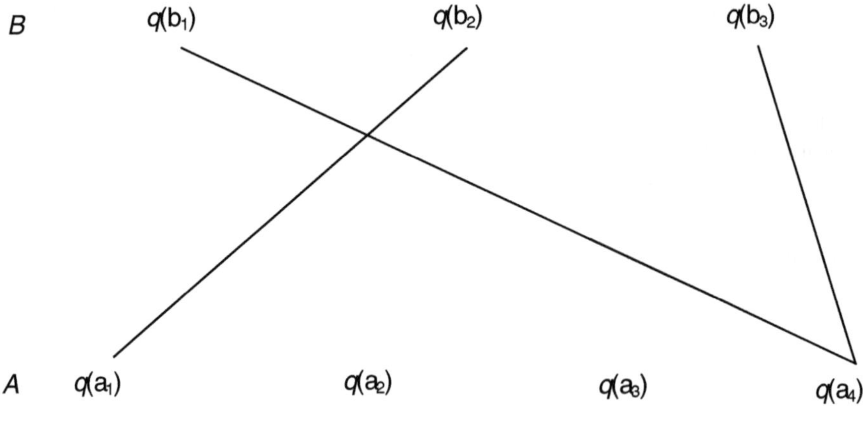

Fig. 2 After the selection of the three best in the above sequence (3) a recommendation within the project iBaMs could be given. Combine employees with the ability tuples $q(b_1)$ and $q(b_3)$ with the user interface $q(a_4)$ and those characterized by $q(b_2)$ with the user interface $q(a_1)$

4 Discussion

The ranking of the tuples of A and those of B was thought of as being performed by tools of partial order theory. Why do we not apply partial order once again to find a ranking for the pairs ($q(a)$,$q(b)$)? The problem is that we are faced with many components, namely $2 \times ma \times mb$. As an increasing number of indicators will increase the number of incomparabilities, it is to be assumed that the result would be an antichain. This kind of situation appears fairly often in applications of partial order tools, and much research was done to improve such a situation (Bruggemann and Voigt 2012; Bruggemann and Carlsen 2014, 2015a, b). Nevertheless, the recommended analysis is time-consuming and still troublesome. Therefore, an outranking method such as that proposed by Copeland (Al-Sharrah 2010) was applied. However, when we leave the concept of partial order then clearly other MCDA methods could be used too. The crucial points, however, are as to how far we can find numerical values for all the needed additional parameters and any weighting would be extremely difficult.

Another critical point is the construction of the matrix M. Within the project iBaMs, it was thought of as an outcome of an expert decision. In that specific situation, the scientists involved in that project knew both the background information for A and that for B. However, in general, neither A nor B is well known. The only premise which must be fulfilled is that the side delivering information about A can define an MIS, as well as that side, which is responsible for B. The indicators qa_i and qb_j may be formulated on a linguistic basis. Therefore, the construction of a matrix such as M needs identification as to how far concepts of A cover those of B and vice versa. A scientific project is planned to find details in this case. Beyond this, in the matching procedure discussed above, the entries of M can be replaced by fractions and these fractions can be used to feed a generalized Copeland procedure, as was suggested by Al-Sharrah (2010).

Finally, the selection principles as formulated in Sect. 3 may still have to fulfill some additional constraints. For example, when most tuples $q(a)$ can be related to just one tuple $q(b)$, then it is a matter of economics whether or not it is necessary to include a second tuple $q(b)$. For example, in the project iBaMs it seems to be improbable that a second user interface will be installed, as was found in Fig. 2.

5 Conclusion

A first concept for a matching was discussed and as the discussion (Sect. 4) has shown there are still many open points whose clarification needs urgently new scientific projects. Nevertheless, when

- two MIS can be established and
- a matrix M can be formulated,

a procedure can be suggested to find a good matching, which is flexible enough to take additional constraints into account.

Another important point is that, albeit that the calculations can easily be done manually, the number $2 \times ma \times mb$ and n_A and n_B can be large and the calculations could be prone to errors and will be tedious. Therefore, a software is needed and within the project iBaMs a software was developed which is accessible by the internet and which will be now continuously improved and extended. This software is based on PyHasse (Bruggemann et al. 2014) and some modules are already available. The interested reader may visit www.phyhasse.org

References

Al-Sharrah G (2010) Ranking using the Copeland score: a comparison with the Hasse diagram. J Chem Inf Model 50(5):785–791

Bruggemann R, Carlsen L (2014) Incomparable-What now? MATCH Commun Math Comput Chem 71:699–714

Bruggemann R, Carlsen L (2015a) Incomparable: what now II? Absorption of incomparabilities by a cluster method. Qual Quant 49:1633–1645

Bruggemann R, Carlsen L (2015b) Incomparable – what now III. Incomparabilities, elucidated by a simple version of ELECTRE III and a fuzzy partial order approach. MATCH Commun Math Comput Chem 73:277–302

Bruggemann R, Voigt K (2012) Antichains in partial order, example: pollution in a German region by Lead, Cadmium, Zinc and Sulfur in the herb layer. MATCH Commun Math Comput Chem 67:731–744

Bruggemann R, Carlsen L, Voigt K, Wieland R (2014) PyHasse Software for partial order analysis: scientific background and description of selected modules. In: Bruggemann R, Carlsen L, Wittmann J (eds) Multi-indicator systems and modelling in partial order. Springer, New York, pp 389–423

Bruggemann R, Edich D, Fuhrmann F, Koppatz P, Scholl M, Teske A, Wiesner-Steiner A (2015) Kann (einfache) Diskrete Mathematik die Lebensqualität von Menschen mit geistiger Behinderung an computergesteuerten Maschinen verbessern helfen? In: Wittmann J, Wieland R (eds) Simulation in Umwelt- und Geowissenschaften, 239–251, Workshop Müncheberg 2015. Shaker, Aachen

Büschenfeldt M, Scholl M (2013) Das Organisationsmodell im Anwendungsmodell: Zauber und Herausforderung des Einsatzes von Web 2.0 Anwendungen in der öffentlichen Verwaltung. In: Schweighofer E, Kummer F, Hötzendorfer W (eds) Transformation juristischer Sprachen, Tagungsband des 16. Internationalen Rechtsinformatik Symposions (IRIS 2013), 21.–23. Februar 2013, Universität Salzburg. books@org.at. Band 292. Österreichische Computer Gesellschaft, Wien, pp 187–194

Clark J, Holton DA (1994) Graphentheorie: Grundlagen und Anwendungen. Spektrum Akademischer Verlag, Heidelberg

Fuhrmann F, Scholl M, Bruggemann R (2016), submitted. How can the empowerment of employees with intellectual disabilities be supported? Special Issue in Social Indicator Research: "Dealing with Complexity in Society: from Data to Synthetic Indicators"

Saari D, Merlin V (1996) The Copeland method I. Relationships and the dictionary. J Econ Theory 8:51–76

Wiesner-Steiner A, Teske A, Fuhrmann F, Scholl M (2014) Preconditions and design requirements for touch-screen control panels for intellectually disabled persons working with CNC machines. In: MCCSIS, Multiconference on Computer Science and Information Systems, 15–19, Juli Lissabon, Portugal, 2014, Proceedings of the international conferences interfaces and human computer interaction 2014, Game and Entertainment Technologies 2014, and Computer Graphics, Visualization, Computer Vision and Image Processing 2014, pp 54–61

Application of the Mixing Partial Order to Genes

William Seitz, Krunoslav Brčić-Kostić, Petar T. Mitrikeski, and Patricia Seitz

1 Introduction

The genes are basic functional units of heredity. In a narrow sense these are the stretches of DNA which code for proteins. Each gene product (protein or peptide) exhibits a biological function that is reflected in one or more phenotypic traits. Some genes affect a single trait, whereas others simultaneously affect many traits. The phenomenon that a single mutation within a gene affects two or more traits is called pleiotropy. Some traits are affected by a single gene (Mendelian genetics); whereas, others are affected by many genes (quantitative genetics). The most important phenotypic trait concerning evolution is fitness since it is a primary target for natural selection. Fitness is the most complex quantitative trait that is affected by all functional genes within a genome. DNA sequence change (molecular evolution) is always ongoing, though its persistence results when such changes coincide with an increase in fitness by natural selection. The molecular evolution can also be the result of fixation of neutral alleles by random genetic drift. Some genes evolve

W. Seitz (✉)
Department of Marine Sciences, Texas A&M University at Galveston, Galveston, TX, USA
e-mail: seitzw@tamug.edu

K. Brčić-Kostić
Institute Ruđer Bošković, Zagreb, Croatia
e-mail: brcic@irb.hr

P.T. Mitrikeski
Institute Ruđer Bošković & Institute for Research and Development of Sustainable Ecosystems, Zagreb, Croatia
e-mail: yeastsman@yahoo.com

P. Seitz
University of Texas Medical Branch, Galveston, TX, USA
e-mail: pkseitz@hotmail.com

M. Fattore, R. Bruggemann (eds.), *Partial Order Concepts in Applied Sciences*,
DOI 10.1007/978-3-319-45421-4_17

rapidly, whereas the others are more conserved and evolve at a slower rate. The degree of gene conservation (or the selective pressure acting on a gene) can be estimated by d*N*/d*S* ratio, namely, the ratio of nonsynonymous substitution rate and synonymous substitution rate.

Genes can be classified by various criteria based on gene ontology annotations including biochemical activity of their products; their biological function; and the location of their function in cells, tissues, or organs. Also, important aspects of gene classification are their phenotypic effects, genetic interactions (epistasis), pleiotropy, gene conservation, etc. Here we propose a new gene property by applying the concept of mixing, and its corresponding partial order, to genes.

To clarify our notation, we briefly review some well-known facts regarding genes. In DNA, genetic information (code) is made up of codons which in turn consist of triples of nucleotides which are specified by the base they carry: Adenine (A), Cytosine (C), Thymine (T), and Guanine (G). Thus, there are a total of 64 possible triples (codons), three of which are so-called stop codons. Each gene consists of a set of codons contained between a start codon at the beginning and a stop codon at the end of the gene sequence. Thus, a gene is a sequence of codons. A gene can have significantly different total numbers of codons—so that (excluding the three stop codons) there is a distribution of the 61 codons that make up a gene. If a gene has N codons, then its "codon distribution" is a set, vector, or diagram $\{\lambda\} = [\lambda_1, \lambda_2, \lambda_3, \ldots, \lambda_v]$, where λ_i is the number of codons of type i in the gene and the λ_i are ordered from most frequent to least frequent (highest to lowest).

Mixing is a fundamental property of sets of objects. Mixing was introduced as being fundamental by Ernst Ruch in 1975 (Ruch 1975), where he proved that mixing is a partial order, namely, the majorization partial order for integers. Quoting his paper; "When we classify a set of N objects according to some principle, we obtain a subdivision into subsets without objects in common. Independently of the nature of the classification principle, we can represent the set structure of this subdivision by a diagram, if, for example, we consider the rows as representing the numbers of equivalent objects." Here we apply the concept of mixing to gene codon distributions as a new fundamental property of genes, namely, characterization by mixing.

2 A Short Digression on Notation and Diagrams

Following Ruch, we consider vectors $\{\lambda\} = [\lambda_1, \lambda_2, \lambda_3, \ldots, \lambda_v]$ where $\lambda_i \geq \lambda_j$ for $i<j$. Such vectors may be represented as diagrams that consist of rows of boxes, the top row containing λ_1 boxes, the second row containing λ_2 boxes, etc., with the length of the rows decreasing. The Young Diagram (YD) was introduced in 1900 by Cambridge University Mathematician Alfred Young (1900). It consists of a finite collection of boxes arranged in left justified rows such that, proceeding from the top to the bottom of the diagram, successive rows have equal or fewer boxes (Fig. 1). A Young Diagram may be regarded as a *picture with rules*.

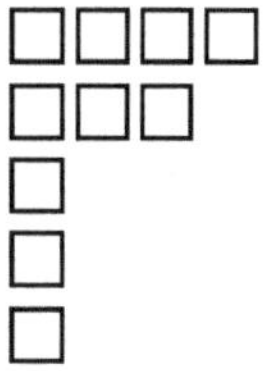

Fig. 1 A Young Diagram for ten objects corresponding to the vector [4,3,1,1,1]

Depending on the rules, diagrams can provide insights into many problems from transportation (Königsberg bridge) to quantum electrodynamics (Feynman diagrams). Young Diagrams are not numbers—they are the "mixing character" of the diagram.

A "diagram" widely used in many chapters in this book is known as a Hasse Diagram. Hasse diagrams depict partial orders as lattices and usually have nodes that are *incomparable* with one another. While the terminology is unfortunate (and confusing) it is important to realize that Young Diagrams and Hasse Diagrams define the word "diagram" completely differently—and in fact the set of all Young Diagrams $\{\lambda\}$ is partially ordered by mixing and a lattice of Young Diagrams represents the mixing partial order. This lattice is, in turn, a Hasse diagram.

To further elaborate, since Young Diagrams are not numbers, it is entirely possible that they cannot be ordered numerically. But there may still be a *partial ordering*. For example, in general, if we consider three Young Diagrams, A, B, and C, it is possible that A and B both "exceed" C, but A and B cannot be compared. As an example, 12 diagrams in Fig. 2 containing stars are incomparable to the diagram containing circles by mixing. The fundamental point in Ruch's seminal paper is that the general concept of mixedness of integer partitions is equivalent to a specific partial order of partitions, namely, the majorization partial order. We will use this theorem to classify the mixing character of genes. Figure 2 shows the full Hasse diagram for all possible mixing characters for ten objects (a "diagram of diagrams").

3 The Young Diagram Lattice and Mixing

A lattice is a set of nodes with connections between them (here we won't consider lattices with isolated nodes).[1] In the Young Diagram Lattice there is only one connection between nodes, and the connection is directed.[2] Using such diagrams, mathematicians have developed formulas for the dimensions of the irreducible

[1] Diagrams are generally rendered in two dimensions, and while they can be "viewed" in higher dimensions they are only properly described as abstract mathematical objects.

[2] People familiar with Geographic Information Systems will recognize that directed graphs underlie *ArcInfo* files while *shapefiles* are undirected.

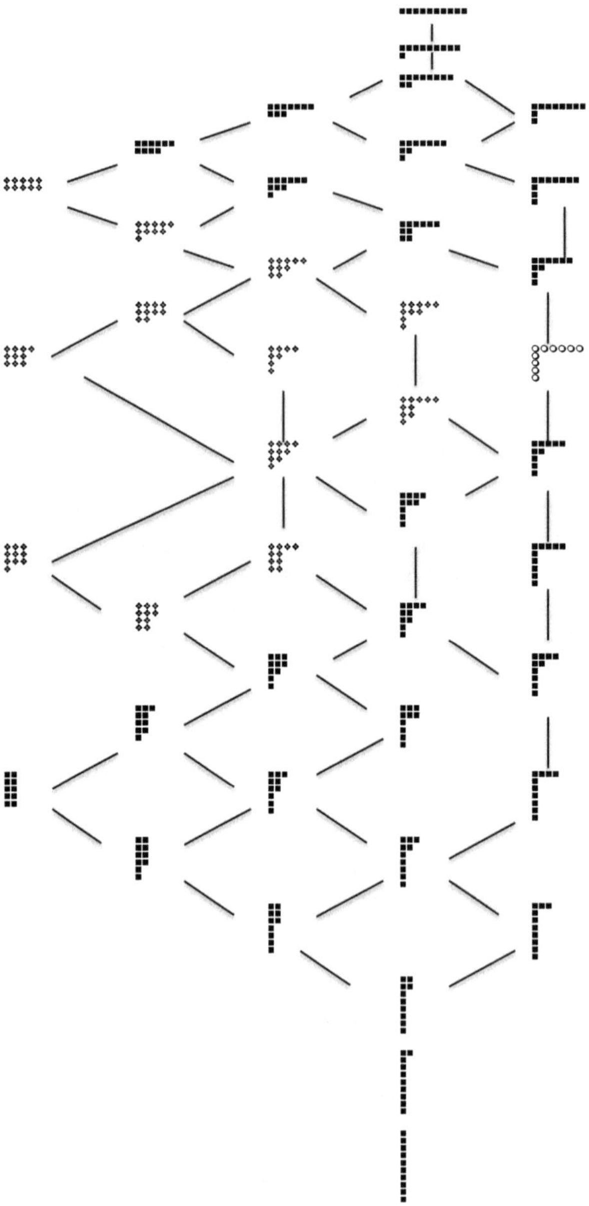

Fig. 2 The Young Diagram Lattice for ten objects. It is a Hasse diagram where each node is a Young Diagram

representations of the symmetric group S_N (Rutherford 1948) and Matsen (1975) also related the Young Diagrams to the Gelfand states of the unitary group U_N. However, these applications are not developed further here.

A key observation is that Young Diagrams are in one-to-one correspondence with the partitions of integers. For $N = 5$ the partitions are

$$\{[5];\ [4,1];\ [3,2];\ [3,1,1];\ [2,2,1];\ [2,1,1,1];\ [1,1,1,1,1]\}$$

(Notes on notation: (a) the partition represents the partition of 5 into [5,0,0,0,0]; (b) the partitions [3,1,1] and [1,1,1,1,1] can be written $[3,1^2]$ and $[1^5]$. Another common representation (Ferrers' Diagram|Mathematics (n.d.)), called Ferrer's Diagrams, substitutes dots for boxes in Fig. 1.)

The number of partitions, IP(N), grows rapidly with N. For example, there are 3,972,999,029,388 partitions for $N = 200$ and more than 10^{31} for $N = 1000$. In the limit of large N, Hardy and Ramanujan obtained an asymptotic formula for the number of partitions of the integer N (Hardy and Ramanujan 1918)

$$\mathrm{IP}(N) \approx \frac{1}{4N\sqrt{3}}\, \mathrm{e}^{\pi\sqrt{\frac{2N}{3}}} \tag{1}$$

The set of partitions of integers can be ordered by various means and also can be partially ordered. An example of an ordering of partitions by their (Boltzmann) entropy is defined as follows. Let $[\lambda_1, \lambda_2, \lambda_3, \ldots, \lambda_N]$ be a partition, then for an isolated system the Boltzmann–Gibbs–Shannon (Wikipedia 2015) (BCS) entropy is given by:

$$S = -\sum_{i=1}^{N} p_i \ln\, p_i \tag{2}$$

where p_i is the probability that the system is in state i, namely, λ_i/N. The entropy, so defined, provides an order for YDs, i.e., the diagrams are ordered according to increasing entropy. (Following Shiner et al. (1999) we have normalized the entropy in (2) by dividing by $k \ln N$ where k is Boltzmann's constant.) For smaller N, the normalized entropy of a macrostate $[\lambda]$ is directly computed by the Boltzmann equation

$$S_{[\lambda]} = \frac{1}{\ln\, N!} \ln\left(\frac{N!}{\prod_i \lambda_i!}\right) \tag{3}$$

Consequently, the partition $[N]$ has entropy 0 and the partition $[1^N]$ has entropy 1. Clearly, if the set $[\lambda_1, \lambda_2, \lambda_3, \ldots, \lambda_v]$ is the codon distribution for a gene, then $S_{[\lambda]}$ would represent the Boltzmann entropy of the gene.

The Young Diagram Lattice is a Hasse diagram that reflects the majorization or mixing partial order of the integer partitions. The lattice for $N = 10$ is shown in Fig. 1. The majorization partial order for integers can be stated mathematically as follows (Marshall and Olkin 1979). Consider the partition $\lambda = [\lambda_1, \lambda_2, \lambda_3, \ldots, \lambda_v]$ of N objects where λ_i is the number of objects of type i (or class i). In general, we can take $\lambda_i \geq \lambda_j$ if $i < j$ and clearly

$$\sum_{i=1} \lambda_i = N \tag{4}$$

The statement of the majorization (or dominance) partial order is that a partition λ is placed above (or exceeds) another partition μ if and only if

$$\sum_{i=1}^{m} \lambda_i \geq \sum_{i=1}^{m} \mu_i \quad \forall \ m = 1 \text{ to } N \tag{5}$$

Ruch (1975) proved that the majorization partial order is the same as the partial ordering of sets by mixing.[3] The fundamental insight in Ruch's paper is that *mixing character* has no numerical value; rather it can only be represented by a diagram that represents an integer partition. [4]

As noted earlier, the Young Diagram Lattice is a specific and fundamental example of a Hasse partial order diagram. Later in the discussion section of this paper we will argue that sets of nodes in any Hasse diagram (such as those studied by Bruggemann and others) that are fundamentally incomparable to others have much science (and economics, political science, etc.) to reveal.

Following we show how codon distributions, when viewed as integer partitions, can lead to a new characterization of genes according to mixing. In earlier work, Seitz and Kirwan (2014) have argued that such a characterization complements entropy, so that mixing is complementary to disorder. Here we will only develop an example for a small number of genes of—in this case—the rainbow trout. Since the purpose here is to illustrate how genes can be characterized by mixing, we consider only 15 genes rather than attempting a full study of the genome. An exhaustive study on the subject is ongoing in our laboratories.

---

[3]Recent attention has been drawn to mixing with regard to molecular genetics (Wan and Wootton 2000) and recently to quantum entanglement (Datta and Vidal 2007) as well.

[4]Ruch also suggested that the approach to equilibrium in physical systems proceeds according to increasing mixing character (as determined by the majorization partial order)—this is a stronger requirement than increase of entropy alone, though entirely consistent with the physics of thermodynamics. As Ruch points out "…evolution toward equilibrium is not determined by a single function such as the entropy, but rather by a "quality"—the mixing character—that is determined by the majorization partial ordering. It should be noted that second law of thermodynamics discussions using arguments that refer to entropy are unaffected by the principle of increasing mixing character."

4 Rainbow Trout Gene Example

Table 1 shows the random sample of 15 genes in the rainbow trout investigated here. The first column is the ID number (arbitrarily assigned to the genes studied), and the rest present additional useful information.[5]

Note that each gene can be viewed as a vector of 64 numbers the values of which are the frequency of occurrence of the codons in the gene. The codons are ordered

CGA CGC CGG CGT AGA AGG CTA CTC CTG CTT TTA TTG TCA TCC TCG TCT AGC AGT ACA ACC ACG ACT CCA CCC CCG CCT GCA GCC GCG GCT GGA GGC GGG GGT GTA GTC GTG GTT AAA AAG AAC AAT CAA CAG CAC CAT GAA GAG GAC GAT TAC TAT TGC TGT TTC TTT ATA ATC ATT ATG TGG TAA TAG TGA

Table 1 Gene products for 15 genes of the rainbow trout (see: http://www.uniprot.org/uniprot/)

ID	Swiss-Prot identifier	Protein name(s)	Gene name(s)	Length of the canonical sequence (amino-acids)
1.	O73631	TFZR1	/	447
2.	O57395	SOX9	/	488
3.	Q9YGF8	Retinal arylalkylamine *N*-acetyltransferase	/	240
4.	O57396	RtSox23	/	585
5.	O57397	Nuclear pore complex glycoprotein p62	/	579
6.	O57399	RtSox24*	GSONMT00079394001	367
7.	Q9TNN8	MHC class I (MHC class I heavy chain)	OnmyUAA-OSU	351
		(MHC class IA antigen)	Onmy-UBA	
		(MHC class Ia heavy chain)	Onmy-UBA*1	
8.	Q9TNX1	MHC class I	Onmy-UAA*11	362
9.	O93244	Androgen receptor alpha	AR-alpha	854
10.	O93245	Androgen receptor beta*	AR-beta	853
11.	Q9YGF7	Embryonic alpha-type globin*	GSONMT00041146001	143
12.	Q9YGF6	Embryonic alpha-type globin2*	GSONMT00009487001	143
13.	Q9YGF5	Embryonic beta-type globin*	GSONMT00009490001	147
14.	Q9YGF4	Embryonic beta-type globin2*	GSONMT00056958001	147
15.	Q9YGF3	Dynein light chain 2*	DLC2	169

*Uncharacterized protein

[5] The codon distribution was obtained from the international DNA sequence databases: status for the year 2000. Nakamura, Y., Gojobori, T. and Ikemura, T. (2000) Nucl. Acids Res. 28, 292.

Table 2 Distributions of codons for the first six genes

• Product = "TFZR1"
6 5 4 1 2 10 2 12 7 2 2 3 3 2 1 2 2 4 7 13 6 7 7 5 6 6 6 10 3 8 8 14 7 3 0 4 10 5 17 35 5 8 2 21 5 4 8 26 23 14 11 7 2 11 9 8 1 6 4 11 4 1 0 0
• Product = "SOX9"
2 5 1 2 6 6 1 10 13 1 3 2 4 17 4 9 18 6 4 15 4 2 7 25 6 5 2 9 4 8 6 27 6 9 2 7 13 0 8 16 11 4 7 33 10 10 8 17 21 7 11 8 1 2 11 1 0 13 2 11 5 0 0 1
• Product = "Retinal arylalkylamine *N*-acetyltransferase"
1 7 3 2 4 3 5 5 12 2 3 3 2 5 0 6 2 2 3 3 3 4 2 3 4 8 5 3 1 4 1 7 3 2 1 1 12 3 3 4 2 0 3 11 4 4 4 14 6 2 4 1 3 2 11 3 0 7 3 6 3 1 0 0
• Product = "RtSox23"
6 7 7 2 12 12 3 5 30 4 0 9 5 12 0 10 15 6 3 12 1 7 14 26 3 10 4 18 4 3 13 13 17 7 0 9 11 3 12 14 19 1 5 50 14 8 6 37 12 7 10 4 4 2 5 5 4 12 2 23 6 0 1 0
• Product = "Nuclear pore complex glycoprotein p62"
1 4 0 0 2 6 4 9 21 7 3 2 9 8 1 12 10 11 23 23 3 22 13 12 4 15 19 30 3 28 24 12 18 13 7 7 8 4 11 8 12 3 13 22 3 2 5 19 12 3 2 1 0 2 13 16 1 7 9 14 3 0 0 1
• Product = "RtSox24"
2 0 4 1 5 4 2 2 8 4 3 6 6 10 4 10 10 10 8 8 1 6 6 5 8 4 7 6 7 7 3 2 1 3 1 2 11 2 17 15 7 6 3 3 7 0 16 19 19 7 8 4 3 1 9 6 1 7 4 12 4 0 1 0

with the last three being the so-called stop codons. Traditionally, the genetic code is represented by the RNA codon table. However, since genes are usually discovered at the DNA level, DNA codon table is more useful and, therefore, we prefer to use it here.

Table 2 presents the distribution for the first six of these genes.

The distributions in Table 2 can be reorganized and expressed as the partitions of N for each gene where N is the sum of the codon frequencies, which is generally different for each gene. Thus to consistently apply the majorization partial order to codon distributions, we must normalize them to $N = 61$; namely, the total number of codons less the 3 stop codons. We choose 61 in order to later compute the incomparability number for each gene with all of the integer partitions for $N = 61$.

5 Numerical Results

While the discussion up until now has involved only integer values, the majorization concept [in (5)] can be applied to vectors with real valued coordinates. The integer partitions in Table 2 can be "normalized" by multiplication by $61/N$ where N is the number of codons in the gene. (In this particular case N is representing the total codon number within a gene rather than the number of codon classes, i.e., 61. So, N here can be, and usually is, different than 61.) For illustration, the normalized first two vectors in Table 2 are presented in Table 3 (where in the table we have rounded off to three significant digits) and sorted them from largest to smallest.

Table 3 Normalized codon distributions for TFZR1 and SOX9

4.78	3.55	3.14	2.87	2.32	1.91	1.91
1.77	1.64	1.50	1.50	1.50	1.36	1.36
1.36	1.23	1.09	1.09	1.09	1.09	1.09
0.96	0.96	0.96	0.96	0.96	0.96	0.82
0.82	0.82	0.82	0.82	0.82	0.68	0.68
0.68	0.68	0.68	0.55	0.55	0.55	0.55
0.55	0.55	0.41	0.41	0.41	0.41	0.27
0.27	0.27	0.27	0.27	0.27	0.27	0.27
0.27	0.14	0.14	0.14	0.00		
4.13	3.38	3.13	2.63	2.25	2.13	2.13
2.00	1.88	1.63	1.63	1.63	1.38	1.38
1.38	1.38	1.25	1.25	1.25	1.13	1.13
1.13	1.00	1.00	1.00	1.00	0.88	0.88
0.88	0.88	0.75	0.75	0.75	0.75	0.75
0.75	0.63	0.63	0.63	0.50	0.50	0.50
0.50	0.50	0.50	0.38	0.25	0.25	0.25
0.25	0.25	0.25	0.25	0.25	0.13	0.13
0.13	0.13	0.13	0.00	0.00		

These vectors represent the "mixing character" of the genes being studied. The partial sums in (5) are then computed and compared. We have done so for the 15 example genes.

The results can be discussed in several ways. First, we list the "incomparabilities" for the genes, where incomparability of a gene is the number of genes in the set of 15 with which it is incomparable. For example, the first gene, whose product is TFZR1, is incomparable to 10 of the 15 genes. The list of incomparabilities for all genes in the example is $\{10\ 7\ 7\ 9\ 8\ 5\ 8\ 8\ 1\ 1\ 4\ 1\ 4\ 4\ 7\}$. Thus, we see that genes 1 and 4 are highly incomparable, while genes 9, 10, and 12 are comparable to all but one of the 15. Further by inspection of the partial sums in (5), we find that genes 9 and 10 are more mixed than the others while 12 is the least mixed. The second way to discuss the results is in terms of which genes a given gene is incomparable to (i.e., the ones that are neither more nor less mixed). These results are given in Table 4.

6 Conclusions and Discussion

We have shown how sets of genes can be compared with one another with regard to their mixing character and that this results in a partial order. But mixing is a fundamental characteristic of any set of objects (in this case genes) characterized by type (in this case codons). It seems worthwhile, therefore, to explore implications of this concept in other areas as well as in future genetic studies. Thus, in general, when any set of objects is partitioned into subsets without objects in common according to any principle, one obtains integer partitions that may be compared via the majorization (or mixing) partial order.

Table 4 Incomparability of genes with one another

Gene ID	IDs of the genes that are incomparable with this gene									
1	2	3	5	6	7	8	11	13	14	15
2	1	3	4	5	7	8	15			
3	1	2	5	6	7	8	15			
4	2	5	7	8	11	12	13	14	15	
5	1	2	3	4	6	7	8	15		
6	1	3	5	7	8					
7	1	2	3	4	5	6	8	15		
8	1	2	3	4	5	6	7	15		
9	10									
10	9									
11	1	4	13	14						
12	4									
13	1	4	11	14						
14	1	4	11	13						
15	1	2	3	4	5	7	8			

Even more generally, since most partial orders (including many in this volume) often (usually) have incomparable members, it seems reasonable to explore the extent of incomparability more deeply. Seitz and Kirwan (2014) have suggested this may be related to complexity of members of the partial order. They also discussed the evolution of systems from low entropy (ordered) to high entropy (disordered) under the assumption that such evolution follows paths on the YDL. This view suggests that complexity is maximized at mid entropy levels. This idea was argued to have qualitative implications in such fields as political science (evolutions of societies and civilizations), economics (managed vs. unmanaged economic systems), etc.

The characterization of genes by mixing is subtle and it remains to be seen if it proves to be of significant biological value. Even considering this small set of genes; we nevertheless do observe that the most mixed genes are those for the androgen receptors which consequently can be compared to all of the other 13 genes in the set. Noting this we also looked at the estrogen receptor genes (not in our random sample) and found that they were less mixed than the androgen receptors. Future work to investigate the set of all genes that comprise a genome is underway in our laboratories.

References

Datta A, Vidal G (2007) Role of entanglement and correlations in mixed-state quantum computation. Phys Rev A 75:042310. doi:10.1103/PhysRevA.75.042310

Ferrers' Diagram|Mathematics (n.d.) Encyclopedia Britannica online. Encyclopedia Britannica. Web. 17 Nov 2015

Hardy GH, Ramanujan S (1918) Asymptotic formulae in combinatory analysis. Proc Lond Math Soc 17:75–115

Marshall AW, Olkin I (1979) Inequalities: theory of majorization and its applications. Academic, New York

Matsen FA (1975) The unitary group formulation of the N-particle problem. Int J Quantum Chem 8:379–388

Ruch E (1975) The diagram lattice as structural principle. Theor Chim Acta (Berl) 38:167–183

Rutherford DE (1948) Substitutional analysis. University Press, Edinburgh

Seitz W, Kirwan AD (2014) Entropy vs. majorization, what determines complexity. Entropy 16:3793–3807

Shiner JS, Davison M, Landsberg PT (1999) Simple measure for complexity. Phys Rev A 59:459–1464

Wan H, Wootton J (2000) A global compositional complexity measure for biological sequences: AT-rich and GC-rich genomes encode less complex proteins. Comput Chem 24:71–94

Wikipedia (2015) Entropy (order and disorder) Wikipedia, the free encyclopedia [Online]. Accessed 11 Feb 2015

Young A (1900) On quantitative substitutional analysis. Proc Lond Math Soc 33:97–146

Analyzing Ethnopharmacological Data Matrices on Traditional Uses of Medicinal Plants with the Contribution of Partial Order Techniques

Stergios Pirintsos, Michael Bariotakis, Danae Laina, Christos Lionis, Elias Castanas, and Rainer Bruggemann

1 Introduction

Ethnobotany is the study of complex relationships between cultures and uses of plants, focusing primary on how plants are used, managed and perceived across human societies, while ethnopharmacology is defined as the interdisciplinary scientific exploration of biologically active agents traditionally employed or observed by man. On one side, ethnobotany is a very broad discipline, interested not only in medicinal plants but also in most natural products derived from plants, such as food, coloring agents, and fibers, while ethnopharmacology is a narrower field. On the other side, however, ethnobotany is only focused on plants, while ethnopharmacology has a broader focus exploring biologically active agents from plants, minerals, animals, fungi, and microbes. The two fields have many interconnections and overlaps and both share the concern on traditional knowledge of human societies.

S. Pirintsos (✉)
Department of Biology, University of Crete, P.O.B. 2208, 71409 Heraklion, Greece

Botanical Garden, University of Crete, Gallos Campus, 74100 Rethymnon, Greece
e-mail: pirintsos@biology.uoc.gr

M. Bariotakis • D. Laina
Department of Biology, University of Crete, P.O.B. 2208, 71409 Heraklion, Greece

C. Lionis
Clinic of Social and Family Medicine, School of Medicine, University of Crete, P.O.B. 2208, 71003 Heraklion, Greece

E. Castanas
Laboratory of Experimental Endocrinology, School of Medicine, University of Crete, P.O.B. 2208, 71003 Heraklion, Greece

R. Bruggemann
Department of Ecohydrology, Leibniz—Institute of Freshwater Ecology and Inland Fisheries, Berlin, Germany

M. Fattore, R. Bruggemann (eds.), *Partial Order Concepts in Applied Sciences*,
DOI 10.1007/978-3-319-45421-4_18

The medicinal uses of plants are better understood if studied as a domain of knowledge embedded in the large body of cultural knowledge, practices, and beliefs of a group (Reyes-Garcia 2010). An important task for ethnopharmacology is to contextualize uses and cultural perceptions of plants. Documentation of the indigenous knowledge through ethnobotanical studies has extensively been attempted on traditional uses of medicinal plants. Several data matrices have been published in many areas all over the world ranging from regional to continental scale (e.g., for Europe see references in Pardo-de-Santayana et al. 2015).

Most ethnopharmacological projects in this direction, whether basic or applied, follow a similar structure. They start from field research among a local community and aim at finding consensus on local disease concepts and the use of specific medicinal plants (Leonti and Weckerle 2015). At this stage, the emic perception, i.e., the perspective of the local people of disease and illness, is of high interest. In the next stage the documented ethnomedical views are usually translated into biomedical disease concepts of other medical and knowledge systems (the etic perception). The emic perspective is generally understood to come from within a culture and is opposed to the etic point of view, which is that of an outsider (Staub et al. 2015).

Although this scheme has repeatedly been followed in many studies, a basic methodological standardization has not yet been achieved since, in practice, there is no consensus on data collection, data analysis, and contextualization process. A notable example could be the botanical aspects (Bennett and Balick 2014; Bussmann 2015), where all the plants mentioned in the data set need to be documented and validated with taxonomic keys and herbarium specimens and every plant name in the data needs to be connected to a specific voucher, which needs to be deposited at an internationally recognized herbarium. Rivera et al. (2014) reported that among the 428 reviewed articles, 308 articles cited plant names incorrectly.

Another example is the disease classification system. Different standard disease classification systems exist, such as the International Classification of Diseases (ICD) by WHO, the International Classification of Primary Care (ICPC), also accepted by WHO or the Economic Botany Data Collection Standard (EBDCS) and their suitability in the different stages of the ethnopharmacological research is highly differentiated (Staub et al. 2015).

Even though the field of ethnopharmacology experiences a shift from the raw compilation of data to a greater methodological and conceptual reorientation, a lot of ethnopharmacological knowledge is still unprocessed and there is plenty of room for handling this knowledge through several methodologies. For example, the fact that some families or some plant groups clearly hold more medicinal species than predicted by chance has been repeatedly demonstrated following different methodologies (e.g., Kapur et al. 1992; Phillips and Gentry 1993a, b; Moerman 1996; Shepard 2004; Bourbonnais-Spear et al. 2005; Amiguet et al. 2006; Bennett and Husby 2008; Douwes et al. 2008; Leonti et al. 2009; Thomas et al. 2009; Saslis-Lagoudakis et al. 2011; Weckerle et al. 2011).

Here we attempt to demonstrate for the first time an implementation of Partial Order Techniques, processing ethnopharmacological information, with the purpose to reveal inner structures and characteristics of raw data, which could potentially

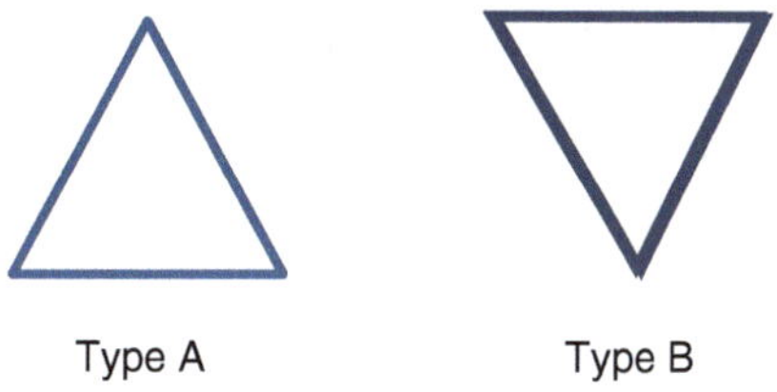

Fig. 1 Type A/B Hasse diagram similar to a partial order having greatest/least element

contribute in the conceptualization and management of ethnopharmacological knowledge.

Specifically, we try to investigate the following queries:

(a) Whether the expected type of Hasse diagram (Bruggemann and Patil 2011) from ethnopharmacology data sets on traditional uses of medicinal plants follows a type A or B (see Fig. 1). A Hasse diagram similar to a partial order having a greatest element is defined here as Type A in contrast to the reversed Type B (Fig. 1).
(b) Classifying the traditional uses of medicinal plants in broader medicinal categories, whether the expected type of Hasse diagram from the derived data set follows a type A or B or neither.
(c) Whether we can classify plant species based on both the use value of each species and its rank order on Hasse diagram.
(d) Whether the use of Hasse Diagram Technique (HDT) can reveal trends in medicinal uses of closely related plant taxa.
(e) Whether we can use successfully HDT in ethnopharmacological comparisons

The previous queries are investigated with actual data, using published data sets on traditional uses of medicinal plants.

2 Materials and Methods

2.1 Partial Order

Here we only give a short description. For more details Bruggemann and Patil (2011) are recommended. Given a finite set of objects it is possible to define partial order relations among them in a multitude of ways, whence partial order theory became a powerful technique in many applied sciences (compare in that context Wolski 2008). Even, when a data matrix is at hand and the data matrix can be considered as suitable for a ranking there are many different possibilities to define a partial order relation.

2.1.1 Hasse Diagram technique

Let X be a finite set of objects (often also called 'elements') and IB the set of indicators q_j ($j = 1, \cdots, m$), then one of the most simple ways to define a partial order is (1):

$$x, y \in X,\ x \leq y : \Longleftrightarrow\ q_j(x) \leq q_j(y) \text{ for all } q_j \in \text{IB} \tag{1}$$

$$\text{When } q_j(x) = q_j(y) \text{ for all } q_j \in \text{IB then } x \cong y \tag{2}$$

With $\cong$ indicating an equivalence.

In most practical applications based on (1) and (2), only a representative element of an equivalence class is considered and (1) is specified as follows:

$$x,\ y \in X,\ x \leq y\ : \Longleftrightarrow\ q_j(x) \leq q_j(y) \text{for all } q_j \in \text{IB with at least one } q_{j*} \text{for which a strict inequality holds.} \tag{3}$$

Equations (1), (3), and (2) are a very specific realization of a partial order, whence in the literature the resulting analysis technique is called Hasse Diagram technique (HDT). Elements which obey (1) are called comparable. If (1) does not hold, then x and y are incomparable, in sign: $x \,||\, y$.

2.1.2 Cover Relation and Drawing a Hasse Diagram

Equation (1) fulfills the axioms of partial order, especially the transitivity:

$$x, y, z \in X, \quad \text{if } x \leq y \text{ and } y \leq z, \text{ then : } x \leq z \tag{4}$$

The situation, where $x \leq z$ and there is no element y for which (4) holds is of special importance, because x and z can be considered as immediate neighbors. A pair $x < y$ where no third element of X is between x and y is called a cover relation and designed as $x <: y$. Cover relations are the basis to draw a Hasse diagram:

(a) $x <: y$ is drawn in a plane such that y is located vertically above x
(b) The location of the elements of X (to be more exact: of the representative elements) is done to get as far as possible a symmetric graph.
(c) The vertical arrangement of the objects should be done with the least number of different heights (taking from the bottom). Hence subsets of objects are arranged in the same vertical level. Technically formulated: the concept "level" is derived from an equivalence relation: Objects x and y are equivalent (are belonging to the same level) if their graph-theoretical distance suitable selected to measure the vertical distance is equal.

(d) In cases objects can be assigned to different levels, they will be located in the highest one. This rule is not justified by any order theoretical argument but by convenience (see c).

2.1.3 Chains and Levels

A chain is a subset $X' \subseteq X$, where all elements are mutually comparable. If $X' = X$ then the complete object set can be linearly ordered. In that case the partial order provides a complete ranking.

If X' is a proper subset of X, then nevertheless X' may be a chain. Interpreted on the basis of (1) the presence of chains indicates that the indicator values are not countercurrent. Therefore, a "partial ranking" is possible. At least for the elements of a chain a ranking can be performed, without applying knowledge beyond the data matrix or the need of sophisticated multicriteria decision support systems (MCDSs—see for instance Munda 2008; Huang et al. 2011; Bruggemann and Carlsen 2012; Wittmann and Bruggemann 2013).

Chains can have different lengths, i.e., different number of elements. Chains of maximal length are of special interest. A maximal chain which has a maximum length defines the number of possible levels:

$$\text{Number of levels} = \text{Length of maximum of maximal chain.} \tag{5}$$

Partial order theory shows that levels constitute an order, i.e., level1 < level 2 < level 3. The order relation between any two levels I and k is based on the order relation among the elements of the level i and level k. Two levels are neighbored if order relations between the elements of both levels are cover relations.

Hence the system of increasing levels indicates a weak order among the elements of X as follows:

$$\begin{aligned} &x \cong y \text{ if } x, y \text{ belong the same level} \\ &x < y \text{ if } x \text{ belongs to levels } i \text{ and } y \text{ belongs to level } k \text{ with } i < k \end{aligned} \tag{6}$$

A total or weak (including ties) order for all objects can be obtained without the need of defining attribute weights or knowledge beyond the data matrices. It is the method of calculation average heights which is described in detail in Bruggemann and Carlsen (2011) and Bruggemann and Annoni (2014). The average heights can be used to derive a ranking. Therefore, the notion Rk(av) is used here.

2.1.4 Comparing Posets

In complex data sets it is often necessary to compare different sets of criteria (attributes). In the similarity analysis we intend to calculate the similarity of different posets (partially ordered sets). This similarity analysis is an important feature of PyHasse (http://www.pyhasse.org). For details see Bruggemann and Patil (2011).

In the similarity analysis we calculate the proximity of different posets (partially ordered sets) based on the same ground set X: $P_1 = (X, \text{IB}_1)$ and $P_2 = (X, \text{IB}_2)$. The outcome of a partial order for two objects a, b may be

$$\begin{aligned} &a < b, \\ &a > b, \\ &a \,||\, b, \\ &a \cong b. \end{aligned}$$

When two partial orders P_1 and P_2 are to be compared, then the different combinations are counted, say for objects a, b, c, d, e, f such as:

$$a{<_1}b,\ a{<_2}b \text{ or } c{<_1}d,\ c{>_2}\,d \ \text{ or}$$

for which we use the shorthand notation ≪, <>, ||<, etc., and write, for instance, $a <_1 b$ instead of $a <_{P1} b$. Most important are the entries like ≫ or ≪, which are counting the 'isotone' character of both partial orders (ISO) and the entries like ><, <> which contribute to the »antitone« character, i.e., to the conflicts between the two partial orders (ANTI). There are still more combinations to look upon: <||, >||, ||>,||<, =||, ||= or || || are considered as indifferent (IND), combinations such as > =, < =, = <, = > are called weak isotone (WISO). Finally the entry of type = = contributes to equivalence relations (IDE). The appropriately normalized contributions of ISO, ANTI, IND, WISO, and IDE are the final result.

2.1.5 Partial Orders Based on Binary Indicator Values

The partial orders which will be studied here are based on indicators having the values 0 or 1 (medicinal code not present/present for a certain plant species). If any possible profile based on the indicators is realized then the width of each level, i.e., the number of elements in each level is determined by the Sperner number:

$$N_k = \binom{m}{k} \tag{7}$$

where m is the number of indicators (of medicinal codes) and $k = 0,1,\ldots$ enumerates the levels starting with the bottom level. A Hasse diagram therefore has the shape of Fig. 2.

Now, for the number of representative elements n most often the following inequality holds:

$$n < \sum_{k=0\ldots m} \binom{m}{k} = 2^m \tag{8}$$

Therefore, the realized Hasse diagram with a limited number of representative numbers can often have a shape similar to the lower part in Fig. 2.

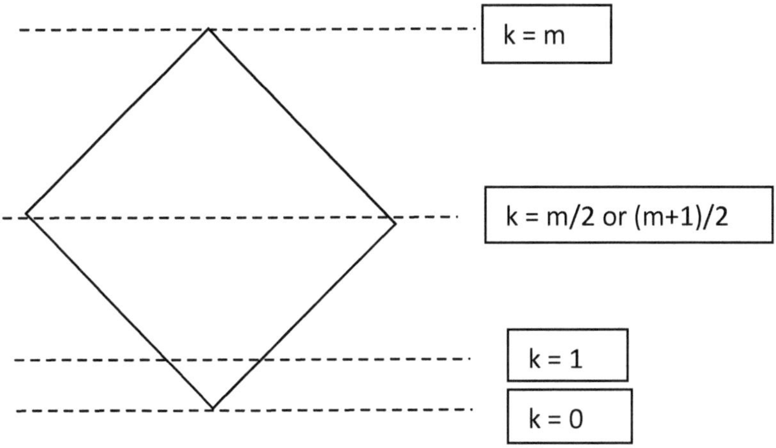

Fig. 2 "Diamond type" of Hasse diagram (Bruggemann and Patil 2011)

2.2 Data

The queries are investigated using the data set published in Polat et al. (2013). This study has identified not only the wild plants collected for medical purposes by local people of Solhan District in the Eastern Anatolia Region but also the uses and local names of these plants. It is a complete and well-done study which provides a source for researchers in ethnobotany, pharmacology, and chemistry by comparing the information obtained from traditionally used herbs with previous laboratory studies.

The medicinal uses (see Appendix 2) reported in Polat et al. (2013) have been classified using the following 15 medicinal categories: digestive system disorders, circulatory system disorders, respiratory system disorders, inflammation, endocrine system disorders, mascular-skeletal system disorders, infections-infestations, genitourinary system disorders, nutritional disorders, metabolic system disorders, injuries, skin/subcutaneous cellular tissue disorders, mental disorders, pain, and blood system disorders. The medicinal categories follow the standardized descriptors and terms of Kew.org Economic Botany Data Standard (http://www.kew.org/tdwguses/).

The use value (UV) of the species has also been used in data processing. The UV (Trotter and Logan 1986) in the data set of Polat et al. (2013) was calculated according to the following formula: UV = U/N, where UV refers to the use value of a species, U to the number of citations per species, and N to the number of informants.

2.3 *Data Analysis*

The analysis of the data was performed by the PyHasse package (http://www.pyhasse.org/). For more information, see also Bruggemann et al. (2014). The software package can be obtained as a CD from the R.B. This software contains more than 100 modules, solving partially pretty specific tasks. For a first insight the interested reader should visit the web page, especially the module spyout.

3 Results and Discussion

3.1 *In the Absence of Panacea*

A plant species ordered as the greatest element of a Hasse diagram of species on the basis of their medicinal uses would be a panacea for all the corresponding diseases. The resulting Hasse diagram of the species reported in Polat et al. (2013) is presented in Fig. 3. The species list and the list of their medicinal uses employed in the Partial Order Techniques are presented in Appendices 1 and 2, respectively, with their corresponding codes.

It is very clear that the scheme of Hasse diagram in Fig. 3 is consistent with what has been described as Hasse diagram Type B (see Fig. 1). In other words, it is a case of absence of greatest element in the poset, which indicates that plants differ in their medicinal uses and none is a panacea for all diseases. Mathematically, a structure like type A or type B is often to be expected, because of the evolution of the binary profiles (see Sect. 2.1.5), but does the absence of a greatest element in the poset sound biologically meaningful?

The answer is positive. The bioactivity of plant in their medicinal uses is based on specialized (secondary) metabolites produced in their tissues. First of all, plants synthesize an immensely rich diversity of specialized metabolites comprising more than 200,000 compounds (Hartmann 2007). Assuming that in plants 10–20 % of the genes in the genome encode enzymes for secondary metabolism (Somerville and Somerville 1999; Pichersky and Gang 2000) and a roughly 1:1 ratio between the number of enzymes and the number of compounds produced in the cell, as

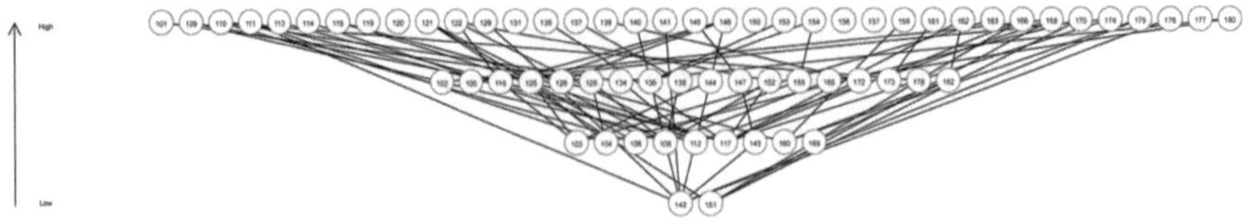

Fig. 3 Hasse diagram of wild plant species based on their uses for medical purposes by local people (data matrix 82 plant species × 54 medical uses). Codes are presented in Appendix 1 and equivalent classes in Appendix 3

well as that, for example, the poplar genome codes for 45,000 genes (Sterck et al. 2007; Michael and Jackson 2013), it becomes clear that each plant species can synthesize only a small fraction of the total number of specialized metabolites found throughout the plant kingdom (Pichersky and Lewinsohn 2011). On the other side, specialized metabolites serve cellular functions to the particular plant in which they occur and are assigned as indispensable components of the survival strategy of plants. Therefore, different plants have independently evolved the ability to make compounds already present in other plant lineages or to make different compounds that fulfill the same role (Moilanen et al. 2015). In the former case, the chemical solutions to a common problem are often different in different plant lineages (different specialized metabolites can have identical function).

According to Weng (2014), the study of specialized metabolic enzymes and pathways across major phyla of the plant kingdom reveals several general patterns, all consistent with the stepwise evolutionary processes underlying the emergence and evolution of the metabolic traits observed in extant plants. First, diverse plant specialized metabolic pathways branch from core primary metabolism at different nodes, second, in general, the taxonomic distribution of plant specialized metabolic traits correlates with the gradual evolutionary development of specialized tissue types, organs, and/or lifestyles observed in land plants as they underwent extensive divergence over the last 500 million and third, the expansion of the specialized metabolism in plants did not involve the emergence of new protein folds, but rather the extensive exploitation of the sequence space in the preexisting protein folds by natural selection.

Hence, identical metabolic traits often arise independently in disparate lineages because a number of plant species derived from divergent lineages often co-occupy a common environment. The repeated evolution of common metabolic traits likely resulted from natural selection driven by similar selective pressures associated with particular environments. Repeated evolution is a special form of convergent evolution in which new enzymes with the same function evolve independently in separate plant lineages from shared pool of related enzymes with similar but not identical functions (Pichersky and Lewinsohn 2011).

Nevertheless, alternative explanations reported, such as an explanation based on the differential gene expression. According to this explanation, in several cases ancestral members of a group evolved the biosynthetic capacity to produce a certain specialized metabolite and the absence of such a trait in phylogenetically derived groups is probably due to differential gene expression, so the corresponding genes are not lost but switched off (Wink 2003). Moreover, evolutionary diversification of the genes involved in the pathways of specialized metabolites has been linked with gene duplication (Flagel and Wendel 2009), a major driving force for gene diversification and recruitment in plants. Gene duplication and new functionalization of the duplicate that no longer underlies the stringent functional demands of the mother gene appears to be the key to explain the evolution of metabolic diversity (Pichersky and Gang 2000; Ober 2005; Hartmann 2007; Kroymann 2011).

Thus, a Hasse diagram Type B seems to be reasonable and biologically meaningful. Plants differ in their medical uses; none is a panacea for all diseases. Identical

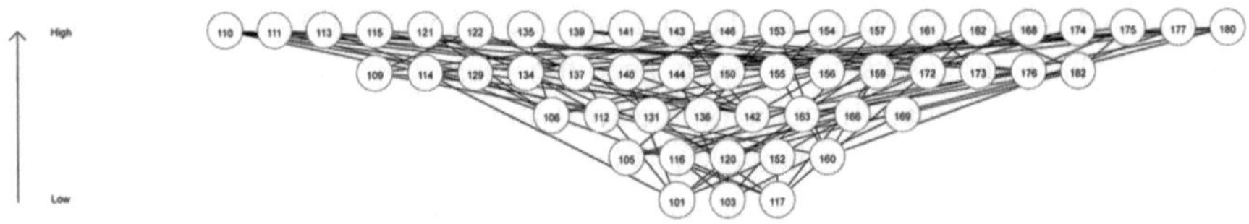

Fig. 4 Hasse diagram of wild plant species based on medical categories of their uses for medical purposes by local people (data matrix 82 plant species × 15 medical categories). Codes follow Appendix 1 and equivalent classes are presented in Appendix 4

specialized metabolites often arise independently in disparate plant lineages, e.g., linalool, an acyclic monoterpene alcohol, is a compound found in many plant species (Pichersky and Lewinsohn 2011). Different compounds in divergent plant lineages can have identical action, e.g., phenolics, terpenoids, and alkaloids among others, widely spread in diverse plant lineages, have recorded for their antimicrobial action (Cowan 1999). Moreover, different plant parts and extraction methods of the same plant may result in opposed effects and uses, e.g., in the same plant, coexisting protein fractions may induce both immunostimulatory and immunosuppressive activities (Karali et al. 2015).

3.2 From Uses of Medicinal Plants to Medical Categories

The 54 medicinal uses (see Appendix 2) have been classified in the 15 medicinal categories which reported in Sect. 2.2 and the outcome has been used for the Partial Order Analyses. The Hasse diagram of the species on the basis of the resulted medicinal categories is presented in Fig. 4. Again, the scheme of the Hasse diagram (Fig. 4) is clearly consistent with what has been described as Hasse diagram Type B (see Fig. 1). It is a case of absence of a greatest element in the poset.

Although the implemented classification of the traditional uses of medicinal plants is part of the etic perception, the earlier result, in the absence of panacea reported previously, indicates a social maturity. It has long been recognized that culture influences the content and practical application of medical knowledge. Even the current state of knowledge about pathology and treatment is not simply the outcome of a neutral process of scientific investigation and discovery, but is shaped by changing theoretical frameworks affected by more general cultural perspectives (Hughes 2015).

The social maturity concerns the process at the societal level, which is not restricted in the causes of diseases or etiology like wrath of deity, evil spirits or sorcery, black eye, violation of taboos, etc., but involves identification of disease by examining the symptoms followed by management using different medicinal plants (Gupta et al. 2014).

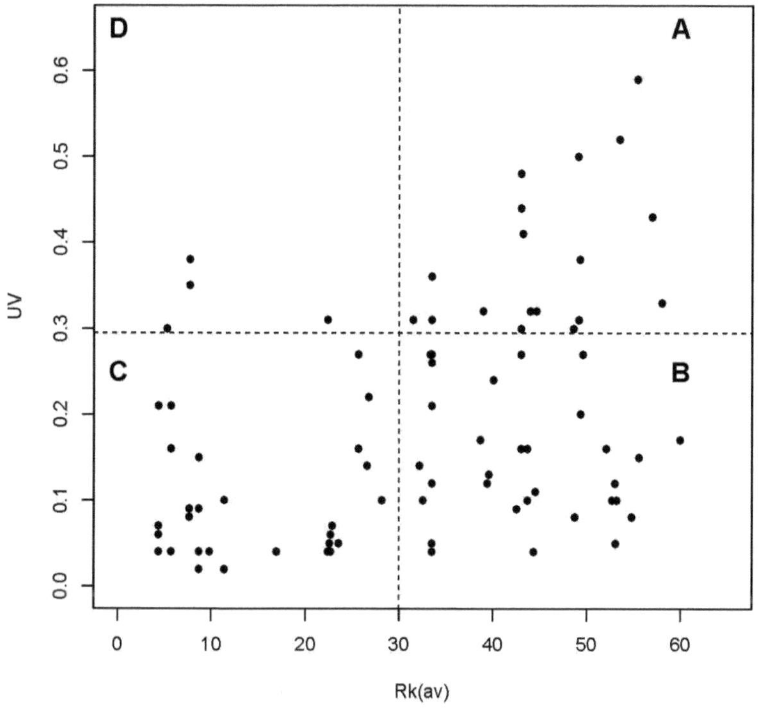

Fig. 5 Scatterplot of plant species. UV = use value of each species, Rk(av) = Plant species averaged height in Hasse diagram of Fig. 3. *Vertical dashed lines* on each axis split points according to the values of *U* and Rk(av), respectively, using the 50 % of the maximum value as threshold

3.3 *Use Value of Plants and Rank Order*

Although there is no panacea for all diseases, some plants are widely used in many different medical categories. *Urtica doica* L. is an example of the data set published in Polat et al. (2013) as it refers to the plants with most use reports. The *Urtica dioica* L. is used in Solhan to treat analgesic, arthritis, digestive, diuretic, genital disorders, hemorrhoids, hepatitis, lipsotrichia, rheumatism. Previous studies showed that *Urtica dioica* L. results in analgesic and antimicrobial and antihyperglycemic activity (Bnouham et al. 2003; Gülçin et al. 2004). *Urtica dioica* L. were reported to be of the highest use value.

The use value (UV) demonstrates the relative importance of species known locally and the classification of species in different categories based on UV and Rk(av) is graphically presented in Fig. 5.

Following Fig. 5 and using as arbitrary threshold the 50 % of the maximum value for each axis the plant species are classified in four groups. In *Group A* plant species of high use value and high Rk(av) values are included (17 species). We can describe this group of species, as species well known locally with multiple medicinal uses.

Urtica dioica L., *Malva neglecta* Wallr., *Anchusa azurea* Mill., *Rosa canina* L., *Juglans regia* L., and *Rubus sanctus* Schreber are included among others.

In *Group B* plant species of low use value and high Rk(av) values are included (32 species). We can describe this group of species, as species with multiple medicinal uses, but not well known locally. *Teucrium polium* L., *Achillea biebersteinii* Afan., *Arum elongatum* Steven subsp. *detruncatum* (C.A. Meyer ex Schott) H.Riedl, *Crataegus aronia* (L.) Bosc. Ex DC. var. *aronia* (L.) Bosc. ex DC., *Stachys lavandulifolia* Vahl. var. *lavandulifolia* Vahl., and *Xanthium spinosum* L. are included among others. In *Group C* plant species of low use value and low Rk(av) values are included (29 species). We can describe this group of species, as species not well known locally with restricted medicinal uses. *Cotoneaster nummularia* Fisch. & Mey., *Scorzonera cinerea* Boiss., *Astragalus brachycalyx* Fischer, *Scorzonera mollis* M. Bieb. subsp. *szovitzii* (DC.) Chamberlain, *Hordeum bulbosum* L., and *Anchusa strigosa* Labill are included among others. In *Group D* plant species of high use value and low Rk(av) values are included (4 species). We can describe this group of species, as species well known locally with restricted medicinal uses. *Rhus coriaria* L., *Allium sativum* L., *Ocimum basilicum* L., and *Vitis vinifera* L. are included.

This classification scheme can further improve the evaluation and consequently the management of the ethnopharmacological knowledge on traditional medical uses of plants. It is worthy to further explore this scheme both in the social (Gale 2014; Gupta et al. 2014; Hughes 2015) and in the biomedical context (Raut and Karuppayil 2014; Ziani et al. 2015). It is also worthy to explore this scheme in comparative studies and to relate this with other classification systems of medicinal flora (Molares and Ladio 2009; Pardo-de-Santayana et al. 2015).

The same approach has also been followed using the use value (UV) of each species and its Rank Order Rk(av) value in Hasse diagram of Fig. 4 and the same thresholds. The plant species have been classified again in four groups based on medical categories of their uses for medical purposes. The number of species classified in each group for both classification efforts (uses and medicinal categories) is presented in Fig. 6.

According to the classification results, in the case of the second classification (medicinal categories) the highest number of species is recorded in Group C, which incorporates species not well known locally with uses restricted in one or few different medicinal categories. This shifting to high participation in Groups C and D (groups with low Rk(av)) is strongly related to the fact that different traditional uses of medicinal plants are classified in the same medicinal category.

3.4 Systematics, Phylogeny, and Evolution

By means of Hasse Diagram Techniques we have tried to reveal trends in medicinal uses of closely related plant taxa selecting the species of two families Lamiaceae and Asteraceae of the Polat et al. (2013) data set. The Hasse diagrams of the medicinal

Fig. 6 The number of species classified in each group using the use value (UV) of each species and its Rank Order Rk(av) values. *Bars* in different colors represent different classifications efforts

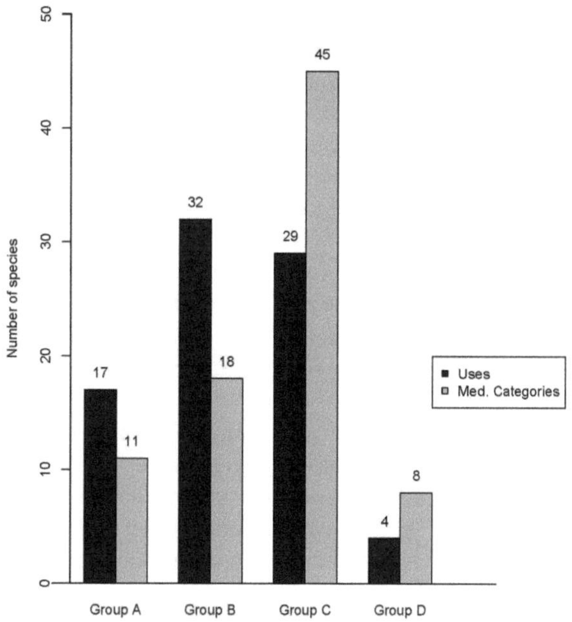

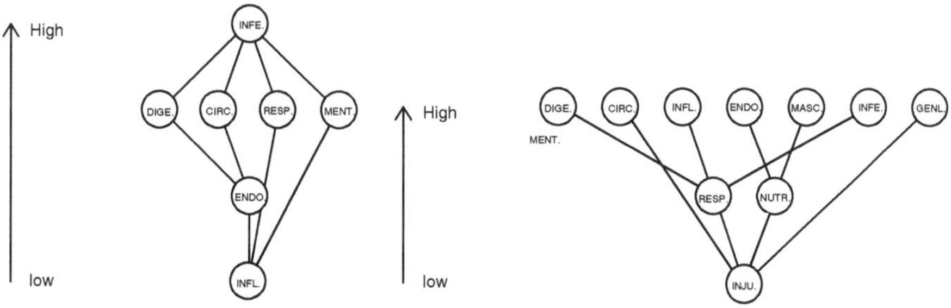

Fig. 7 Hasse diagrams of medicinal categories of the families Lamiaceae (*left*) and Asteraceae (*right*). Abbreviations use the first four letters of the medicinal categories described in Sect. 2.2

categories for these two families are presented in Fig. 7. For the construction of each of the Hasse diagrams only the species of each family have been used.

It is very clear that the species of the data set from these families are used in traditional medicine to treat different medicinal categories of diseases. Despite their similarities, the rank order profile of medicinal categories derived from the Hasse Diagram Technique reveals a marked difference. In the case of Asteraceae the Hasse diagram follows type B (see Fig. 1). It is a case of absence of a greatest element in the poset. In the case of Lamiaceae, all the species have been reported for traditional uses in the medicinal category infections–infestations, so this medicinal category is the greatest element of the Hasse diagram. This evidence is also supported by the fact that major antimicrobial specialized metabolites (Cowan 1999) have been identified in most studied members of this family (Chorianopoulos et al. 2007).

It is generally accepted that in ethnopharmacology certain plant families and higher taxa are used more extensively than others (Moerman et al. 1999; Leonti et al. 2009; Weckerle et al. 2011; de Medeiros et al. 2013). As it has been mentioned earlier, this argument has been tested several times using different methodologies. For example, Asteraceae and Lamiaceae from the Euasterids tend to be overused for medicinal purposes in the holarctic region in contrast to other families such as Poaceae and Cyperaceae and Orchidaceae (Moerman et al. 1999; Leonti and Weckerle 2015).

Recently, the phylogenetic dimension of these findings was emphasized, especially with and after the publication of Saslis-Lagoudakis et al. (2012a), revitalizing what we call *phylo-chemical paradox*, where on one side medicinal plants and their uses show strong phylogenetic clustering while on the other side similar or identical specialized (secondary) metabolites with an apparently shared biosynthetic origin appear in clearly disparate higher plant taxa, impoverishing the relationship between phylogeny and chemotaxonomy. However, as Yessoufou et al. (2015) reported, the relevance of phylogenetic approach in ethnomedicine has recently been questioned (see Gertsch 2012; Saslis-Lagoudakis et al. 2012a debate), and the finding that biochemistry is not always phylogenetically clustered (e.g., Rønsted et al. 2012; Savo et al. 2015; Yessoufou et al. 2015) indicates that this approach requires further exploration. Considering specialized metabolites as plant traits (Hadacek 2002; Wink 2003), further caution is required in their relationships with phylogeny as the theoretical concept in this view is interdisciplinary broader (Rønsted et al. 2012; Winter et al. 2013; Kelly et al. 2014).

The suggested approach, the implementation of Hasse Diagram Techniques, can reveal trends in medicinal uses of closely related plant taxa independently of whether the close relatedness is based on phylogeny or not. Selecting the group of interest (we can use several criteria of grouping species and taxa) and analyzing their medicinal uses or the medicinal categories of their uses, we can reveal the rank order profile of their medicinal applications, which is valuable for the analysis and management of ethnopharmacological data in several fields.

3.5 Comparing Ethnopharmacological Data Matrices

A second data matrix can be produced from the data set of Polat et al. (2013), this concerning the traditional uses of medicinal plants of the area reported in the scientific literature. Assuming that it is worthy to compare the two data sets, the first data matrix derived from the interview records and the second data matrix derived from the literature records, and beyond many other used methodologies for comparisons, we suggest the use of Hasse Diagram Techniques.

For this purpose we have incorporated in the second data matrix only the same uses already recorded from the interviews. Moreover classifying the medical uses of the second matrix in medical categories we have produced a new matrix for comparisons on the basis of medical categories. The comparisons by means of the

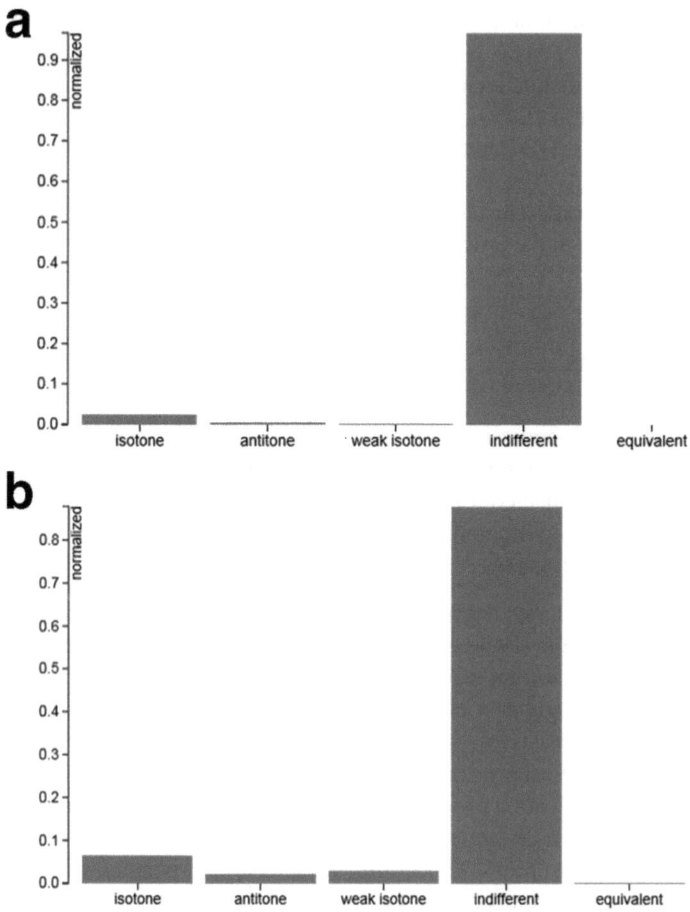

Fig. 8 Proximity of the recorded (first) and literature (second) data matrices based on HDT. Data set derived from Polat et al. (2013), while (**a**) and (**b**) correspond to medicinal uses and medical categories, respectively

similarity modules of PyHasse show in both cases that IND (see Sect. 2.1.4) is overwhelming followed by small values for ISO, whereas all other contributions to proximity have less values (Fig. 8).

This finding is not surprising for this case because we refer in different spatial scales, and it is well known that there is spatial variation in the production of specialized metabolites and their biological activity (e.g., Homer et al. 2000; Lee et al. 2002; Viljoen et al. 2005; Chen et al. 2015; Georgescu et al. 2016).

Comparisons of ethnobotanical data have been attempted many times waiting on different targets and using different methodologies. For example, cross-cultural comparisons in a phylogenetic context (Saslis-Lagoudakis et al. 2011, 2012b; Yessoufou et al. 2015) took place among others, using methodologies such as regression and binomial analyses, the net relatedness index (NRI) and the net taxon index (NTI) (Webb et al. 2002), the D statistic (Fritz and Purvis 2010), the evolutionarily distinctiveness metric (ED) (Isaac et al. 2007) and phylogenetic distances (Webb et al. 2008).

Here we suggest, HDT and especially the tool deriving the proximity of different partially ordered sets. The advantage of this comparison is related to the fact that we compare partially ordered sets and not primary data and therefore this comparison reveals similarities in the ordered structure of the data, i.e., is relationally based.

4 Conclusions

Here we attempt to demonstrate for the first time an implementation of Partial Order Techniques in processing ethnopharmacological information. A published data set of the wild plants collected for medical purposes by local people of Solhan District in the Eastern Anatolia Region and their uses have been utilized.

According to the results:

(a) The scheme of Hasse diagram of species on the basis of their medicinal uses was consistent with the absence of greatest element in the poset, which indicates that plants differ in their medicinal uses and none is a panacea for all diseases.
(b) The absence of greatest element in the poset also was resulted analyzing the species on the basis of the medicinal categories (produced classifying their uses), which indicates a social maturity.
(c) A further classification of species in different categories based on their use value (UV) and their rank order [Rk(av)] has been achieved, which is valuable in decision-making process concerning the selection of plants for medicinal use.
(d) Results revealed trends in medicinal uses of closely related plant taxa independently of whether the close relatedness is based on phylogeny or not, indicating that selecting the plant group of interest and analyzing their medicinal uses we can reveal the rank order profile of their medicinal applications.
(e) The process has achieved a meaningful comparison of ethnobotanical data using the proximity of different partially order sets, suggesting that a comparison of the ordered structure of the ethnobotanical data can reveal relationally based similarities.

Thus, the suggested methodologies reveal inner structures and characteristics of raw data, which could potentially contribute in the conceptualization and management of ethnopharmacological knowledge. Nevertheless, further effort is required in this direction in order to explore and document the potentialities of the suggested methodologies.

Appendix 1: Species List and Codes Used in the Partial Order Techniques

Amaranthus retroflexus L. (**101**), Pistacia terebinthus L. subsp. palaestina (Boiss.) Engler. (**102**), Rhus coriaria L. (**103**), Anthriscus cerefolium (L.) Hoffm. (**104**), Heracleum trachyloma Fisch & Mey. (**105**), Pastinaca armena Fisch. & Mey. (**106**), Pimpinella olivieroides Boiss. & Hausskn. (**107**), Sium sisarum L. var. lancifolium (M. Bieb.) Thell. (**108**), Arum elongatum Steven subsp. detruncatum (C.A. Meyer ex Schott) H. Riedl (**109**), Achillea biebersteinii Afan. (**110**), Anthemis wiedemanniana Fisch. & Mey. Endemic, Least concern (LC). (**111**), Centaurea kurdica Reichardt. Endemic, Near threatened (NT). (**112**), Chaerophyllum bulbosum L. (**113**), Gundelia tournefortii L. var. tournefortii (**114**), Helichrysum plicatum DC. (**115**), Helianthus tuberosus L. (**116**), Scorzonera mollis M. Bieb. subsp. szovitzii (DC.) Chamberlain (**117**), Scorzonera cinerea Boiss. (**118**), Tragopogon reticulatus Boiss. & Huet (**119**), Onopordum acanthium L. (**120**), Xanthium spinosum L. (**121**), Anchusa azurea Mill. (**122**), Anchusa strigosa Labill. (**123**), Alyssum pateri Nyar. subsp. pateri Endemic, Least concern (LC). (**124**), Nasturtium officinale R.Br. (**125**), Capsella bursa-pastoris (L.) Medik. (**126**), Sambucus nigra L. (**127**), Silene vulgaris (Moench) Garcke var. vulgaris (**128**), Cornus mas L. (**129**), Chenopodium album L. subsp. album var. album (**130**), Equisetum ramosissimum Desf. (**131**), Astragalus brachycalyx Fischer (**132**), Astragalus lamarckii Boiss. Endemic, Least concern (LC) (**133**), Trifolium pratense L. var. pretense (**134**), Quercus petraea (Mattuschka) Liebl. subsp. pinnatiloba (C. Koch) Menitsky (**135**), Quercus libani Olivier (**136**), Hypericum perforatum L. (**137**), Hypericum scabrum L. (**138**), Juglans regia L. (**139**), Mentha longifolia (L.) Hudson subsp. typhoides (Briq.) Harley var. typhoides (**140**), Mentha spicata L. subsp. spicata (**141**), Ocimum basilicum L. (**142**), Origanum vulgare L. (**143**), Satureja hortensis L. (**144**), Stachys lavandulifolia Vahl. var. lavandulifolia Vahl. (**145**), Teucrium polium L. (**146**), Teucrium chamaedrys L. subsp. sinuatum (Celak.) Rech. f. (**147**), Thymus kotschyanus Boiss. & Hohen (**148**), Allium sativum L. (**149**), Allium cepa L. (**150**), Asparagus acutifolius L. (**151**), Eremurus spectabilis Bieb. (**152**), Alcea pallida Waldst. & Kit. (**153**), Malva neglecta Wallr. (**154**), Malva sylvestris L. (**155**), Morus nigra L. (**156**), Plantago lanceolata L. (**157**), Plantago major L. subsp. major (**158**), Zea mays L. (**159**), Hordeum bulbosum L. (**160**), Polygonum cognatum Meissn. (**161**), Rheum ribes L. (162), Rumex acetosella L. (**163**), Rumex tuberosus L. (**164**), Polygonum cognatum Meissn. (**165**), Portulaca oleracea L. (**166**), Crataegus atrosanguinea Pojark. (**167**), Crataegus aronia (L.) Bosc. Ex DC. var. aronia (L.) Bosc. ex DC. (**168**), Crataegus szovitsii Pojark. (**169**), Crataegus orientalis (Mill.) M. Bieb. (**170**), Cotoneaster nummularia Fisch. & Mey. (**171**), Cydonia oblonga Miller (**172**), Prunus kurdica Fenzl ex Fritsch (**173**), Rubus sanctus Schreber (**174**), Rosa canina L. (**175**), Rosa dumalis Bechst. (**176**), Salix alba L. (**177**), Hyoscyamus niger L. (**178**), Celtis tournefortii Lam. (**179**), Urtica dioica L. (**180**), Vitis vinifera L. (**181**), Tribulus terrestris L. (**182**)

Appendix 2: Uses and Codes Used in the Partial Order Techniques

Diarrhea (**1**), digestive (**2**), stomach-ache (**3**), ulcer (**4**), antihypertensive (**5**), asthma (**6**), bronchitis (**7**), anti-inflammatory (**8**), abdominal pain (**9**), diabetes disease (**10**), guatr (**11**), rheumatism (**12**), hepatitis (**13**), sinusitis (**14**), toothache (**15**), menstruation pain (**16**), antitussive (**17**), colds and flu (**18**), appetite (**19**), cholesterol (**20**), aphrodisiac (**21**), kidney stones (**22**), intestinal inflammation (**23**), hepatic diseases (**24**), hemorrhoids (**25**), cardiac disorder (**26**), embolism (**27**), carminative (**28**), wound healing (**29**), headache (**30**), urinary inflammations (**31**), antilipsotrichia (**32**), inflammatory (**33**), wounds (**34**), antifungal (**35**), antilubrication (**36**), psoriasis (**37**), for hair (**38**), respiratory tract problem (**39**), sedative (**40**), anypnia (**41**), insomnia (**42**), analgesic (**43**), infertility (**44**), eczema (**45**), abscess (**46**), anemia (**47**), diuretic (**48**), constipation (**49**), joint pain (**50**), stress (**51**), expectorant (**52**), arthritis (**53**), lipsotrichia (**54**)

Appendix 3: Equivalence Classes of Fig. 3

Count	Representant	Equivalence classes		
2	102	133		
2	103	149		
2	106	107		
3	108	123	130	
3	112	124	127	
4	117	118	132	171
2	140	179		
2	143	148		
2	151	167		
2	157	158		
2	163	164		
2	173	181		

Appendix 4: Equivalence Classes of Fig. 4

Count	Representant	Equivalence classes								
10	101	102	104	108	123	128	130	133	138	178
4	103	149	151	167						
3	106	107	119							
4	112	124	126	127						
2	114	165								
4	117	118	132	171						
3	120	125	170							
2	140	179								
4	143	145	147	148						
2	157	158								
2	163	164								
2	173	181								

References

Amiguet VT, Arnason JT, Maquin P, Cal V, Sánchez-Vindas P, Poveda Alvarez L (2006) A regression analysis of Q'eqchi' maya medicinal plants from Southern Belize. Econ Bot 60:24–38

Bennett BC, Balick MJ (2014) Does the name really matter? The importance of botanical nomenclature and plant taxonomy in biomedical research. J Ethnopharmacol 152:387–392

Bennett BC, Husby CE (2008) Patterns of medicinal plant use: an examination of the Ecuadorian Shuar medicinal flora using contingency table and binomial analysis. J Ethnopharmacol 116:422–430

Bnouham M, Merhfour FM, Ziyyat A, Mekhfi H, Aziz M, Legssyer A (2003) Antihyperglycemic activity of the aqueous extract of *Urtica dioica*. Fitoterapia 74:677–681

Bourbonnais-Spear N, Awad R, Maquin P, Cal V, Vindas PS, Poveda L, Arnason JT (2005) Plant use by the Q'eqchi' Maya of Belize in ethnopsychiatry and neurological pathology. Econ Bot 59:326–336

Bruggemann R, Annoni P (2014) Average heights in Partially Ordered Sets. MATCH Commun Math Comput Chem 71:101–126

Bruggemann R, Carlsen L (2011) An improved estimation of averaged ranks of partially orders. MATCH Commun Math Comput Chem 65:383–414

Bruggemann R, Carlsen L (2012) Multi-criteria decision analyses. Viewing MCDA in terms of both process and aggregation methods: some thoughts, motivated by the paper of Huang, Keisler and Linkov. Sci Total Environ 425:293–295

Bruggemann R, Patil GP (2011) Ranking and prioritization for multi-indicator systems - introduction to partial order applications. Springer, New York

Bruggemann R, Carlsen L, Voigt K, Wieland R (2014) PyHasse software for partial order analysis: scientific background and description of selected modules. In: Bruggemann R, Carlsen L, Wittmann J (eds) Multi-indicator systems and modelling in partial order. Springer, New York

Bussmann RW (2015) Taxonomy – an irreplaceable tool for validation of herbal medicine. In: Mukherjee P (ed) Evidence-based validation of herbal medicines farm to pharma. Elsevier, Amsterdam

Chen J, Xu Y, Wei G, Liao S, Zhang Y, Huang W et al (2015) Chemotypic and genetic diversity in *Epimedium sagittatum* from different geographical regions of China. Phytochemistry 116: 180–187

Chorianopoulos N, Nychas GJE, Haroutounian SA (2007) Essential oils of Lamiaceae family Taxa as natural preservatives of food preparations. Food 1:202–215

Cowan MM (1999) Plant products as antimicrobial agents. Clin Microbiol Rev 12:564–582

de Medeiros PM, Ladio AH, Santos AMM, de Albuquerque UP (2013) Does the selection of medicinal plants by Brazilian local populations suffer taxonomic influence? J Ethnopharmacol 146:842–852

Douwes E, Crouch NR, Edwards TJ, Mulholland DA (2008) Regression analyses of southern African ethnomedicinal plants: informing the targeted selection of bioprospecting and pharmacological screening subjects. J Ethnopharmacol 119:356–364

Flagel LE, Wendel JF (2009) Gene duplication and evolutionary novelty in plants. New Phytol 183:557–564

Fritz SA, Purvis A (2010) Selectivity in mammalian extinction risk and threat types: a new measure of phylogenetic signal strength in binary traits. Conserv Biol 24:1042–1051

Gale N (2014) The sociology of traditional, complementary and alternative medicine. Sociol Compass 8:805–822

Georgescu L, Stefanakis MK, Kokkini S, Katerinopoulos HE, Pirintsos SA (2016) Chemical and genetic characterization of *Phlomis* species and wild hybrids in Crete. Phytochemistry 122: 91–102

Gertsch J (2012) Cross-cultural comparisons of medicinal floras—what are the implications for bioprospecting? J Ethnopharmacol 139:685–687

Gülçin İ, Küfrevioğlu OI, Oktay M, Büyükokuroğlu ME (2004) Antioxidant, antimicrobial, antiulcer and analgesic activities of nettle (*Urtica dioica* L.). J Ethnopharmacol 90:205–215

Gupta P, Sharma VK, Sharma S (2014) Health and folk medicine. In: Gupta P, Sharma VK, Sharma S (eds) Healing traditions of the Northwestern Himalayas. Springer, New Delhi

Hadacek F (2002) Secondary metabolites as plant traits: current assessment and future perspectives. Crit Rev Plant Sci 21:273–322

Hartmann T (2007) From waste products to ecochemicals: fifty years research of plant secondary metabolism. Phytochemistry 68:2831–2846

Homer LE, Leach DN, Lea D, Lee LS, Henry RJ, Baverstock PR (2000) Natural variation in the essential oil content of *Melaleuca alternifolia* Cheel (Myrtaceae). Biochem Syst Ecol 28: 367–382

Huang IB, Keisler J, Linkov I (2011) Multi-criteria decision analysis in environmental sciences: ten years of applications and trends. Sci Total Environ 409:3578–3594

Hughes D (2015) Cultural influences on medical knowledge. In: Schramme T, Edwards S (eds) Handbook of the philosophy of medicine. Springer, New York

Isaac NJB, Turvey ST, Collen B, Waterman C, Baillie JEM (2007) Mammals on the EDGE: conservation priorities based on threat and phylogeny. PLoS One 2:e296

Kapur SK, Shahi AK, Sarin YK, Moerman DE (1992) The medicinal flora of Majouri-Kirchi forests (Jammu and Kashmir State), India. J Ethnopharmacol 36:87–90

Karali D, Georgescu L, Pirintsos S, Athanassakis I (2016) T cell regulation by *Phlomis lanata* protein extracts in mice. Pharm Biol 54(2):207–214

Kelly S, Grenyer R, Scotland RW (2014) Phylogenetic trees do not reliably predict feature diversity. Divers Distrib 20:600–612

Kroymann J (2011) Natural diversity and adaptation in plant secondary metabolism. Curr Opin Plant Biol 14:246–251

Lee LS, Brooks LO, Homer LE, Rossetto M, Henry RJ, Baverstock PR (2002) Geographic variation in the essential oils and morphology of natural populations of *Melaleuca alternifolia* (Myrtaceae). Biochem Syst Ecol 30:343–360

Leonti M, Weckerle CS (2015) Quantitative and comparative methods in ethnopharmacology. In: Heinrich M, Jäger AK (eds) Ethnopharmacology. Wiley, New York

Leonti M, Casu L, Sanna F, Bonsignore L (2009) A comparison of medicinal plant use in Sardinia and Sicily – De Materia Medica revisited? J Ethnopharmacol 121:255–267
Michael TP, Jackson S (2013) The first 50 plant genomes. Plant Genome 6:1–7
Moerman DE (1996) An analysis of the food plants and drug plants of native North America. J Ethnopharmacol 52:1–22
Moerman DE, Pemberton RW, Kiefer D, Berlin B (1999) A comparative analysis of five medicinal floras. J Ethnobiol 19:49–70
Moilanen J, Koskinen P, Salminen JP (2015) Distribution and content of ellagitannins in Finnish plant species. Phytochemistry 116:188–197
Molares S, Ladio A (2009) Ethnobotanical review of the Mapuche medicinal flora: use patterns on a regional scale. J Ethnopharmacol 122:251–260
Munda G (2008) Social multi-criteria evaluation for a sustainable economy. Springer, Berlin
Ober D (2005) Seeing double: gene duplication and diversification in plant secondary metabolism. Trends Plant Sci 10:444–449
Pardo-de-Santayana M, Quave CL, Sõukand R, Pieroni A (2015) Medical ethnobotany and ethnopharmacology of Europe. In: Heinrich M, Jäger AK (eds) Ethnopharmacology. Wiley, New York
Phillips OL, Gentry AH (1993a) The useful plants of Tambopata, Peru: I. Statistical hypotheses tests with a new quantitative technique. Econ Bot 47:15–32
Phillips OL, Gentry AH (1993b) The useful plants of Tambopata, Peru: II. Additional hypotheses testing in quantitative ethnobotany. Econ Bot 47:33–43
Pichersky E, Gang DR (2000) Genetics and biochemistry of secondary metabolites in plants: an evolutionary perspective. Trends Plant Sci 5:439–445
Pichersky E, Lewinsohn E (2011) Convergent evolution in plant specialized metabolism. Annu Rev Plant Biol 62:549–566
Polat R, Cakilcioglu U, Satıl F (2013) Traditional uses of medicinal plants in Solhan (Bingöl—Turkey). J Ethnopharmacol 148:951–963
Raut JS, Karuppayil SM (2014) Bioprospecting of plant essential oils for medicinal uses. In: Fulekar MH, Pathak B, Kale RK (eds) Environment and sustainable development. Springer, Delhi
Reyes-Garcia V (2010) The relevance of traditional knowledge systems for ethnopharmacological research: theoretical and methodological contributions. J Ethnobiol Ethnomed 6:32
Rivera D, Allkin R, Obón C, Alcaraz F, Verpoorte R, Heinrich M (2014) What is in a name? The need for accurate scientific nomenclature for plants. J Ethnopharmacol 152:393–402
Rønsted N, Symonds MRE, Birkholm T et al (2012) Can phylogeny predict chemical diversity and potential medicinal activity of plants? A case study of Amaryllidaceae. BMC Evol Biol 12:182–193
Saslis-Lagoudakis CH, Williamson EM, Savolainen V, Hawkins JA (2011) Cross-cultural comparison of three medicinal floras and implications for bioprospecting strategies. J Ethnopharmacol 135:476–487
Saslis-Lagoudakis CH, Savolainen V, Williamson EM et al (2012a) Phylogenies reveal predictive power of traditional medicine in bioprospecting. Proc Natl Acad Sci U S A 109:15835–15840
Saslis-Lagoudakis CH, Williamson EM, Savolainen V, Hawkins JA (2012b) Cross-cultural comparisons of medicinal floras and bioprospecting revisited. J Ethnopharmacol 139:688–690
Savo V, Joy R, Caneva G, McClatchey WC (2015) Plant selection for ethnobotanical uses on the Amalfi Coast (Southern Italy). J Ethnobiol Ethnomed 11:58
Shepard G (2004) A sensory ecology of medicinal plant therapy in two Amazonian societies. Am Anthropol 106:252–266
Somerville C, Somerville S (1999) Plant functional genomics. Science 285:380–383
Staub PO, Geck MS, Weckerle CS, Casu L, Leonti M (2015) Classifying diseases and remedies in ethnomedicine and ethnopharmacology. J Ethnopharmacol 174:514–519
Sterck L, Rombauts S, Vandepoele K, Rouze P, Van de Peer Y (2007) How many genes are there in plants (... why are they there)? Curr Opin Plant Biol 10:199–203

Thomas E, Vandebroek I, Sanca S, Van Damme P (2009) Cultural significance of medicinal plant families and species among Quechua farmers in Apillapampa, Bolivia. J Ethnopharmacol 22:60–67

Trotter RT, Logan MH (1986) Informant consensus: a new approach for identifying potentially effective medicinal plants. In: Etkin NL (ed) Plants in indigenous medicine and diet, behavioural approaches. Redgrave Publishing Company, Bedford Hills, NY

Viljoen AM, Subramoney S, Van Vuuren SF, Başer KHC, Demirci B (2005) The composition, geographical variation and antimicrobial activity of *Lippia javanica* (Verbenaceae) leaf essential oils. J Ethnopharmacol 96:271–277

Webb CO, Ackerly DD, McPeek MA, Donoghue MJ (2002) Phylogenies and community ecology. Annu Rev Ecol Evol Syst 33:475–505

Webb CO, Ackerly DD, Kembel SW (2008) Phylocom: software for the analysis of phylogenetic community structure and trait evolution. Bioinformatics 24:2098–2100

Weckerle CS, Cabras S, Castellanos ME, Leonti M (2011) Quantitative methods in ethnobotany and ethnopharmacology: considering the overall flora—hypothesis testing for over- and underused plant families with the Bayesian approach. J Ethnopharmacol 137:837–843

Weng JK (2014) The evolutionary paths towards complexity: a metabolic perspective. New Phytol 201:1141–1149

Wink M (2003) Evolution of secondary metabolites from an ecological and molecular phylogenetic perspective. Phytochemistry 64:3–19

Winter M, Devictor V, Schweiger O (2013) Phylogenetic diversity and nature conservation: where are we? Trends Ecol Evol 28:199–204

Wittmann J, Bruggemann R (2013) Comparison of six simple multicriteria decision methods. In: Wittmann J, Müller M (eds) Simulation in Umwelt- und Geowissenschaften; Workshop Leipzig 2013. Shaker, Aachen

Wolski M (2008) Information quanta and approximation operators: once more around the track. In: Transactions in rough sets VIII, vol 5084, Lecture notes in computer science. Springer, New York, pp 237–250

Yessoufou K, Daru BH, Muasya AM (2015) Phylogenetic exploration of commonly used medicinal plants in South Africa. Mol Ecol Resour 15:405–441

Ziani BE, Calhelha RC, Barreira JC, Barros L, Hazzit M, Ferreira IC (2015) Bioactive properties of medicinal plants from the Algerian flora: selecting the species with the highest potential in view of application purposes. Ind Crop Prod 77:582–589

Part V
Software Developments

PARSEC: An R Package for Partial Orders in Socio-Economics

Alberto Arcagni

1 Introduction

The R (R Core Team 2015) package parsec provides tools for multidimensional evaluation with ordinal variables. In particular its functions allow the application of the poset-based approach (Fattore et al. 2011a,b, 2012) and of the counting approach (Alkire and Foster 2011a,b). The package is available on CRAN (the Comprehensive R Archive Network).

Recently, Fattore published an article where he describes in detail the poset-based approach and specifies how to include external information on attribute relevance (Fattore 2015). Therefore parsec has been upgraded by introducing functions to generate more elaborated posets so that they represent such external informations.

The basic functions of the package are described in a chapter of the book "Multi-indicator Systems and Modelling in Partial Order" (Fattore and Arcagni 2014) therefore this work focuses on the new functionalities. In this work it is assumed that the reader has a basic knowledge of the methodology, of the R language and of the object oriented programming. The work is divided into two parts.

Section 2 describes how to create incidence matrices that are the main structures used in parsec to model posets. Observe that some methods to generate incidence matrices require to distinguish between *achievement posets* and *attribute posets* (Fattore 2015). The first ones are the posets used to define the comparabilities between profiles, whereas attribute posets are necessary to include the external information on attribute relevance. The package provides various functions to create incidence matrices, and they are exposed in the following sections. Section 2.1 illustrates how to check if a generic boolean matrix can be utilized as incidence

A. Arcagni (✉)
Department of Statistics and Quantitative Methods, University of Milano-Bicocca, Milan, Italy
e-mail: alberto.arcagni@unimib.it

M. Fattore, R. Bruggemann (eds.), *Partial Order Concepts in Applied Sciences*,
DOI 10.1007/978-3-319-45421-4_19

matrix. Section 2.2 describes the functions useful to generate a *basic* achievement poset. Section 2.3 focuses on functions for the creation of attribute posets and for the generation of achievement posets derived by them.

Section 3 represents the second part of this work. Once the achievement poset is realized it can be used to evaluate the indicators provided by the poset-based approach. Such section describes the basic use of the `evaluation` function. Functions are exemplified by four achievement poset, then Sect. 3.1 concludes the examples by proposing a comparison of the results.

Section 4 is devoted to conclusions.

2 Incidence Matrices and Posets

A partial order set, *poset*, is a set X equipped with a partial order relation $\leq$ (for details see Fattore 2015). In `parsec` posets are represented through *incidence matrices*. They are objects of class `incidence` that are boolean matrices whose rows and columns are named in order to list the elements of the set. Boolean values assumed by the incidence matrix at line `i` and column `j` indicate if the `i`th element of the set is lower or equal to the `j`th one.

Through examples, it is illustrated how to construct an incidence matrix. In the package there are functions:

- to check if a boolean matrix can be used as incidence matrix;
- to generate incidence matrices of basic achievement posets;
- to generate incidence matrices of attribute posets;
- to generate incidence matrices of achievement posets starting from an attribute poset.

2.1 Check Boolean Matrices

First of all, we show how to check if a boolean matrix can be used as incidence matrix. By definition, a partial order relation satisfies the properties of reflexivity, antisymmetry and transitivity. Let the set $X = \{x, y, z\}$ represent it through a character vector and create the named boolean matrix `I` that will store the partial order relation $\leq$.

```
X <- c("x", "y", "z")
I <- matrix(FALSE, 3, 3)
rownames(I) <- colnames(I) <- X
I
#       x     y     z
# x FALSE FALSE FALSE
# y FALSE FALSE FALSE
# z FALSE FALSE FALSE
```

Reflexivity property states that $x \leq x$ for all $x \in X$. It can be obtained by setting the diagonal of matrix `I` all equal to `TRUE`. Function `reflexivity` checks if the matrix satisfies such property.

```
diag(I) <- TRUE; I
#       x     y     z
# x  TRUE FALSE FALSE
# y FALSE  TRUE FALSE
# z FALSE FALSE  TRUE
reflexivity(I)
# [1] TRUE
```

Antisymmetry property states that if $x \leq y$ and $y \leq x$ then $x = y$, for all $x, y \in X$. In the following example we set `x` lower or equal to `y` and `y` lower or equal to `x`. Since they are represented in two different rows and columns of matrix `I` they cannot be the same element of the set, therefore the function `antisymmetry` returns `FALSE`. Then we have to set `y` greater than `x` in order to get an antisymmetric relation.

```
I["x", "y"] <- TRUE; I["y", "x"] <- TRUE
antisymmetry(I)
# [1] FALSE
I["y", "x"] <- FALSE; I
#       x     y     z
# x  TRUE  TRUE FALSE
# y FALSE  TRUE FALSE
# z FALSE FALSE  TRUE
antisymmetry(I)
# [1] TRUE
```

Transitivity states that if $x \leq y$ and $y \leq z$, then $x \leq z$, $x, y, z \in X$. Therefore, if `I["y", "z"]` is set `TRUE` the function `transitivity(I)` returns `FALSE` because, in the previous step, `I["x", "y"]` is `TRUE`. Starting from matrix `I`, there are two choices to obtain boolean matrices `I1` and `I2` that can be used as incidence matrices where z and y can be compared:

- matrix `I1`: set `I1["z", "y"]` equal to `TRUE`, instead of `I1["y", "z"]`;
- matrix `I2`: set `TRUE` both `I2["y", "z"]` and `I2["x", "z"]`.

```
I1 <- I
I1["z", "y"] <- TRUE
transitivity(I1)
# [1] TRUE
I2 <- I
I2["y", "z"] <- TRUE
I2["x", "z"] <- TRUE
transitivity(I2)
# [1] TRUE
```

Once the boolean matrix satisfies all the required properties (function `is.partialorder` checks all of them) its class can be set to `incidence`. Then the matrix represents a poset and it can be used with methods and functions of the package. For instance, the method of function `plot` associated by the package to the class `incidence` returns a Hasse-diagram instead of a scatter-plot. Hasse-diagram is an oriented graph where the edge directions are from top to bottom and represents the cover relation.

```
is.partialorder(I)
# [1] TRUE
class(I) <- "incidence"; I
#        x     y     z
# x  TRUE  TRUE FALSE
# y FALSE  TRUE FALSE
# z FALSE FALSE  TRUE
# attr(,"class")
# [1] "incidence"
plot(I, main = "I")

is.partialorder(I1)
# [1] TRUE
class(I1) <- "incidence"; I1
#        x    y     z
# x  TRUE TRUE FALSE
# y FALSE TRUE FALSE
# z FALSE TRUE  TRUE
# attr(,"class")
# [1] "incidence"
plot(I1, main = "I1")

is.partialorder(I2)
# [1] TRUE
class(I2) <- "incidence"; I2
#        x     y    z
# x  TRUE  TRUE TRUE
# y FALSE  TRUE TRUE
# z FALSE FALSE TRUE
# attr(,"class")
# [1] "incidence"
plot(I2, main = "I2")
```

The images generated by the `plot` commands are shown in Fig. 1. From Fig. 1 the partial order relations described in this section can be summarized. Apart reflexivity, matrix `I` only states that $x \leq y$, i.e., only `I["x", "y"]` is TRUE. This can be observed in the corresponding Hasse-diagram where there is only the edge from y to x. Observe that z cannot be compared with any other element of the set X.

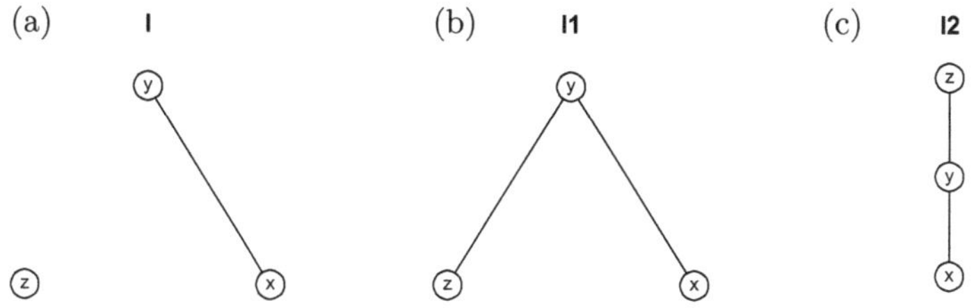

Fig. 1 Hasse-diagrams of partial orders described by incidence matrices `I`, `I1` and `I2`

On the contrary, in the Hasse-diagram of `I1` matrix it can be observed that z is lower than y because `I1["z", "y"]` has been set `TRUE`. Matrix `I2` is similar to `I` but `I2["y", "z"]` is `TRUE`. From the corresponding Hasse-diagram it can be observed the reason why by transitivity $x \leq z$ and `I2["x", "z"]` has to be `TRUE`. Matrix `I2` represents a complete order.

2.2 *The Basic Achievement Poset*

In multidimensional measurement, a set of k ordinal variables (they are also called *attributes*) $v_1, \ldots, v_k$ may be identified to describe a complex phenomena. Let the variable v_j have m_j degrees, $j = 1, \ldots, k$. All the combinations of their modalities represent the set of *achievement profiles*, denoted by Π. Without introducing any assumption, the *basic achievement poset* can be obtained simply comparing the scores that each variable assumes in a profile. The following example shows how to obtain in `parsec` the basic achievement poset from a set of $k = 3$ variables with $m_1 = 3, m_2 = 2$ and $m_3 = 4$.

Variables can be simply defined through the vector of their degrees. It can be named a vector in order to assign at each variable a label. Then the function `var2prof` generates the set of achievement profiles Π.

```
m <- c(x=3, y=2, z=4)
PI <- var2prof(varlen = m); PI
# $profiles
#     x y z
# 111 1 1 1
# 211 2 1 1
# ...
# 224 2 2 4
# 324 3 2 4
#
# $freq
# 111 211 311 121 221 321 112 212 312 122 222 322 113 213
# 1   1   1   1   1   1   1   1   1   1   1   1   1   1
```

```
# 313 123 223 323 114 214 314 124 224 324
# 1   1   1   1   1   1   1   1   1   1
#
# attr(,"class")
# [1] "wprof"
```

Function `var2prof` returns an object of class `wprof` composed by a `data.frame` representing the set of achievement profiles Π and a vector of frequencies associated with each profile, that can be modified and it is useful to evaluate synthetic measures whenever the set of achievement profiles is associated with a population.

An object of class `wprof` can be used as argument of function `getzeta` which returns the incidence matrix of the basic achievement poset (generally identified with letter *Z*). Such function returns an object of class `incidence` that can be used as shown before.

```
Z <- getzeta(PI)
plot(Z)
```

The output of `plot` command is shown in Fig. 2.

2.3 External Information on Attribute Relevance

How to obtain the basic achievement poset is described in Sect. 2.2 and we pointed out that it is the achievement poset that is defined simply by comparing the scores that each attribute assumes in profiles. But it is also possible to introduce external informations on attribute relevance. It can be done through an *attribute poset*,

Fig. 2 Basic achievement poset

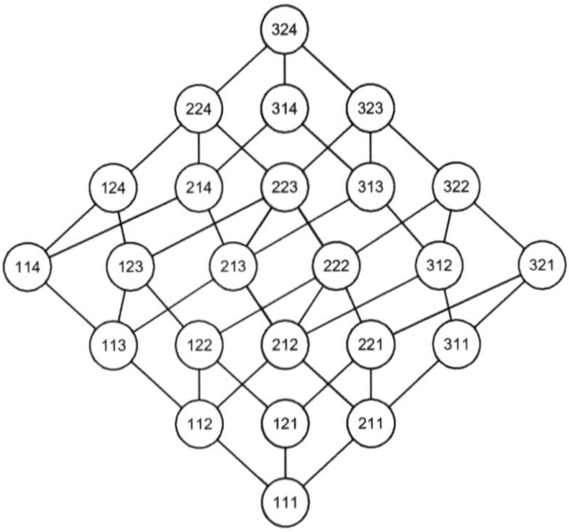

generally identified as Λ. The function `getlambda` has been recently introduced in the package to define the attribute poset and it generates an incidence matrix simply by the definition of comparabilities through attributes. For instance, the incidence matrices defined in Sect. 2.1 can be obtained with function `getlambda` simply by declaring comparabilities and by listing the attributes that are not comparable with the others.

```
I   <- getlambda(x < y, z)
I1 <- getlambda(x < y, z < y)
I2 <- getlambda(x < y, z > y)
```

Once the attribute poset is defined it is necessary to get its set of linear extensions $\Omega(\Lambda)$ (see Fattore 2015). Function `LE` returns the linear extensions of a poset. The output is a list of character vectors. Each vector represents the attribute names from the lowest to the highest one.

```
Omega_I <- LE(I);
# $LE1
# [1] "x" "y" "z"
#
# $LE2
# [1] "x" "z" "y"
#
# $LE3
# [1] "z" "x" "y"

Omega_I1 <- LE(I1); Omega_I1
# $LE1
# [1] "x" "z" "y"
#
# $LE2
# [1] "z" "x" "y"

Omega_I2 <- LE(I2); Omega_I2
# $LE1
# [1] "x" "y" "z"
```

Observe that matrix `I2` represents a complete order therefore it has only a linear extension. Function `mrg` generates the incidence matrix of the achievement poset from $\Omega(\Lambda)$ and from the attribute definitions. Consider, for instance, the attributes defined in Sect. 2.2. The achievement posets corresponding to each attribute poset `I`, `I1` and `I2` are shown in Fig. 3 and the code to get them follows:

```
m <- c(x=3, y=2, z=4)

Z_I <- mrg(Omega_I, varlen = m)
plot(Z_I, shape = "equispaced")
```

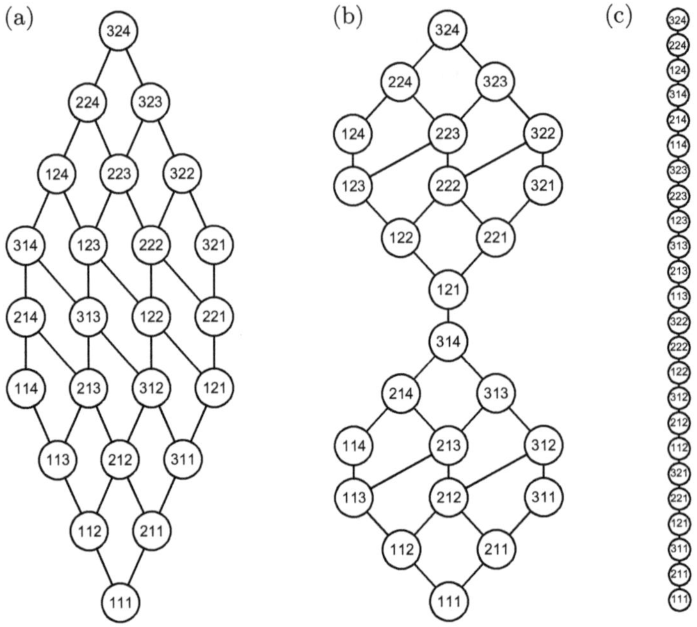

Fig. 3 Achievement poset obtained from attribute posets (**a**) `I`, (**b**) `I1` and (**c**) `I2`

```
Z_I1 <- mrg(Omega_I1, varlen = m)
plot(Z_I1, shape = "equispaced")

Z_I2 <- mrg(Omega_I2, varlen = m)
plot(Z_I2)
```

In conclusion of this section, observe that the basic achievement poset shown in Fig. 2 can also be obtained through the `getlambda` function by listing the attributes names without introducing any relation between them. In this case, the set $\Omega(\Lambda)$ is represented by all the permutations of attributes names.

```
I0 <- getlambda(x, y, z)
Omega_I0 <- LE(I0);Omega_I0
# $LE1
# [1] "x" "y" "z"
#
# $LE2
# [1] "x" "z" "y"
#
# $LE3
# [1] "y" "x" "z"
```

```
#
# $LE4
# [1] "y" "z" "x"
#
# $LE5
# [1] "z" "x" "y"
#
# $LE6
# [1] "z" "y" "x"
m <- c(x=3, y=2, z=4)
Z <- mrg(Omega_I0, varlen = m)
plot(Z)
```

3 From the Achievement Poset to the Incidence Function

In Sect. 2, four achievement posets were proposed for attributes x, y and z: `Z`, `Z_I`, `Z_I1` and `Z_I2`. Once an achievement poset is available, the choice of a *threshold* allows the application of the poset-based methodology in order to obtain an *incidence function*. Details about thresholds selections are available in Fattore et al. (2011a, 2012) and Fattore (2015), here we recall only that the threshold is an antichain of the poset whose downset identifies the profiles that certainly represent a condition of deprivation.

In `parsec` a threshold is a vector whose elements allow to identify rows/-columns of the incidence matrix. Rows and columns of the incidence matrix can be identified through a boolean vector or a vector of profiles. Function `evaluation` applies the poset-based methodology starting from a threshold and an object of class `incidence` representing the achievement poset. It is also possible to add an object of class `wprof` in order to define the profile frequency distribution to get the synthetic measures in relation to a population.

In the following the results of function `evaluation` are shown through examples starting from the incidence matrices `Z`, `Z_I`, `Z_I1` and `Z_I2` defined before.

Profiles `122`, `221` and `113` are used as threshold proposal. The function `downset` returns the downset of a subset of profiles. If the subset of profiles is not an antichain it means that one or more profiles are not necessary to generate such downset. The function `gen.downset` returns the antichain that generates the input downset. Observe that antichains depend on the incidence relation, therefore the threshold proposal cannot be an antichain even if such profiles are not comparable in the basic achievement poset. Therefore, for each example the threshold proposal is analysed.

First of all, consider the basic achievement poset.

```
proposal <- c("122", "221", "113")

dwn <- downset(Z, proposal); which(dwn)
# 111 211 121 221 112 122 113
#   1   2   4   5   7  10  13

threshold <- gen.downset(Z, dwn); which(threshold)
# 221 122 113
#   5  10  13
```

The downset of the threshold proposal in this case is represented by profiles `111`, `211`, `121`, `221`, `112`, `122` and `113`. The antichain that generates this downset is equal to the proposal, therefore it is used as threshold.

```
results <- evaluation(threshold = threshold, zeta = Z)
```

The `evaluation` function returns an object of class `parsec`. Such class of objects is a list summarizing the input informations and the results. In particular, in the results are provided the distributions of different indices computed by uniform sampling of the linear extensions of the poset through a `C` implementation of the Bubley–Dyer algorithm (Bubley and Dyer 1999).

Methods for the class `parsec` has been associated with functions `summary` and `plot`. The `evaluation` function returns a large variety of results useful to get synthetic measures, plots and to develop further analysis. Therefore it may be useful to summarize its results. Function `summary` returns a data-frame that for each profile provides

weights its weight in population;
threshold the boolean variable indicating if it belongs to the threshold;
id. function the corresponding value of the identification function, i.e., the fraction of sampled linear extensions where the profile is in the downset of the threshold;
average rank the average of the ranks that the profile assumes in each of the sampled linear extension;
abs. severity (absolute severity) the average graph distance of the profile from the first one above all threshold elements (the distance is set equal to 0 for profiles above the threshold);
rel. severity (relative severity) equal to the absolute severity divided by its maximum, that is, the absolute severity of the minimal element in the linear extension;
abs. wealth gap (absolute wealth gap) the average graph distance from the maximum threshold element (the distance is set equal to 0 for profiles in the downset of threshold elements);
rel. wealth gap (relative wealth gap) equal to the absolute wealth gap divided by its maximum, that is, the absolute distance of the threshold from the maximal element in the linear extension.

The `evaluation` function returns the synthetic indicators `poverty_gap` and `wealth_gap` that are, respectively, the weighted means of the relative severity and of the relative wealth gap. The code to get the summary of the results and the plots follows:

```
sry_results <- summary(results); str(sry_results)
# 'data.frame': 24 obs. of  8 variables:
# $ weights          : num  1 1 1 1 1 1 1 1 1 1 ...
# $ threshold        : logi  FALSE FALSE FALSE TRUE ...
# $ id. function     : num  1 1 1 1 1 ...
# $ average rank     : num  1 4.54 6.2 7.36 8.84 ...
# $ abs. severity    : num  9.95 7.73 7.06 2.45 8 ...
# $ rel. severity    : num  1 0.769 0.708 0.251 0.797 ...
# $ abs. wealth gap: num  0 0 0 0 0 ...
# $ rel. wealth gap: num  0 0 0 0 0 ...

results$poverty_gap
# [1] 0.2736463
results$wealth_gap
# [1] 0.4427691

plot(results)
```

Figure 4 shows the images provided by function plot. Figure 4a highlights threshold profiles in the Hasse-diagram of the achievement poset. Figure 4b shows the relative frequencies of the times a profile is threshold in the sampled linear extensions of the achievement poset, therefore it is a representation of the relevance of the threshold profiles. The identification function is shown in Fig. 4c. Finally, Fig. 4d shows a representation of the rank distributions of each profile in the sampled linear extensions. The grey scale used to fill the rectangles represents the rank value (white means rank equal to 1, black means maximum rank). The height of the rectangles represents the corresponding relative frequency. Therefore, to each profile corresponds a bar of height 1, and a dark bar means that the profile tends to assume higher ranks.

3.1 Final Comparisons

In conclusion a comparison of the results obtained with the different achievement posets is proposed.

It may be of interest a comparison on the threshold variations induced by the introduction of external informations on attribute relevance. Previously we defined a threshold proposal that in the basic achievement poset is an antichain therefore it is suitable to be used as threshold. We also observed that the introduction of external informations on attribute relevance changes the achievement poset, consequentially the threshold proposal may not be an antichain. Therefore we proposed to evaluate

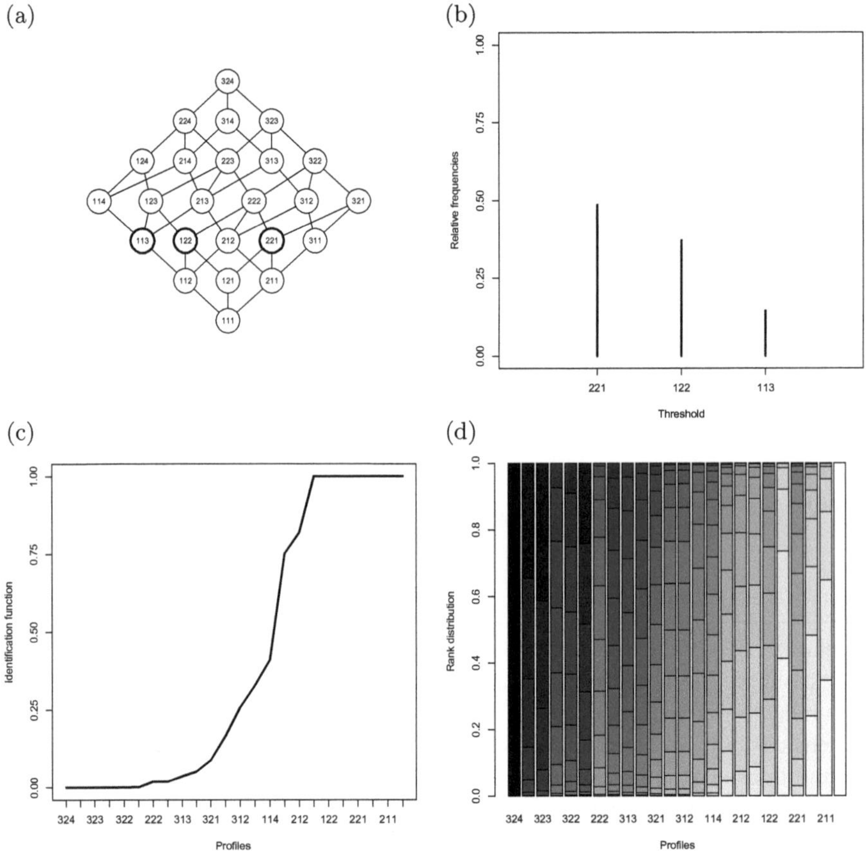

Fig. 4 Plots of the results provided by the `evaluation` function. (**a**) Threshold in the Hasse-diagram. (**b**) Relevance of threshold profiles. (**c**) Identification function. (**d**) Profiles rank distributions

the proposal downset and then to identify the antichain that generates such downset. Here below it is proposed the code to apply this procedure to the achievement posets generated by the attribute posets `I`, `I1` and `I2`.

```
dwn <- downset(Z_I, proposal); which(dwn)
# 111 211 311 121 221 112 212 312 122 113
#   1   2   3   4   5   7   8   9  10  13
threshold_I <- gen.downset(Z_I, dwn);
which(threshold_I)
# 221 122 113
#   5  10  13

dwn <- downset(Z_I1, proposal); which(dwn)
# 111 211 311 112 212 312 113 213 313 114 214 314
#   1   2   3   4   5   6   7   8   9  10  11  12
# 121 221 122
#  13  14  16
threshold_I1 <- gen.downset(Z_I1, dwn);
which(threshold_I1)
# 221 122
#  14  16

dwn <- downset(Z_I2, proposal); which(dwn)
# 111 211 311 121 221 321 112 212 312 122 222 322
#   1   2   3   4   5   6   7   8   9  10  11  12
# 113
#  13
threshold_I2 <- gen.downset(Z_I2, dwn);
which(threshold_I2)
# 113
#  13
```

In the case of the achievement poset generated by the attribute poset `I`, Figs. 3a and 1a, the threshold proposal still remains an antichain.

When the attribute y is more important than attributes x and z, attribute poset `I1` shown in Fig. 1b, the profile `113` is lower than profiles `122` and `221`, Fig. 3b, because, even if its score on attribute z is 3, it is less important than the score it assumes on attribute $y = 1$, which is lower than the scores of y in the other two profiles. Therefore it is not necessary to introduce profile `113` into the threshold, as it can be observed in the commented results of the code proposed before.

On the contrary, in the achievement poset (Fig. 3c) obtained by the attribute poset `I2` (Fig. 1c) profile `113` is the only one of the proposed one that is useful as threshold because attribute z is the most relevant.

For the sake of brevity, only the comparison between the synthetic measures, Table 1, and the inequality curves, Fig. 5, are shown.

For all the examples the poverty gap is lower than the wealth gap, this is consequence of the fact that profiles in the threshold proposal are closer to the bottoms of the different achievement posets than to their tops. Moreover the tendency of

Table 1 Comparison of synthetic measures between different achievement posets obtained through different attribute posets and with same threshold proposal

Attribute poset	Poverty gap	Wealth gap
Basic	0.2736	0.4428
`I`	0.3425	0.4393
`I1`	0.4528	0.5388
`I2`	0.5385	0.5455

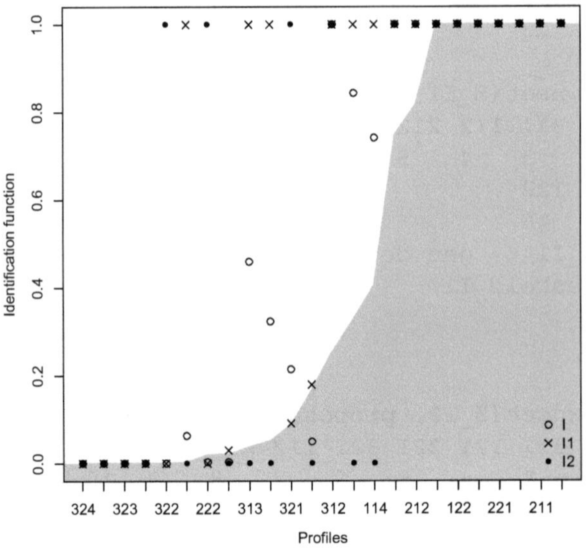

Fig. 5 Comparison of identification functions between different achievement posets obtained through different attribute posets and with same threshold proposal

such gaps to increase from the basic achievement poset to the achievement poset based on `I2` may be a consequence of the evolution of the examples, obtained by the progressive introduction of external information. This procedure increased the achievement posets levels and reduced the threshold size increasing the relative distances of the profiles from the threshold.

Generally, identification functions are represented in increasing order. In Fig. 5 we preferred to maintain the profile order obtained with the identification function generated by the basic poset (the gray background) and to show the other identification functions through different point characters. In the last example, based on `I2`, the achievement poset is a complete order, therefore the identification function assumes only two values: 1 when the profile is lower or equal than the threshold `113`, 0 otherwise.

4 Conclusion

Some examples were presented in order to expose the new functions of `parsec`, from poset definitions by incidence matrices to the analysis of the results provided by the main functions of the package. In particular, this work focuses on the introduction of external informations on attribute relevance, it shows that function `getlambda` simplifies the creation of posets and that function `LE` generates their linear extensions. Then, from the linear extensions and from the attribute definitions, it has been shown how to get achievement poset with function `mrg`. In the second part, we suggested how to choose a threshold, starting from a proposal and by its analysis in a specific achievement poset. Then we compared the main results of the function `evaluation`.

In conclusion we recall that the package `parsec` is available on CRAN and that it is still in development in order to add new functionalities and to optimize its procedures.

References

Alkire S, Foster J (2011a) Counting and multidimensional poverty measurement. J Public Econ 95(7):476–487

Alkire S, Foster J (2011b) Understandings and misunderstandings of multidimensional poverty measurement. J Econ Inequal 9(2):289–314

Bubley R, Dyer M (1999) Faster random generation of linear extensions. Discret Math 201(1): 81–88

Fattore M (2015) Partially ordered sets and the measurement of multidimensional ordinal deprivation. Soc Indic Res 1–24

Fattore M, Arcagni A (2014) Parsec: an R package for poset-based evaluation of multidimensional poverty. In: Multi-indicator systems and modelling in partial order. Springer, Berlin, pp 317–330

Fattore M, Bruggemann R, Owsiński J (2011a) Using poset theory to compare fuzzy multidimensional material deprivation across regions. In: New perspectives in statistical modeling and data analysis. Springer, Berlin, pp 49–56

Fattore M, Maggino F, Greselin F (2011b) Socio-economic evaluation with ordinal variables: Integrating counting and poset approaches. Statistica & Applicazioni 1(2011):31–42

Fattore M, Maggino F, Colombo E (2012) From composite indicators to partial orders: evaluating socio-economic phenomena through ordinal data. In: Quality of life in Italy. Springer, Berlin, pp 41–68

R Core Team (2015) R: a language and environment for statistical computing. R Foundation for Statistical Computing, Vienna, Austria, 2015

»PyHasse« and Cloud Computing

❖ Partial Order Concepts in Applied Sciences ❖

Peter Koppatz and Rainer Bruggemann

1 Introduction

This article is describing the new software package »PyHasse«. For a clear separation of the old and the newer version, we name the older one PyHasse 2 and the newer one PyHasse 3. Whereas PyHasse 2 started in 2007, a rapid development on the Internet made it necessary to follow the new development and react with a new strategy for PyHasse. Our answer is version PyHasse 3. The development process and first results are described in this article. The development of the new PyHasse 3 was strongly accelerated by the needs of the project iBaMs. Within this project ranking will be a necessary step to find best panels of CNC machines, suitable for intellectually disabled people. For more details, see Bruggemann et al. (2015).

2 PyHasse 2

Now reporting about a new way to use multi-indicator systems, we should mention in more detail, what was the PyHasse development until today and why we started a new development for »PyHasse and cloud computing«. PyHasse 2 is a software with the following properties:

P. Koppatz (✉)
Technical University of Applied Sciences Wildau, Hochschulring 1, 15745 Wildau, Germany
e-mail: pkoppatz@web.de

R. Bruggemann
Department of Ecohydrology, Leibniz—Institute of Fresh Water Ecology and Inland Fisheries, Berlin, Germany

M. Fattore, R. Bruggemann (eds.), *Partial Order Concepts in Applied Sciences*,
DOI 10.1007/978-3-319-45421-4_20

- written in Python 2.6
- using Tcl/Tk for user interaction (graphical user interface)
- more than 100 modules
- individual distribution via CD and mail
- Developer: Only one (R.Bruggemann). Details, see Bruggemann et al. (2014).

3 Development Rethought

3.1 What Are the Missing Points?

First question: What are the missing points which in the end are the starting point for a new development?

The following list shows some important and critical remarks:

- Python has reached the version 3.x and is not always compatible with Python 2.x.
- Presentation of a science projects in the web is nowadays standard (missing so far).
- Distribution of software via web technologies is standard (not available so far).
- Missing and/or unsatisfactory development steps:
  - automatic testing
  - documentation
  - quality checks
  - no public source repositories.

Thinking more about the gaps we developed a strategy to fill these gaps. We are describing the used solutions as an example for other projects or developer/scientists who wants to support the further development.

3.2 Python Version

There is no doubt Python development is an ongoing process and using Python 3 is a good starting point. New projects should not use Python 2.x any more! A second reason to start the development from ground up was the mix of code for presentation—the GUI—and scientific parts. This was acceptable for a rapid development but it prevented a reuse of code and a modular concept. With PyHasse 3 presentation layer and core calculations are separated and reusable by other modules.

3.3 Presentation

If you want to get your Software usable in the World Wide Web, using a browser for this task is the easiest and most accepted solution. While generating output as text or presenting results as table is an easy task, producing graphical output is not so easy. But in this field also many things are changing. So we are using a JavaScript library to draw, interact and present Hasse diagrams in the browser. The Tcl/Tk GUI was replaced by a Client–Server solution whereas the browser is the GUI for the user. The main advantage of this solution is the availability of a browser on every modern internet device! A nice side effect is that now different scientists or students are able to use the same installation simultaneously.

3.4 Client–Server

All software used is supporting Python version 3 and is a must. For the web framework we are using a one file solution named »Bottle«. Beside its simplicity it is satisfactory for our project. This component to generate HTML output is complemented by other components:

- waitress—as server
- beaker—session management
- docutils—generate HTML from simple text
- jinja2—templates for structuring the layout.

On the client side all modern browsers are supported if JavaScript is not deactivated.

3.5 Modular Concept

Beside of solving mathematical problems we decided to start with a modular concept and to focus on the core functionality used for all modules. The result was a tripartite namespace.

mis.pyhasse.xxx

in which the first part »mis« acts as an umbrella for multi-indicator systems, allowing to add Python programmed modules not necessarily using the PyHasse realization. The second part represents the »pyhasse« specific solutions and the last part (here as a placeholder »xxx«) represents a specific module.

Today we have finished three core modules and seven modules based on the core modules (Fig. 1 and Table 1).

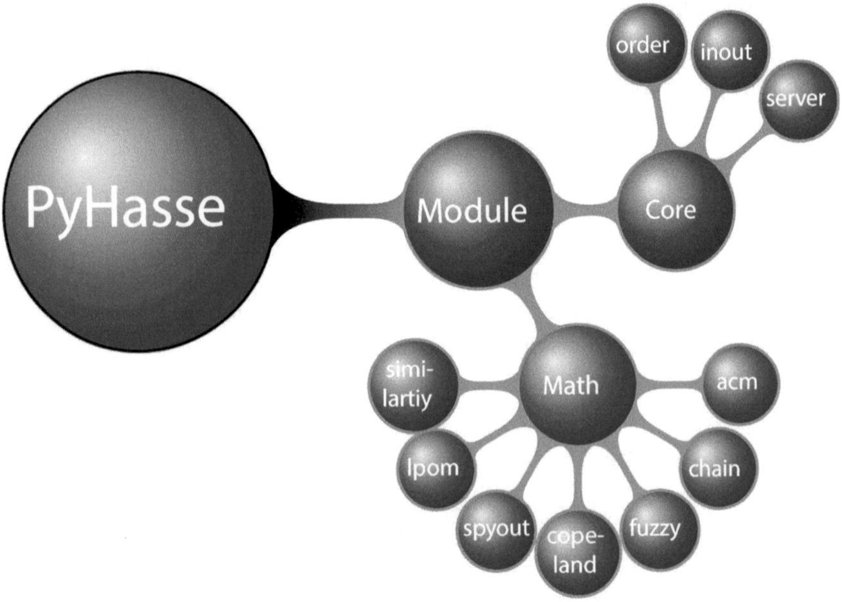

Fig. 1 Module in PyHasse

Table 1 Available modules

Module	Short description
Spyout	Getting some experience
Antichain	Incomparable objects
Chain	Comparable objects
Copeland	Classical construction of a ranking index
LPOM	Posetic construction of ranking indices
Fuzzy	Greater than/Less than in the fuzzy framework
Similarity	Comparing two posets

4 Development Process/Tools

The development process in PyHasse 2 was simple—only one developer (R. Bruggemann). Now the process is more complex because at the moment three developers and one designer are involved and more developers would be highly welcomed in the future.

The question is: How to organize the development process? This question is not new and there are many solutions available. So we tried different variants and picked up some best practices, which are listed upon below:

Mercurial a distributed code revision system
PyFlakes measures test coverage
PyTest a testing framework

Pylint variable usage, syntax checks
Pep8 checking the syntax
Tox part of PyTest.

4.1 *Mercurial*

If a developer is interested in a protocol of changes and at the same time have a backup of his work, a version control system is the best solution. Most used today are systems like git, mercurial, bazaar and others. We selected mercurial[1] and made good experiences, because there are only five commands to use the software effective: status, pull, push, add and commit. Of course also a GUI is available, however, we are using the command-line. In contrast to a GUI the command-line is more flexible, albeit without any comfort with respect to editing texts.

4.2 *PyFlakes*

Is a passive checker of Python programs which checks Python source files for errors, e.g.,

- unused import statements
- wrong naming conventions
- wrong number of arguments
- undefined names
- 'return' outside of a function.

4.3 *PyTest*

PyTest: helps you write better programs[2]. Testing is a very important task and without test, a software is marked as »It is not working«. Not all code can be tested, but the corner cases. The feature list of this tool is very long. The output after a test run is shown as an example. Only parts of the output after a test run are shown:

```
> py.test
======== test session starts ================================
platform darwin -- Python 3.4.2 -- py-1.4.25 -- pytest-2.6.3
collected 299 items
...
```

[1]https://mercurial.selenic.com.
[2]https://pypi.python.org/pypi/pytest.

```
spyout/tests/test\_class\_spyout.py ...
spyout/tests/test\_class\_spyout\_houses.py ...
...
========= 291 passed, 8 skipped in 4.55 seconds =============
```

This gives you a good feeling after changing the code and run the tests again.

4.4 Pylint

Pylint[3] is a Python source code analyzer which looks for programming errors and helps to enforce coding standards (see also: Martin Fowler's Refactoring book). According to the website[4] the following aspects are under control:

- »checking line-code's length,«
- »checking if variable names are well-formed according to your coding standard«
- »checking if imported modules are used«

It generates reports similar to PyFlakes.

4.5 Pep8

PEP 0008—Style Guide for Python Code.[5] If an other developer has to read your code, it is nice to have some formal rules with respect to the programming style and layout. In the Python community this rules are defined in a pep (Python Enhancement Proposal) and a standard nowadays. A sample output is shown:

```
> pep8 mis/pyhasse/spyout/spyout.py

spyout/spyout.py:25:24: W291 trailing whitespace
spyout/spyout.py:42:21: E231 missing whitespace after ','
spyout/spyout.py:54:9: E265 block comment should start with
```

4.6 Tox

Tox[6]—from the website—is a generic virtualenv management and test command-line tool you can use for:

[3] http://www.pylint.org.

[4] http://www.pylint.org.

[5] https://www.python.org/dev/peps/pep-0008.

[6] https://pypi.python.org/pypi/tox.

1. checking your package installs correctly with different Python versions and interpreters
2. running your tests in each of the environments, configuring your test tool of choice
3. acting as a frontend to Continuous Integration servers, greatly reducing boilerplate and merging CI and shell-based testing.

4.7 Pyvenv

The venv module provides support for creating lightweight »virtual environments« with their own site directories, optionally isolated from system site directories. Each virtual environment has its own Python binary (allowing creation of environments with various Python versions) and can have its own independent set of installed Python packages.

Now with Python 3.5 all tools to create new »virtual environments« are available. No extra installation steps are necessary any more.

5 Deployment

The old way to deploy the newest version of »PyHasse« was a brand new CD delivered by mail. Today delivering software via download is state of the art. So a website has to be available: *pyhasse.org*[7] is the result of this idea. We are supporting three versions to get and use the software package:

1. Online demo (no installation effort).
2. VirtualBox: no knowledge about Python necessary. Only the handling of VirtualBox, most of the time a click and run tool.
3. Local installation via PyPi, which is thought of for developers and scientists interested in discovering the software or to catch a bug.

There is not only a set of software packages, but the source code is also available. It is hosted at bitbucket[8] a platform hosting open source projects. New scientists and developer are welcome to develop new modules or add new functionality to available modules. To make it easy we have selected the Mozilla Public License (MPL). The documentation extracted from the code and a short version of the mathematical background (for the full set of literature have a look at our collection of references[9]).

---

[7] https://pyhasse.org.

[8] https://bitbucket.org.

[9] http://pyhasse.org/references.

6 Results

Lets us summarize the development process of »PyHasse«. The results are a huge step to bring a useful software to more people in the most easiest way of today, i.e., by the Internet. The new flexibility introduces new workflows, easy to adapt for new developers. The software itself is easy to install and to apply for end users.

The main entry point is now the website https://pyhasse.org (Please see Figs. 2 and 3).

7 Future

PyHasse 2 has a long history and PyHasse 3 is a very young project. There are many variants of math and calculation to migrate from PyHasse 2 to the new PyHasse 3. At the moment there are also more requirements to export results in order to publish the data. The corresponding activities will be continued. But there are also new components we are thinking about. For example, we are working on a first prototype to relax the limitations of the 2D-Hasse diagrams or supplement the Hasse diagram view. A new 3D-Hasse diagram-Editor could be a step forward (Fig. 4).

Fig. 2 homepage of website pyhasse.org

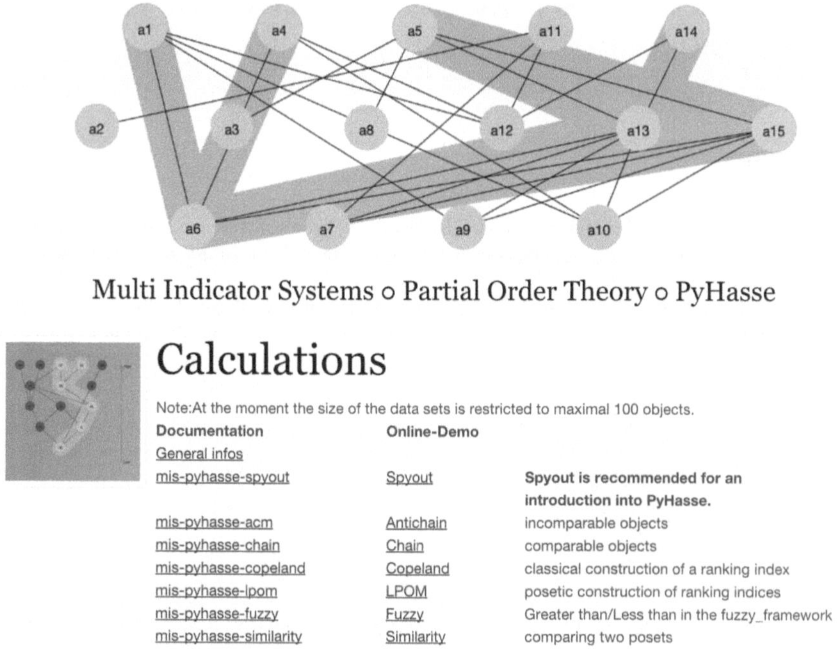

Fig. 3 Website with list of module and Hasse diagram

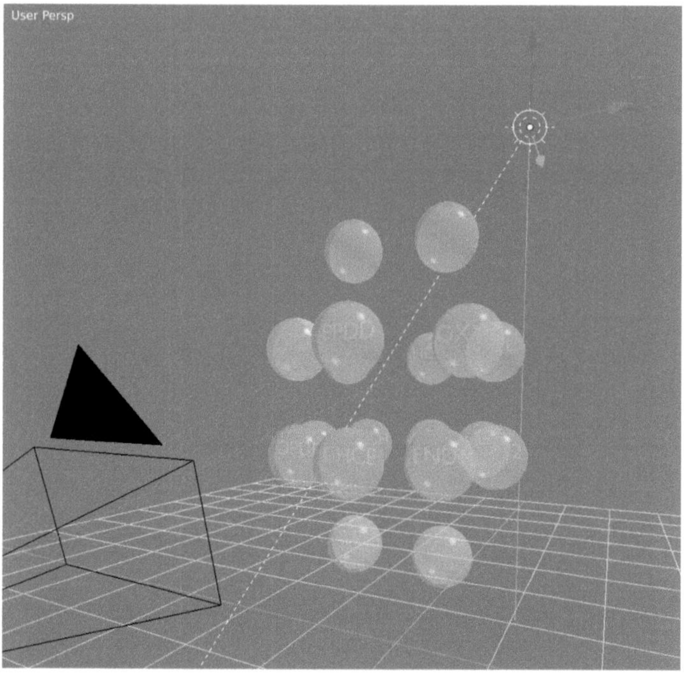

Fig. 4 Prototyp: HD3D-Viewer

We are looking forward to new users, developers and new features for the PyHasse-Project.

References

Bruggemann R, Carlsen L, Voigt K, Wieland R (2014) PyHasse software for partial order analysis: scientific background and description of selected modules. In: Multi-indicator systems and modelling in partial order, pp 389–423

Bruggemann R, Edich D, Fuhrmann F, Koppatz P, Scholl M, Teske A, Wiesner-Steiner A (2015) Kann (einfache) Diskrete Mathematik die Lebensqualität-von Menschen mit geistiger Behinderung an computergesteuerten Maschinen verbessern helfen? In: Wittmann J, Wieland R (eds) Simulation in Umwelt- und Geowissenschaften, workshop MÜncheberg, pp 239–251

Index

M. Fattore, R. Bruggemann (eds.), *Partial Order Concepts in Applied Sciences*,
DOI 10.1007/978-3-319-45421-4

Q

MIX
Papier aus verantwortungsvollen Quellen
Paper from responsible sources
FSC® C105338

If you have any concerns about our products,
you can contact us on
ProductSafety@springernature.com

In case Publisher is established outside the EU,
the EU authorized representative is:
Springer Nature Customer Service Center GmbH
Europaplatz 3, 69115 Heidelberg, Germany

Printed by Libri Plureos GmbH
in Hamburg, Germany